WITHDRAWN
From the
Dean B. Ellis Library
Arkansas State University

The Mathematical Theory of Coding

The Mathematical Theory of Coding

IAN F. BLAKE

Department of Electrical Engineering
University of Waterloo
Waterloo, Ontario, Canada

RONALD C. MULLIN

Department of Combinatorics and Optimization
University of Waterloo
Waterloo, Ontario, Canada

ACADEMIC PRESS   New York   San Francisco   London   1975
A Subsidiary of Harcourt Brace Jovanovich, Publishers

To our parents

COPYRIGHT © 1975, BY ACADEMIC PRESS, INC.
ALL RIGHTS RESERVED.
NO PART OF THIS PUBLICATION MAY BE REPRODUCED OR
TRANSMITTED IN ANY FORM OR BY ANY MEANS, ELECTRONIC
OR MECHANICAL, INCLUDING PHOTOCOPY, RECORDING, OR ANY
INFORMATION STORAGE AND RETRIEVAL SYSTEM, WITHOUT
PERMISSION IN WRITING FROM THE PUBLISHER.

ACADEMIC PRESS, INC.
111 Fifth Avenue, New York, New York 10003

United Kingdom Edition published by
ACADEMIC PRESS, INC. (LONDON) LTD.
24/28 Oval Road, London NW1

Library of Congress Cataloging in Publication Data

Blake, Ian F
    The mathematical theory of coding.

    Bibliography: p.
    Includes index.
    1. Coding theory.   I. Mullin, Ronald Cleveland,
joint author.   II. Title.
QA268.B56   519.4   74-10218
ISBN 0-12-103550-6

PRINTED IN THE UNITED STATES OF AMERICA

Contents

PREFACE ix
ACKNOWLEDGMENTS xi

1. Finite Fields and Coding Theory

1.1	Introduction	1
1.2	Fields, Extensions, and Polynomials	2
1.3	Fundamental Properties of Finite Fields	7
1.4	Vector Spaces over Finite Fields	10
1.5	Linear Codes	15
1.6	Polynomials over Finite Fields	29
1.7	Cyclic Codes	40
1.8	Linear Transformations of Vector Spaces over Finite Fields	58
1.9	Code Invariance under Permutation Groups	63
1.10	The Polynomial Approach to Coding	66
1.11	Bounds on Code Dictionaries	83
1.12	Comments	89
	Exercises	90

2. Combinatorial Constructions and Coding

2.1	Introduction	95
2.2	Finite Geometries: Their Collineation Groups and Codes	95
2.3	Balanced Incomplete Block Designs and Codes	119
2.4	Latin Squares and Steiner Triple Systems	132
2.5	Quadratic Residues and Codes	139

2.6	Hadamard Matrices, Difference Sets, and Their Codes	144
2.7	Self-Dual and Quasicyclic Codes	152
2.8	Perfect Codes	161
2.9	Comments	165
	Exercises	166

3. Coding and Combinatorics

3.1	Introduction	170
3.2	General t Designs	170
3.3	Matroids	173
3.4	Chains and Chain Groups	175
3.5	Dual Chain Groups	178
3.6	Matroids, Graphs, and Coding	179
3.7	Perfect Codes and t Designs	183
3.8	Nearly Perfect Codes and t Designs	186
3.9	Balanced Codes and t Designs	189
3.10	Equidistant Codes	192
3.11	Comments	201
	Exercises	201

4. The Structure of Semisimple Rings

4.1	Introduction	204
4.2	Rings, Ideals, and the Minimum Condition	205
4.3	Nilpotent Ideals and the Radical	207
4.4	The Structure of Semisimple Rings	210
4.5	The Structure of Simple Rings	215
4.6	The Group Algebra and Group Characters	217
4.7	The Structure of Cyclic Codes	222
4.8	Abelian Codes	227
4.9	Comments	244
	Exercises	244

5. Group Representations

5.1	Introduction	246
5.2	Representation of Groups	246
5.3	Group Characters	253
5.4	Orthogonality Relationships and Properties of Group Characters	259
5.5	Subduced and Induced Representations	269
5.6	Direct and Semidirect Products	273
5.7	Real Representations	276
5.8	Modules, Group Algebras, and Representations	284
5.9	Comments	290
	Exercises	290

6. Group Codes for the Gaussian Channel

6.1	Introduction	292
6.2	Codes for the Gaussian Channel	293
6.3	Group Codes for the Gaussian Channel	297
6.4	The Configuration Matrix	304
6.5	Distance Properties of Group Codes	314
6.6	The Initial Vector Problem	318
6.7	Comments	326
	Exercises	326

Appendix A. The Möbius Inversion Formula   329

Appendix B. Lucas's Theorem   330

Appendix C. The Mathieu Groups   331

References   339

INDEX   349

Preface

The subject of coding theory, for both discrete and continuous channels, has developed rapidly over the past twenty-five years with the application of more and more diverse algebraic and combinatoric methods. The aim of this book is to present a unified treatment of these mathematical ideas and their use in the construction of codes. It is not at all concerned with the practical matters of code implementation, and the subject of decoding is considered only insofar as it relates to the mathematical ideas involved. In many instances we have purposely chosen for a problem an approach that is mathematically more advanced than required in order to expose the reader to as wide a scope of concepts as possible, within the context of coding. Such an approach is not designed to achieve a given result with maximum efficiency.

In many ways this book complements the currently available texts on algebraic coding theory. It is assumed that the reader has an introductory knowledge of modern algebra since it would otherwise have been impossible to cover the contents in a single volume. Many readers will want to skip sections containing familiar mathematical ideas, and the book has been written with this in mind.

The first three chapters deal with coding for the discrete channel, i.e., algebraic coding theory, while the last two deal with codes obtainable from group representations for the continuous channel. Chapter 4 contains algebraic material that is common to both discrete and continuous coding. In the rather lengthy first chapter we have presented an exposition on finite fields together with properties of polynomials over finite fields, the linear groups of transformations of a vector space over a finite field, and elementary geometric properties of such vector spaces. These ideas are applied to the analysis and construction of several classes of linear and cyclic codes. Codes that are best studied from a combinatorial point of view are given in Chapters 2 and 3. Thus, for example, the finite geometry codes, while they can be considered as a subclass of cyclic codes or extensions of cyclic codes and their duals, are in our opinion best considered from the geometrical point of view, where their essential geometric flavor is retained and the importance of their geometric structure is demonstrated. The results of Chapter 3 are all stated and proven with the use of matroids. This was not necessary but was in keeping with our aim of providing maximum exposure to mathematical concepts that either have proved useful or have the potential of being so. In Chapter 4, the structure of semisimple rings is explored. This material is common to coding for both discrete and continuous channels. For the discrete case, of course, it is not necessary to have a knowledge of semisimple rings. However, study of the manifestations of this assumption and the resulting structure of the codes, viewed as ideals in the appropriate algebra, is a rewarding exercise.

Elements of the theory of group representations are contained in Chapter 5. A modern approach to this subject would use module-theoretic arguments. The structure of semisimple rings plays a vital role in such an approach. Initially it was our intention to build on the structure theorems of Chapter 4 to give this approach. This was abandoned, however, since it would have required a considerably expanded fourth chapter. Thus our presentation on group representations is from the matrix point of view and the two approaches are reconciled in Section 5.8.

The final chapter utilizes group representations to construct codes for the Gaussian channel. While this theory is still in its infancy, there are sufficiently interesting results available to warrant inclusion. Indeed, it would be rewarding to the authors if the account of group representations and their codes given here encouraged further consideration of this problem.

Reference to a section within the same chapter is given by the section number only. Reference to a section in another chapter includes the chapter number. The same is true for theorems, the theorem numbering beginning anew for each section. At the end of each chapter are sections for comments and problems. An attempt is made in the comment section to cite sources that for one reason or another were not given in the text. The problems are used both to provide exercises and to indicate results that were not considered in the chapter. Several problems in this last category are quite difficult and for these references are quoted.

Acknowledgments

The authors are indebted to many individuals for their assistance in the preparation of this book. The most significant contribution was provided by Dr. Phillipe Delsarte of MBLE in Belgium, who painstakingly read a rough draft and provided us with many constructive suggestions. In many cases he saved us from making false or inconclusive statements, and we gratefully acknowledge his assistance. The authors are also grateful to Professor W. Tutte for his permission to make such liberal use of his lucid treatment of matroid theory in Chapter 3. Thanks are also due Mrs. Janet Schell for her patient and skillful typing and retyping of the manuscript. Finally, we cannot omit mention of our wives for their understanding and forbearance throughout the project. Their support was greatly appreciated.

1 Finite Fields and Coding Theory

1.1 Introduction

Many introductory books on algebra contain a section on finite fields and prove some of their basic properties. Often their interest is to discuss Wedderburn's theorem on finite division rings or to consider the structure of groups of transformations of vector spaces over fields. For these ends only the elementary properties of finite fields are needed. In this chapter we attempt to give a more detailed account of the theory of finite fields, including material on polynomials over finite fields and linear transformations of vector spaces over finite fields. For this purpose it is convenient to assume of the reader an introductory course of algebra. This allows an efficient development of the properties of finite fields and places them in their proper algebraic perspective as particular and interesting examples of a more general theory. The next section reviews some of the theory of fields, extensions of fields, and polynomials over fields. No proofs are given in this section since these are readily available elsewhere.

The remainder of the chapter is concerned with the use of finite field theory in the construction and analysis of codes. Many of the important classes of codes and approaches to coding are included. Those that either arise from combinatorial constructions or else are best treated from a combinatorial point of view are considered in the next chapter, along with the required combinatorial background.

1.2 Fields, Extensions, and Polynomials

We shall denote fields by italic capitals, usually E and F or K, L, M, etc. Let L be a field containing a subset K, which is itself a field under the operations inherited from L. Then L is called an extension of K, and K a subfield of L. Every field has a smallest subfield, called the prime subfield, which is isomorphic to either the rationals Q, in which case we say it has characteristic zero, or to the integers mod p, p a prime, in which case we say it has characteristic p. We shall denote the characteristic of an arbitrary field K by char K. For elements of an arbitrary field we shall use lowercase Greek letters, α, β, etc. If $\alpha \in L \supset K$ and $\alpha \notin K$, then the smallest field containing both K and α will be denoted by $K(\alpha)$. Similarly, the smallest field containing K and the elements $\alpha_1, \ldots, \alpha_n$, $\alpha_i \in L \supset K$ will be written $K(\alpha_1, \ldots, \alpha_n)$. Any extension field L of K can be viewed as a vector space over K and the dimension of this vector space is called the degree of L over K and is denoted by $[L:K]$. If L is a finite-dimensional extension of K and M a finite extension of L, then $[M:K] = [M:L][L:K]$.

Denote by $K[x]$ the ring of polynomials over K in the variable x. Then $K[x]$ is a principal ideal domain. A polynomial in $K[x]$ is said to be monic if the coefficient of the highest power in x is unity. It is irreducible if it is not the product of two nonscalar polynomials in $K[x]$. If $f(x)$ is an irreducible polynomial in $K[x]$, then there exists a smallest extension field L of K, which contains a zero of $f(x)$. Furthermore, L is isomorphic to the quotient field $K[x]/\langle f(x) \rangle$, where $\langle f(x) \rangle$ denotes the ideal in $K[x]$ generated by $f(x)$. If the irreducible polynomial $f(x)$ is of degree n, then $[L:K] = n$. If a is the zero of $f(x)$ in question, then $L = K(a)$ and the elements $1, a, \ldots, a^{n-1}$ form a basis of L over K.

It is often of interest to construct fields containing all the roots of a given polynomial and we introduce the following definition.

Definition Let $f(x) \in K[x]$. The smallest field containing K and all the zeros of $f(x)$ is called the *splitting field* of $f(x)$ over K.

The following theorem justifies our calling it "the" splitting field.

Theorem 2.1 Let $f(x)$ be an irreducible polynomial in $K[x]$. Then a splitting field for $f(x)$ over K exists and any two such splitting fields are isomorphic.

If $f(x)$ is a polynomial in $K[x]$ of degree n, then its splitting field over K is at most of degree $n!$ and this bound may or may not be obtained, depending on the polynomial and the field. We now consider some elementary properties of polynomials over a field. If $f(x) \in K[x]$ is given by

$$f(x) = \beta_n x^n + \beta_{n-1} x^{n-1} + \cdots + \beta_1 x + \beta_0, \qquad \beta_i \in K$$

1.2 Fields, Extensions, and Polynomials

then by the derivative of $f(x)$, denoted by $f'(x)$, will be meant the polynomial

$$f'(x) = n\beta_n x^{n-1} + (n-1)\beta_{n-1} x^{n-2} + \cdots + \beta_1$$

which is a polynomial of degree at most $n - 1$. Notice that the derivative of a polynomial may be zero even though the polynomial is not a constant. Thus, for example, over the field of two elements the polynomial $f(x) = x^2 + 1$ has a zero derivative. A polynomial $f(x) \in K[x]$ is said to have a zero α of multiplicity m in some extension field F of K if m is the largest positive integer for which

$$(x - \alpha)^m | f(x)$$

in $F[x]$, where the vertical bar indicates $(x - \alpha)^m$ divides $f(x)$. The zeros of an irreducible polynomial $f(x) \in K[x]$ in the splitting field for $f(x)$ are called conjugates. Let $f(x), g(x) \in K[x]$ be two polynomials such that $\deg f > \deg g$, where $\deg f$ denotes the degree of $f(x)$. By the Euclidean algorithm there exist two polynomials $q(x)$ and $r(x)$ such that

$$f(x) = q(x)g(x) + r(x), \qquad \deg r < \deg g$$

By repeated application of the algorithm the greatest common divisor (hereafter denoted gcd) $d(x)$ of $f(x)$ and $g(x)$ can be expressed as

$$d(x) = (f(x), g(x)) = a(x)f(x) + b(x)g(x), \qquad a(x), b(x) \in K[x]$$

For further reference we collect some elementary properties of polynomials in the following theorem.

Theorem 2.2 Let $f(x), g(x) \in K[x]$ and let L be any extension of K. Then:

(i) If $(f(x), g(x)) = d(x)$ in $K[x]$, then $(f(x), g(x)) = d(x)$ in $L[x]$.
(ii) $f(x)|g(x)$ in $K[x]$ iff $f(x)|g(x)$ in $L[x]$.
(iii) $f(x)$ has a multiple zero iff $(f(x), f'(x)) \neq 1$.

It is significant that two polynomials have a common root in some extension field if they have a common divisor over the original field.

The question of multiplicities of roots of irreducible polynomials can be settled precisely. If $f(x) \in K[x]$, char $K = 0$, and $f(x)$ is irreducible, then $f(x)$ cannot have multiple zeros. If char $K = p$, then an irreducible polynomial $f(x)$ having multiple zeros must be of the form $f(x) = g(x^p)$ for some polynomial $g(x) \in K[x]$. In this case each zero of $f(x)$ has the same multiplicity. We will show in Section 1.6 that if $f(x)$ is an irreducible polynomial over a finite field, then it has only simple (multiplicity one) zeros. Thus the only fields over which an irreducible polynomial may have a multiple zero are infinite fields with characteristic p. Moreover, irreducible polynomials with multiple roots over such fields do exist [e.g., Herstein (1964, p. 193)]. Recall also the following important and basic theorem.

Theorem 2.3 A polynomial $f(x) \in K[x]$ of degree n has at most n zeros in any extension of K.

We consider now the problem of classifying the various types of extensions of a field and give the basic properties of such extensions. In later sections we will only be interested in finite extensions of finite fields, but the present material will place the problem in better perspective.

An element $\alpha \in L \supset K$ is said to be algebraic of degree n over K if it satisfies an irreducible polynomial of degree n over K. That monic polynomial $m(x)$ in $K[x]$ of lowest degree of which α is a zero is called the minimal polynomial of α over K. The other zeros of $m(x)$ are the conjugates of α, the terminology coming from the more usual concept of the roots of an irreducible quadratic over the reals. If every element is algebraic over K, then L is called an algebraic extension of K. It follows easily that if $\alpha \in L$ is algebraic over K and if $[K(\alpha):K] = n$, then the degree of the minimal polynomial of α over K is n. It can be shown that any finite extension of a field is an algebraic extension. An extension that is not algebraic is called transcendental.

An extension L of K is called simple if there exists an element $\alpha \in L \backslash K = \langle \beta \in L | \beta \notin K \rangle$ such that $L = K(\alpha)$. If K has characteristic zero, then any finite extension is both algebraic and simple. Also, if K is a finite field and L a finite extension of K, then L is again both simple and algebraic.

By a normal extension L of K we mean a finite extension which has the property that any irreducible polynomial $f(x) \in K[x]$ that has one zero in L will have all its zeros in L. It can be shown that if K has characteristic zero, then L is a normal extension of K iff it is the splitting field of some polynomial over K. We will shortly tie the concept of normal extensions to that of normal subgroups of a certain group of automorphisms.

An element $\alpha \in L \supset K$ is separable over K if it satisfies an irreducible polynomial with simple zeros. If every element in L is separable, it is called a separable extension. Every finite-dimensional, separable extension of a field K is a simple extension. We have already noted that if char $K = 0$, then an extension is normal if it is a splitting field for some polynomial. The same is true if K is a finite field. If K is an infinite field with char $K = p$, then an extension of K is normal iff it is separable and a splitting field for some polynomial. The separability condition eliminates the possibility of an irreducible polynomial having multiple zeros.

A field K is perfect if all its finite-dimensional extensions are separable. Any field with characteristic zero is perfect, and it will follow from later work that every finite field is perfect. A field K is said to be algebraically closed iff every $f(x) \in K[x]$ factors into linear factors in $K[x]$. This is equivalent to the statement that K possesses no proper algebraic extensions. Every field possesses an extension that is algebraically closed.

1.2 Fields, Extensions, and Polynomials

We consider now the concept of field isomorphism, which will be useful in the investigation of finite fields. An isomorphism of the field K_1 onto the field K_2 is a one-to-one onto map that preserves both field operations, i.e.,

$$\tau(\alpha + \beta) = \tau(\alpha) + \tau(\beta), \qquad \tau(\alpha\beta) = \tau(\alpha)\tau(\beta), \qquad \alpha, \beta \in K_1$$

An automorphism of K is an isomorphism of K onto itself. The set of all automorphisms of a field forms a group under composition. If $\tau_1, \ldots, \tau_n$ are distinct isomorphisms of K_1 onto K_2, then these isomorphisms are linearly independent over K_2 in the sense that if

$$\sum_{i=1}^{n} a_i \tau_i(b) = 0, \qquad a_i \in K_2$$

for all $b \in K_1$, then $a_i = 0$, $i = 1, \ldots, n$. In particular, distinct automorphisms of a field are linearly independent.

A field automorphism clearly leaves elements of the prime subfield fixed. The automorphism may leave a larger subfield fixed. More generally, if

$$K' = \{a \in K \mid \tau_i(a) = a, i = 1, \ldots, n\}$$

then K' is a subfield of K, called the fixed field of K with respect to the distinct automorphisms τ_i. It can be shown that $[K: K'] \geq n$. We will be interested in subgroups of the group of all automorphisms that fix certain subfields. We shall denote the group of all automorphisms of a field L by $G(L)$ and the subgroup of $G(L)$ that fixes all elements of the subfield $K \subseteq L$ by $G(L/K)$. It is important to note that the fixed field of $G(L/K)$ may properly contain K. It is easily shown that $G(L/K)$ is a subgroup of $G(L)$. Conversely, if H is a subgroup of $G(L)$, then the set of elements of L fixed by H is a subfield of L. This correspondence between subfields and subgroups is fundamental to Galois theory.

We call $G(L/K)$ the Galois group of L over K. If $G(L/K)$ is a finite subgroup of $G(L)$, then $[L: K] \geq |G(L/K)|$ = the order of $G(L/K)$. The inequality arises essentially from the fact that we have already noted, namely, that $G(L/K)$ may fix a larger field than K. If K is the largest subfield fixed by $G(L/K)$, then it can be shown that $[L: K] = |G(L/K)|$. If L is a normal extension of K, however, then K is always the fixed field of $G(L/K)$. Indeed an alternative way of defining a normal extension is to say that L is a normal extension of K if, for any element of $L\setminus K$, there exists an automorphism of $G(L/K)$ moving it.

The original Galois theory was concerned with the solvability by radicals of polynomials and we briefly point out the connection between this problem and the foregoing material. Let $f(x)$ be an irreducible polynomial of degree n in $K[x]$ and L a splitting field for $f(x)$. If the zeros of $f(x)$ are denoted $\alpha_1, \ldots, \alpha_n$, then there exist elements $\tau_i \in G(L/K)$ such that $\alpha_i = \tau_i(\alpha_1)$, $i = 1, \ldots, n$. Of course for any $\tau \in G(L/K)$, $\tau(\alpha_1)$ is a zero of $f(x)$ since $f(\tau(\alpha_1)) = \tau(f(\alpha_1))$

$= 0$. It follows that elements of $G(L/K)$ permute the roots of $f(x)$, and hence $G(L/K)$ is isomorphic to a permutation group on n letters. Since every normal extension is a splitting field for some polynomial, the same statement may be made for any normal extension L of K.

The Galois group of a polynomial $f(x) \in K[x]$ is $G(L/K)$, where L is the splitting field for $f(x)$. The term "solvable group" originates from the famous theorem of Galois stating that a polynomial is solvable by radicals iff its Galois group is solvable (in the group-theoretic sense of the word). The concept of a normal subgroup is connected with normal extensions in the following way. Let L be the splitting field for a polynomial $f(x) \in K[x]$ and let K' be an intermediate field, $K \subset K' \subset L$. Then K' is a normal extension of K iff $G(L/K')$ is a normal subgroup of $G(L/K)$. The extension L of K is cyclic, Abelian, solvable, etc., if $G(L/K)$ is cyclic, Abelian, solvable, etc.

We conclude this section with a brief discussion of the norm and trace of an element. Let L be a normal extension of finite degree n over a subfield K. It follows that there are precisely n distinct elements $\tau_1, \tau_2, \ldots, \tau_n$ of $G(L/K)$. We define the functions on L

$$N(\alpha) = \prod_{i=1}^{n} \tau_i(\alpha), \qquad T(\alpha) = \sum_{i=1}^{n} \tau_i(\alpha)$$

called the norm and trace of α, respectively. The dependence of these functions on the fields L and K is understood. If we view L as a vector space over K, then by a linear functional we mean a K-linear map of L into K. The set of all such maps is called the dual of L over K, denoted by L^*, and it is itself a vector space over K. The trace function is easily seen to be a linear functional by showing that for any element $\alpha \in L$, $T(\alpha) \in K$. Notice that the trace function is additive and the norm function is multiplicative. Suppose the minimum polynomial $m(x)$ of $\alpha \in L\backslash K$ is of degree n. If τ is an arbitrary element of $G(L/K)$, then $m(\tau(\alpha)) = \tau(m(\alpha)) = 0$ and hence $\tau_i(\alpha)$ are all roots of the minimum polynomial. Thus the minimum polynomial can be written as

$$m(x) = \prod_{i=1}^{n} [x - \tau_i(\alpha)] = x^n - \left[\sum_{i=1}^{n} \tau_i(\alpha)\right] x^{n-1} + \cdots + (-1)^n \prod_{i=1}^{n} \tau_i(\alpha)$$

and the second and last coefficients with appropriate sign changes are the trace and norm, respectively. It is a simple calculation to observe that the coefficients of this polynomial are left invariant by elements of $G(L/K)$ and thus lie in K. We shall have some use for these functions in the investigation of finite fields.

From the foregoing we conclude that a finite extension of a finite field is always separable, normal, simple, algebraic, and perfect. It is indeed a well-behaved structure.

1.3 Fundamental Properties of Finite Fields

Much of the material in the previous section was given to simplify some of the proofs presented here and in later sections and to establish the proper algebraic setting in which to discuss finite fields. Many of the theorems discussed can be proved by more elementary methods. However, such a treatment tends to be less satisfying.

Theorem 3.1 Any finite field L with characteristic p has p^n elements for some positive integer n.

Proof Denote by K the prime subfield of L. The vector space of L over K is of some finite dimension, say n, and there exists a basis $\alpha_1, \ldots, \alpha_n$ of L over K. Since every element of L can be expressed as a linear combination of the α_i over K, i.e.,

$$\gamma = \sum_{i=1}^{n} \beta_i \alpha_i, \qquad \beta_i \in K$$

and since K has p elements, L must have p^n elements. $\square$

The uniqueness and existence can be deduced from the following theorem.

Theorem 3.2 Let L be a field with characteristic p and prime subfield K. Then L is the splitting field for $f(x) = x^{p^n} - x$ if L has p^n elements.

Proof Suppose L is the splitting field for $f(x) = x^{p^n} - x$ over K. Since $(f(x), f'(x)) = 1$, the roots of $f(x)$ are distinct and L has at least p^n elements. Consider the subset $E = \{\alpha \in L \,|\, \alpha^{p^n} = \alpha\}$ of L. Clearly E contains p^n elements since it consists of the roots of $f(x)$. Suppose $\alpha, \beta \in E$; then $(\alpha\beta)^{p^n} = \alpha^{p^n} \beta^{p^n} = \alpha\beta$ and hence $\alpha\beta \in E$. Also,

$$(\alpha + \beta)^{p^n} = \sum_i \binom{p^n}{i} \alpha^i \beta^{p^n - i} = \alpha^{p^n} + \beta^{p^n} = \alpha + \beta$$

since $p \,|\, \binom{p^n}{i}$ for $0 < i < p^n$, and hence $(\alpha + \beta) \in E$. The existence of additive and multiplicative inverses in E is trivial and hence E is a subfield of L and also a splitting field for $f(x)$. Thus $E = L$ and L contains p^n elements.

Suppose now that L contains p^n elements. The multiplicative group of L, for which we shall adopt the notation L^*, forms a group of order $p^n - 1$ and hence the order of any element in L^* divides $p^n - 1$. Thus $\alpha^{p^n} = \alpha$ for all $\alpha \in L^*$ and the relation is trivially true for $\alpha = 0$. Thus $f(x)$ splits in L. Notice that

$$f(x) = \prod_{\alpha \in L} (x - \alpha) \quad \square$$

Some important corollaries follow immediately from this theorem.

Corollary 1 Any two finite fields with p^n elements are isomorphic.

Proof Since $f(x) = x^{p^n} - x$ splits over any field with p^n elements and splitting fields are isomorphic (from Theorem 2.1), the result follows. □

We shall denote a field with p^n elements by $GF(p^n)$, the Galois field of order p^n. Where the prime p and integer n play no essential role or are understood, we shall often use the notation $GF(q)$.

Corollary 2 $GF(p^n)$ is the splitting field for $f(x) = x^{p^n} - x$ over $GF(p)$.

Proof This is a restatement of Theorem 3.2 and Corollary 1. □

Corollary 3 For any prime p and integer n, $GF(p^n)$ exists.

Proof By Theorem 2.1 the splitting field of $x^{p^n} - x$ over $GF(p)$ exists and by Corollary 2 it is $GF(p^n)$. □

The following important theorem will be used to establish, among other things, the subfield structure of $GF(p^n)$.

Theorem 3.3 $GF(p^n)^*$ is cyclic.

Proof The multiplicative group $GF(p^n)^*$ is, by definition, Abelian and of order $p^n - 1$. If $p^n - 1 = p_1^{e_1} \cdots p_k^{e_k}$, then, factoring $GF(p^n)^*$ into a direct product of its Sylow subgroups, we have

$$GF(p^n)^* = S(p_1) \times \cdots \times S(p_k)$$

where $S(p_i)$ is the Sylow subgroup of order $p_i^{e_i}$. The order of every element in $S(p_i)$ is a power of p_i and let $a_i \in S(p_i)$ have the maximal order, say $p_i^{e_i'}$, $e_i' \leq e_i$, for $i = 1, \ldots, k$. Since $(p_i, p_j) = 1$, $i \neq j$, the element $a = a_1 a_2 \cdots a_k$ has maximal order $m = p_1^{e_1'} \cdots p_k^{e_k'}$ in $GF(p^n)^*$. Furthermore, every element of $GF(p^n)^*$ satisfies the polynomial $x^m - 1$, implying that $m \geq p^n - 1$. Since $a \in GF(p^n)^*$ has order m, m divides $p^n - 1$ and hence $m = p^n - 1$. The element a, by construction, is a generator of $GF(p^n)^*$. □

To consider the subfield structure of $GF(p^n)$, it is convenient to have the following theorem and a few definitions.

Theorem 3.4 Over any field K, $(x^m - 1) | (x^n - 1)$ iff $m | n$.

1.3 Fundamental Properties of Finite Fields

Proof If $n = qm + r$, $r < m$, then by direct computation

$$x^n - 1 = x^r(\sum_i x^{im})(x^m - 1) + (x^r - 1)$$

It follows that $(x^m - 1)|(x^n - 1)$ iff $x^r - 1 = 0$, i.e., $r = 0$. □

Corollary For any prime integer p, $(p^m - 1)|(p^n - 1)$ iff $m|n$.

Proof The proof parallels that of Theorem 3.4 and is omitted. □

In any group G, if $g \in G$ is of order n, then g^k is of order $n/(n, k)$, a consequence of the fact that $kn/(k, n)$ is the least common multiple of k and n.

Definition The Euler totient function $\varphi(n)$ denotes the number of integers less than n and relatively prime to n.

Definition A primitive element of $GF(p^n)$ is a generator of $GF(p^n)^*$.

It follows from these two definitions that the number of primitive elements of $GF(p^n)$ is simply $\varphi(p^n - 1)$. The basic subfield structure of finite fields is contained in the following theorem.

Theorem 3.5 $GF(p^m) \subset GF(p^n)$ iff $m|n$.

Proof Suppose $GF(p^m) \subset GF(p^n)$; then $GF(p^n)$ may be interpreted as a vector space over $GF(p^m)$ with dimension, say, k. Hence $p^n = p^{km}$ and $m|n$. Now suppose $m|n$, which, from the previous theorem and its corollary, implies that $(x^{p^m-1} - 1)|(x^{p^n-1} - 1)$. Thus every zero of $x^{p^m} - x$ that is in $GF(p^m)$ is also a zero of $x^{p^n} - x$ and hence in $GF(p^n)$. It follows that $GF(p^m) \subset GF(p^n)$. Notice that there is precisely one subfield of $GF(p^n)$ of order p^m, since otherwise $x^{p^m} - x$ would have more than p^m roots. □

Example As an example of this theorem, the inclusion relationships of $GF(2^{12})$ can be demonstrated as shown in Fig. 1.

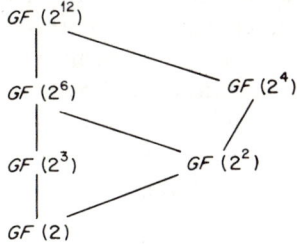

Fig. 1

When considering an arbitrary finite field, we shall often denote it by $GF(q)$, where $q = p^n$ for some prime p and integer n. An extension field of dimension k over $GF(q)$ will then be denoted $GF(q^k)$. It follows from the proof of the previous theorem that the elements of $GF(q^k)$ that are also in $GF(q)$ are precisely those that satisfy the equation $x^q = x$. This is often a useful characterization of a subfield.

Let us now examine the group of automorphisms of $GF(q^n)$ that fix $GF(q)$, i.e., the Galois group of the extension $GF(q^n)$ over $GF(q)$, $G(GF(q^n)/GF(q))$, which we shall denote by $G(q^n/q)$ for convenience.

Theorem 3.6 $G(q^n/q)$ is cyclic.

Proof Since any finite extension of $GF(q)$ is normal, it follows that $|G(q^n/q)| = n$. Consider the map $\tau: \alpha \to \alpha^q$ which takes $GF(q^n)$ into itself. It is readily checked that τ is an automorphism of $GF(q^n)$ that fixes $GF(q)$. The map τ^n must be the identity map i since $x^{q^n} = x$ for all $x \in GF(q^n)$. Furthermore, if $\tau^k = i$, for $k < n$, then $x^{q^k} = x$ for all $x \in GF(q^n)$, which is clearly impossible. Hence $G(q^n/q)$ is the set of maps $\{\tau^j, j = 1, \ldots, n\}$ and is cyclic. $\square$

Most of the fundamental properties of finite fields have been established. These will be used extensively in the rest of this chapter.

1.4 Vector Spaces over Finite Fields

Perhaps one of the first problems that should be considered in any discussion of vector spaces is the existence of certain types of bases, and once again the particular nature of finite fields provides some interesting examples of the more general theory.

The problem of proving the existence of an orthogonal basis, with respect to some form that is yet to be defined, does not depend on the finiteness of the field so much as on the characteristic of the field. A field of characteristic two must be treated separately. In light of this we give only an informal discussion of the problem and for the present will consider the general case of a vector space V over an arbitrary field K. If V is of dimension n over K, we shall sometimes refer to it as $V_n(K)$.

Definition A bilinear form f on V is a mapping from $V \times V$ to K such that

$$f(\alpha x + \beta y, z) = \alpha f(x, z) + \beta f(y, z)$$
$$f(x, \alpha y + \beta z) = \alpha f(x, y) + \beta f(x, z)$$

1.4 Vector Spaces over Finite Fields

for all $x, y, z \in V$ and $\alpha, \beta \in K$. If, in addition,

$$f(x, y) = f(y, x)$$

for all $x, y \in V$, then f is said to be symmetric.

The particular symmetric bilinear form of interest to us will be the usual inner product obtained by fixing a basis in V. Thus if $v_1, v_2, \ldots, v_n$ is a basis of V and

$$x = \sum \alpha_i v_i, \qquad y = \sum \beta_i v_i, \qquad \alpha_i, \beta_i \in K$$

then

$$f(x, y) = \sum \alpha_i \beta_i$$

If $v_1, v_2, \ldots, v_n$ is a fixed basis of V, then the matrix $F = (f(v_i, v_j))$ is called the matrix of the form f with respect to the basis $v_1, v_2, \ldots, v_n$. A change of basis will yield a matrix that is congruent to F. Thus the following definition is consistent.

Definition The bilinear form f is said to be degenerate if the rank of the matrix of f is less then n, the dimension of V. If the rank is n, f is said to be nondegenerate.

Two vectors $x, y \in V$ are said to be orthogonal with respect to f if $f(x,y) = 0$. If V_1 is a subspace of V, we define the subset $V_1^\perp$ to be

$$V_1^\perp = \{x \in V | f(x, y) = 0 \text{ for all } y \in V_1\}$$

It is easily seen to be a subspace. It is also not difficult to show that a symmetric bilinear form f on V is nondegenerate iff $V^\perp = \{0\}$.

For the case of $V = V_n(\mathbb{R})$, $\mathbb{R}$ the reals, if f is the usual inner product, we have that

$$n = \dim V = \dim V_1 + \dim V_1^\perp$$

where V_1 is any subspace of V. For such a vector space it is also true that

$$V = V_1 \oplus V_1^\perp$$

i.e., V is the direct sum of V_1 and $V_1^\perp$ and each element $v \in V$ can be written uniquely as $v = v_1 + v_2$, $v_1 \in V_1$, $v_2 \in V_1^\perp$. For the general case, if f is a nondegenerate bilinear form on V, then, again,

$$\dim V = \dim V_1 + \dim V_1^\perp$$

If f is a nondegenerate form on V, then its restriction to a subspace V_1 may or may not be degenerate. In order for V to be the direct sum of a subspace V_1 and $V_1^\perp$, it is sufficient that the symmetric bilinear form f be nondegenerate on both V and V_1.

As a simple example of these ideas, let V be the space of 5-tuples over $GF(2)$ and let f be the usual inner product. If V_1 is the subspace generated by the vectors $(1, 0, 0, 0, 0)$ and $(0, 1, 0, 0, 0)$, then f is nondegenerate on both V and V_1 and $V_1^\perp$ is spanned by $(0, 0, 1, 0, 0)$, $(0, 0, 0, 1, 0)$, and $(0, 0, 0, 0, 1)$ and $V = V_1 \oplus V_1^\perp$. If V_2 is the subspace spanned by $(1, 1, 0, 0, 0)$ and $(0, 0, 1, 1, 0)$, then $V_2^\perp$ is spanned by $(1, 1, 0, 0, 0)$, $(0, 0, 1, 1, 0)$, and $(0, 0, 0, 0, 1)$ and $V_2 \subset V_2^\perp$. The form f on the subspace V_2 is degenerate and, although $\dim V = \dim V_2 + \dim V_2^\perp$, it is not true that $V = V_2 \oplus V_2^\perp$.

If the bilinear form f on V is such that $f(x, x) = 0$ for all $x \in V$, we call f an *alternating form*. If V is over a field K such that char $K \neq 2$, then f is alternating iff it is skew symmetric, i.e.,

$$f(x + y, x + y) = f(x, x) + f(y, y) + f(x, y) + f(y, x)$$

and hence

$$f(x, y) = -f(y, x)$$

for any $x, y \in V$. On the other hand, if f is skew symmetric, then $f(x, y) = -f(y, x)$ for any $x, y \in V$. In particular if $x = y$, we have $2f(x, x) = 0$, which implies $f(x, x) = 0$ for any $x \in V$. Thus the case of char $K = 2$ is special and is usually treated separately. However, the result that we want, which we state without proof, is valid for either case.

Theorem 4.1 Let V be a nondegenerate vector space with respect to the symmetric, bilinear, nonalternating form f. Then V possesses an orthogonal basis.

If $v_1, v_2, \ldots, v_n$ is an orthogonal basis for $V_n(K)$, then the matrix of the form f is diagonal. The existence of an orthonormal basis then depends on whether or not the elements on the diagonal are squares in K. Thus for vector spaces over finite fields we are always assured of an orthogonal basis if not an orthonormal one.

For some purposes bases other than orthogonal may be of interest. One such basis has already been used, namely, the basis $1, \alpha, \alpha^2, \ldots, \alpha^{n-1}$ for $GF(q^n)$ over $GF(q)$, where $\alpha \in GF(q^n)$ has a minimal polynomial of degree n. For the remainder of the section we confine our attention to the vector space of $GF(q^n)$ over $GF(q)$.

Recall from Theorem 3.6 that the Galois group of $GF(q^n)$ over $GF(q)$, $G(q^n/q)$, is cyclic and is generated by the map $\tau: x \mapsto x^q$. If there exists an element $u \in GF(q^n)$ such that

$$u, \tau(u), \tau^2(u), \ldots, \tau^{n-1}(u), \qquad \tau^i(u) = u^{q^i}$$

is a basis for $GF(q^n)$ over $GF(q)$, then it is said to be a normal basis. We now show that a normal basis always exists for $GF(q^n)$ over $GF(q)$. The automorphisms τ^i, $i = 0, 1, \ldots, n - 1$ are linearly independent, i.e., if $\sum_{i=0}^{n-1} \alpha_i \tau^i(\beta) = 0$ for all $\beta \in GF(q^n)$, then $\alpha_i = 0$ for all i. However $\tau^n = i$, the identity map, implying that the minimal polynomial of τ as a linear transformation of $GF(q^n)$ is $x^n - 1$. However, since the characteristic polynomial of any linear transformation of $GF(q^n)$ over $GF(q)$ is of degree n, we must have that the characteristic polynomial of τ is also $x^n - 1$. Thus τ can have only the one nontrivial similarity invariant and its rational canonical form must be the companion matrix of $x^n - 1$, or

$$\begin{bmatrix} 0 & 1 & 0 & 0 & \cdots & 0 \\ 0 & 0 & 1 & 0 & \cdots & 0 \\ \vdots & \vdots & \vdots & \vdots & & \vdots \\ 0 & 0 & 0 & 0 & \cdots & 1 \\ 1 & 0 & 0 & 0 & \cdots & 0 \end{bmatrix}$$

implying that there exists a basis of $GF(q^n)$ over $GF(q)$ for which τ is represented by this matrix. Suppose $v_1, v_2, \ldots, v_n$ is the given basis; then the action of τ on this basis is $\tau(v_i) = v_{i+1}$, $i = 1, \ldots, n - 1$, and $\tau(v_n) = v_1$. Thus $v_i = \tau^i(v_1)$, $i = 0, 1, \ldots, n - 1$, and v_1 is the desired element which gives the desired normal basis. Note that the basis can also be written as $v_1, v_1^q, \ldots, v_1^{q^{n-1}}$. It has been shown that for $GF(p^n)$ over $GF(p)$ there are at least $(p - 1)^n$ elements u of $GF(p^n)$ such that $u, u^p, \ldots, u^{p^{n-1}}$ is a normal basis. Futhermore, the element u may always be taken to be primitive, in which case we call it a primitive normal basis. It should be mentioned that the notion of a normal basis of an extension field over a given field arises for normal extensions of infinite fields also.

We shall have need of another type of basis, called a dual basis, which is related to any given basis, say $v_1, \ldots, v_n$ of $GF(q^n)$ over $GF(q)$. We first consider the trace function more carefully and recall that the trace of any element $\alpha \in GF(q^n)$ over $GF(q)$ is

$$\text{tr } \alpha = \alpha + \alpha^q + \alpha^{q^2} + \cdots + \alpha^{q^{n-1}}$$

This function may be viewed as a linear map from $GF(q^n)$ to $GF(q)$, i.e., a linear functional, since it is clear that

$$\text{tr}(\alpha + \beta) = \text{tr } \alpha + \text{tr } \beta, \qquad \alpha, \beta \in GF(q^n)$$
$$\text{tr}(\gamma\alpha) = \gamma \text{ tr } \alpha, \qquad \gamma \in GF(q), \quad \alpha \in GF(q^n)$$

and that $\text{tr } \alpha \in GF(q)$ for any $\alpha \in GF(q^n)$. But now a linear functional from $GF(q^n)$ to $GF(q)$ is completely determined by its action on a basis of $GF(q^n)$ over $GF(q)$. Since there are n basis elements and each element may be mapped into any of the q elements of $GF(q)$, there are preciseley q^n distinct linear functions of $GF(q^n)$ into $GF(q)$. Now choose a fixed element $\eta \in GF(q^n)$ and consider the linear function that takes $\alpha \in GF(q^n)$ to $\text{tr}(\eta\alpha) \in GF(q)$ for any $\alpha \in GF(q^n)$. This is clearly a linear function. Moreover, if $\text{tr}(\eta\alpha) = \text{tr}(\delta\alpha)$, $\delta \in GF(q^n)$, for all $\alpha \in GF(q^n)$, then $\text{tr}[(\eta - \delta)\alpha] = 0$, which implies that $\eta - \delta = 0$ and hence the q^n trace functions are distinct. We conclude that any linear map of $GF(q^n)$ into $GF(q)$ can be taken as a trace function.

Now let $\alpha_1, \alpha_2, \ldots, \alpha_n$ be n elements of $GF(q^n)$. Suppose these elements are linearly dependent over $GF(q)$. Then the subspace spanned by them, say V_1, is a proper subspace. This implies that we can construct a nontrivial linear map f such that $f(\alpha_i) = 0$, $i = 1, \ldots, n$, but that there exist elements $\alpha \in V \setminus V_1$ such that $f(\alpha) \neq 0$, i.e., there exists an $\eta \in GF(q^n)$ such that $\eta \neq 0$ and $f(\alpha) = \text{tr}(\eta\alpha)$ such that $\text{tr}(\eta\alpha_i) = 0$, $i = 1, \ldots, n$. Conversely, suppose that $\text{tr}(\eta\alpha_i) = 0$, $i = 1, \ldots, n$. Then $\text{tr}(\eta\alpha) = 0$ for $\alpha \in V_1$, V_1 the subspace spanned by α_i, $i = 1, \ldots, n$. Since the map $\text{tr}(\eta \cdot)$ is assumed nontrivial, the subspace V_1 is proper and thus the α_i are dependent.

Expanding the trace function, this condition yields the set of equations

$$\sum_{j=0}^{n-1} (\eta\alpha_i)^{q^j} = 0$$

which has a solution for the vector $(\eta, \eta^q, \ldots, \eta^{q^{n-1}})$ and hence for η iff the matrix

$$\begin{bmatrix} \alpha_1 & \alpha_1^q & \cdots & \alpha_1^{q^{n-1}} \\ \alpha_2 & \alpha_2^q & \cdots & \alpha_2^{q^{n-1}} \\ \vdots & \vdots & & \vdots \\ \alpha_n & \alpha_n^q & \cdots & \alpha_n^{q^{n-1}} \end{bmatrix}$$

is singular. This is perhaps the more familiar condition.

We return now to the problem of constructing a dual basis for the given basis $\alpha_1, \ldots, \alpha_n$, a concept that we have yet to define. If ε is an arbitrary element of $GF(q^n)$, then

$$\varepsilon = \sum_{i=1}^{n} a_i \alpha_i, \qquad a_i \in GF(q)$$

and the map or projection

$$f(\varepsilon) = a_i$$

for some fixed i is a linear function from $GF(q^n)$ to $GF(q)$. Thus there exist elements $\beta_1, \beta_2, \ldots, \beta_n \in GF(q^n)$ such that

$$a_i = \text{tr}(\beta_i \varepsilon), \qquad i = 1, \ldots, n$$

1.5 Linear Codes

In particular, choosing $\varepsilon = \alpha_j$, we have that

$$\delta_{ij} = \operatorname{tr}(\beta_i \alpha_j)$$

where δ_{ij} is the Kronecker delta function. Expressing this equation in matrix form gives

$$\begin{bmatrix} \beta_1 & \beta_1^q & \cdots & \beta_1^{q^{n-1}} \\ \beta_2 & \beta_2^q & \cdots & \beta_2^{q^{n-1}} \\ \vdots & \vdots & & \vdots \\ \beta_n & \beta_n^q & \cdots & \beta_n^{q^{n-1}} \end{bmatrix} \begin{bmatrix} \alpha_1 & \alpha_2 & \cdots & \alpha_n \\ \alpha_1^q & \alpha_2^q & \cdots & \alpha_n^q \\ \vdots & \vdots & & \vdots \\ \alpha_1^{q^{n-1}} & \alpha_2^{q^{n-1}} & \cdots & \alpha_n^{q^{n-1}} \end{bmatrix} = \begin{bmatrix} 1 & & & 0 \\ & 1 & & \\ & & \ddots & \\ 0 & & & 1 \end{bmatrix}$$

Thus the β matrix is nonsingular, and the elements $\beta_1, \beta_2, \ldots, \beta_n$ are independent and form the dual basis of $\alpha_1, \alpha_2, \ldots, \alpha_n$. From the form of the trace function and the fact that $\alpha^{q^n} = \alpha$ for any $\alpha \in GF(q^n)$ it is readily seen that if $\alpha_1, \alpha_2, \ldots, \alpha_n$ is a normal basis, i.e., $\alpha_i = \alpha_1^{q^{i-1}}$, then $\beta_1, \beta_2, \ldots, \beta_n$ is also a normal basis, i.e., $\beta_i = \beta_1^{q^{i-1}}$.

We consider briefly the problem of computations in a finite field. Essentially the problem is to determine a suitable representation for the elements of a finite field, so that addition and multiplication are as simple as possible. If x is a primitive element in $GF(q^n)$, then, if any two nonzero elements are expressed as powers of x, say x^i and x^j, multiplication becomes trivial. On the other hand, for addition we have $x^i + x^j = x^i(1 + x^{j-i})$ and we require a tabulation of the elements $1 + x^k$. If one represents the elements of $GF(q^n)$ as n-tuples over $GF(q)$ with respect to some basis, then addition is quite trivial, while multiplication tends to be cumbersome.

Suppose we define a function $z(i)$ on the integers i, $0 \leq i \leq q^n - 2$, such that if x is a primitive element of $GF(q^n)$,

$$x^j + 1 = x^{z(j)}$$

With the aid of this function, all arithmetic operations in $GF(q^n)$ become simple. Similar functions can be defined for the vector approach.

1.5 Linear Codes

As in Section 1.4, we denote by $V_n(q)$ a vector space of dimension n over $GF(q)$. We fix a basis in $V_n(q)$ and identify an element of $V_n(q)$ by the n-tuple of its components in the basis.

Definition A linear block code over $GF(q)$ of block length n and dimension k is a k-dimensional subspace of $V_n(q)$. It will be denoted as an (n, k) code.

The term block code refers to the fact that all codewords have the same length, i.e., are n-tuples. The word linear follows since the codewords form a subspace. Very often these codes are referred to as group codes since a subspace forms an Abelian subgroup. This notation is less than desirable, however, since over a nonprime field we can have a collection of vectors that forms an additive subgroup but that does not form a subspace, e.g., if $z \in GF(3^2)\backslash GF(3)$, then the set of vectors $\{(0, 0), (z, 1), (2z, 2)\}$ forms a cyclic additive Abelian group but not a subspace.

Definition The Hamming distance $d(u, v)$ between two n-tuples u and v in $V_n(q)$ will be the number of coordinate places in which they differ. The Hamming weight $\omega(u)$ of the n-tuple $u \in V_n(q)$ is the number of nonzero coordinates of u.

Most work in coding theory is concerned with the Hamming metric and we confine our attention exclusively to this case. Another metric for which some results are available is the Lee metric. The Lee weight of a codeword $u = (x_1, \ldots, x_n)$ is defined as the rational sum of the Lee weights $\omega_L(x_i)$ of its components and

$$\omega_L(x_i) = |x_i|$$

and

$$|x_i| = \begin{cases} x_i & \text{if } 0 \leqslant x_i \leqslant \tfrac{1}{2}p \\ -x_i \pmod{p} & \text{if } \tfrac{1}{2}p < x_i \leqslant p - 1 \end{cases}$$

where the components are interpreted as integers mod p, where p is a prime. It is not hard to verify that this weight function can be used to define a metric. Although in certain physical applications the Lee metric may be a more natural one to use, we shall not mention it further.

In the transmission of a codeword over some medium, errors may occur, where by an error we mean any change in a coordinate of the transmitted n-tuple. The point of coding is that under certain conditions these errors can be corrected. We examine the situation further.

Definition By a sphere $S_r(x)$ of radius r about $x \in V_n(q)$, where r is some nonnegative integer, is meant the set of vectors of $V_n(q)$ of distance r or less from x, i.e.,

$$S_r(x) = \{y \in V_n(q): \ d(x, y) \leqslant r\}$$

In a decoding operation we can elect either to correct or detect errors, where detection implies that we are aware that an error has been made in transmission but that we cannot correct it. When we say that a code is capable

1.5 Linear Codes

of correcting t errors and detecting $t + s$, $s > 0$, errors, we mean that when t or fewer errors are made in transmission, the structure of the code is such that it will permit the correction of these errors, while if $t + j$, $0 < j \leqslant s$, errors occur, we will be able to detect this fact without being able to correct it.

Definition The minimum distance d of a block code $\mathscr{C}$ is

$$d = \min_{\substack{u, v \in \mathscr{C} \\ u \neq v}} d(u, v)$$

Perhaps the single most important reason for considering linear codes is that the minimum distance of such codes is the weight of the minimum weight codeword, which follows since

$$d(u, v) = d(u - v, 0) = \omega(u - v)$$

and, from linearity, $u - v$ is in the code if u and v are in the code. As a matter of notation, when we want to exhibit the minimum distance (or simply distance) of an (n, k) code explicitly we will refer to it as an (n, k, d) code. The field over which the code is defined will be understood. If the code happens to be nonlinear and has M codewords, we refer to it as an (n, M, d) code, where now d is the minimum distance between any two distinct codewords. If M is written as an integer, then it will be clear from the context whether it refers to a dimension or the actual number of codewords. In order to decode a transmitted codeword, suppose we adopt the rule to choose that codeword closest to the received word as the transmitted codeword. For a large class of channels this results in what is called a maximum likelihood decoder, which minimizes the probability of error if each codeword is equally likely to be transmitted. We can now show the following theorem.

Theorem 5.1 A block code with distance d is capable of correcting all patterns of t or fewer errors and detecting all patterns of $t + j$, $0 < j \leqslant s$, errors if $2t + s < d$.

Proof Assume the block code is a subset (not necessarily a subspace) of vectors from $V_n(q)$. We use the "closest codeword" (also called nearest neighbor) decoding rule described earlier and notice that any received codeword with t or fewer errors will lie in the sphere of radius t about the transmitted codeword. Thus, in order to correct t errors, we merely have to verify that the spheres of radius t about the codewords are disjoint. Suppose that they are not and that $u \in V_n(q)$ lies in $S_t(x_1)$ and $S_t(x_2)$, where x_1 and x_2 are codewords. Then

$$d(x_1, x_2) \leqslant d(x_1, u) + d(u, x_2) \leqslant 2t$$

the first inequality due to the triangle inequality of metrics. This contradicts the fact that the code dictionary was assumed to have distance $d > 2t + s$, $s > 0$. In order to detect $t + j$ errors, $0 < j \leqslant s$, it is sufficient to show we can detect $t + s$ errors. To achieve this, we show that a received word with $t + s$ errors will not lie in any sphere of radius t about a codeword and hence be corrected in error. By the same argument as before, this cannot happen if $d > 2t + s$. □

It is possible to have a situation where the vectors lying within and on spheres of radius t about the codewords of a linear (n, k) code account for all vectors in $V_n(q)$. Such a code is called a perfect t-error-correcting code. It has recently been established (Tietäväinen, 1971, 1973) that all perfect codes are known. All of these will be encountered in later sections and the next chapter.

Since an (n, k) code, which at times it is convenient to denote by a script letter, say $\mathscr{C}$, is a vector subspace of $V_n(q)$, we can choose a set of basis vectors $v_1, v_2, \ldots, v_k$ and any linear combination

$$\sum_{i=1}^{k} \alpha_i v_i, \qquad \alpha_i \in GF(q)$$

will be a code vector. It is thus convenient to think of the code as the row space of a generator matrix G whose rows are the basis $v_1, v_2, \ldots, v_k$. Since elementary row operations leave the code space invariant, we may also choose a basis of the subspace, such that the corresponding generator matrix G' is of the form

$$G' = [I_k \vdots A]$$

where I_k is the $k \times k$ identity matrix and A is some $k \times (n - k)$ matrix over $GF(q)$. The matrix G' is just the row reduced form of G. Choosing this basis for the code is attractive since it leads to the following encoding situation. If $\alpha = (\alpha_1, \alpha_2, \ldots, \alpha_k)$ is a row vector of k information symbols, then we can identify the codeword $\alpha G'$ with α. In other words, we map the information sequence α into the codeword $\alpha G'$. However, the first k symbols of $\alpha G'$ are the information symbols, and the remaining $n - k$ symbols are certain $GF(q)$ sums on the first k symbols determined by the matrix A. Such a code is called a systematic code.

The orthogonal complement of the (n, k) code $\mathscr{C}$ is of dimension $n - k$, as discussed in Section 1.4, and is spanned by the vectors $u_1, u_2, \ldots, u_{n-k}$, say, where every vector of the code $\mathscr{C}$ is orthogonal to each u_i, $i = 1, \ldots, n - k$. As a subspace of $V_n(q)$, the orthogonal complement, denoted by $\mathscr{C}^\perp$, is also a code and is called the dual code of $\mathscr{C}$. It is clearly an $(n, n - k)$ code and its generating matrix H can always be written in the form

$$H' = [B \vdots I_{n-k}]$$

1.5 Linear Codes

where B is an $(n - k) \times k$ matrix over $GF(q)$. The orthogonality of the codes $\mathscr{C}$ and $\mathscr{C}^\perp$ can be expressed by the fact that if G is any generating matrix of $\mathscr{C}$ and H any generating matrix of $\mathscr{C}^\perp$, then

$$GH^T = 0$$

Thus, for the two particular forms G' and H' it follows that the matrices A and B must satisfy the relationship $B = -A^T$. It will be convenient at times to think of the original (n, k) code $\mathscr{C}$ as the null space of a matrix H^T, i.e., an element $x \in V_n(q)$ is in $\mathscr{C}$ iff

$$xH^T = 0$$

A code defined in this manner is often referred to as a parity check code.

Example Let $\mathscr{C}$ be a binary linear (n, k) code. To show that either the weight of every codeword is even, or else half of them have even weight and the other half odd weight, consider adding a row of all ones to the parity check matrix H of $\mathscr{C}$, giving

$$H' = \begin{bmatrix} H \\ 1 \quad 1 \quad \cdots \quad 1 \end{bmatrix}$$

If the last row is linearly independent of the first $n - k$ rows, then the null space of H', say $\mathscr{C}'$, consists of the codewords of even weight in $\mathscr{C}$ and the dimension of $\mathscr{C}'$ is $k - 1$, i.e., half the codewords of $\mathscr{C}$ have even weight. If the last row is linearly dependent on the first $n - k$ rows, then every word of $\mathscr{C}$ has even weight.

We shall call two codes $\mathscr{C}_1$ and $\mathscr{C}_2$ equivalent if the words of $\mathscr{C}_1$ can be obtained from the words of $\mathscr{C}_2$ by applying a fixed permutation to the coordinate places of all words in $\mathscr{C}_2$. If $\mathscr{C}_1$ and $\mathscr{C}_2$ are (n, k) codes over $GF(q)$, and D_1 and D_2 are $q^k \times n$ matrices formed by using the q^k codewords of $\mathscr{C}_1$ and $\mathscr{C}_2$ as rows, then $\mathscr{C}_1$ and $\mathscr{C}_2$ are equivalent codes if and only if D_1 can be obtained from D_2 by appropriate permutation of rows and columns of D_2. It is clear in a physical sense that this concept can be expanded upon. Suppose D is the matrix of codewords of an (n, k) code $\mathscr{C}$. Then apart from permuting the columns of D, we can multiply any column of D by a nonzero field element and even interchange the nonzero field elements by some permutation of them, and we will not affect the error-correcting capabilities of the code. Now suppose that G_1 and G_2 are the generator matrices of the two codes $\mathscr{C}_1$ and $\mathscr{C}_2$. Then (MacWilliams, 1961) we can call $\mathscr{C}_1$ and $\mathscr{C}_2$ equivalent iff

$$G_2 = A\phi(G_1)\Pi\Lambda$$

where (i) A is a $k \times k$ nonsingular matrix (which preserves the subspace); (ii) ϕ is an automorphism of the field $GF(q)$ and $G_1 = (g'_{ij})$; then $\phi(G_1) = (\phi(g'_{ij}))$; (iii) Π is an $n \times n$ permutation matrix (permutes the columns); (iv) Λ is an $n \times n$ nonsingular diagonal matrix (multiplies columns by nonzero field elements).

Although this definition of the equivalence of codes is natural, we will, in general, reserve equivalence to mean a permutation of coordinate places unless specified otherwise. It is clear that equivalent codes have precisely the same distance structure, but codes that have the same distance structure may not be equivalent. The matrix H is the parity check matrix of the code $\mathscr{C}$ and the generator matrix of the code $\mathscr{C}^\perp$. It is a simple matter to show that for every linear code there is an equivalent parity check code and conversely, and also that the classes of linear systematic and parity check codes coincide. It is, of course, possible to have systematic nonlinear codes.

The equation

$$xH^T = 0$$

defines a set of parity check equations for the code $\mathscr{C}$. It has been assumed that both x and the rows of H are n-tuples of $V_n(q)$. However, the columns of the $(n-k) \times n$ matrix H are $(n-k)$-tuples over $GF(q)$ and hence we can identify a column of H with an element of $GF(q^{n-k})$ to obtain a $1 \times n$ matrix H' over $GF(q^{n-k})$. The solutions in $V_n(q)$ to

$$xH'^T = 0$$

are again the code $\mathscr{C}$. Similarly, if s is a divisor of $n-k$, then we can identify a column of H with an $[(n-k)/s]$-tuple over $GF(q^s)$. This formulation of the parity check matrix is sometimes useful and convenient.

Let $\mathscr{C}$ be a given (n, k) code in $V_n(q)$, which we treat as a vector subspace. The factor space $V_n(q)/\mathscr{C} = \{x_i + \mathscr{C}\}$, where the x_i are coset representatives, has significance in decoding. We list the n-tuples of cosets in rows of an array, which Slepian (1956) calls the standard array, and which we denote by $S = (s_{ij})$, where each s_{ij} is an n-tuple over $GF(q)$. The first row of S contains the elements of $\mathscr{C}$ and for standardization we assume that s_{11} is the zero vector. The elements of the first column of S are a set of coset representatives and we assume that each representative is a minimal weight vector of its coset. These are, of course, not in general unique. In coding theory such a set of vectors is termed a set of coset leaders. We denote a set of coset leaders by $\{x_1 = 0, x_2, \ldots, x_m\}$ with $m = q^{n-k}$ and the code vectors by $\{c_1 = 0, c_2, \ldots, c_{q^k}\}$. Thus the element s_{ij} of the standard array is given by $s_{ij} = x_i + c_j$. A simple decoding rule that is maximum likelihood for a large class of channels would be to assume that if a received word x is $x_i + c_j$, then the trans-

1.5 Linear Codes

mitted codeword was c_j. It is easy to see that if all possible error patterns with t or fewer nonzero coordinate places over $GF(q)$ occur as coset leaders, then this decoding method will correct all patterns of t or fewer errors. If more than t errors occur, then errors may occur in the decoding. Essentially the coset leaders represent correctable error patterns. Clearly, if the minimal weight n-tuple in a coset is not unique and is of weight t, then the code cannot correct all error patterns with t errors. If x is the received vector, we call the quantity xH^T the syndrome of x. Two elements have the same syndrome if and only if they belong to the same coset. If the syndromes of the coset leaders are known, then the table lookup method of decoding may be simplified as follows. For the received word x, calculate the syndrome S. Find the coset leader x_i with the same syndrome. Then $x - x_i$ is the codeword assumed to have been transmitted. It is easily shown that this standard array decoding is just minimum distance decoding.

From the foregoing discussion it is clear that the concept of a perfect t-error-correcting code introduced earlier can be interpreted as one for which the coset leaders are unique and contain exactly all those vectors of weight t or less. Since there are few classes of such codes, the weaker concept of a quasiperfect code is sometimes of interest. Peterson (1961) defines this as a code that, for some t, has all vectors of weight t or less appearing as coset leaders, some vectors of weight $t + 1$ appearing as coset leaders, and no vectors of weight greater than $t + 1$ appearing as coset leaders.

If we are given an (n, k) linear code over $GF(q)$ with parity check matrix H and generator matrix G, there are many ways G and H can be modified to yield other codes. We will mention only two such methods, which will be of use to us. Suppose that to H we add a row of all ones to form H_1. This has the effect of demanding that all codewords in the new code have coordinates that sum to zero. Thus if the all-ones row is linearly dependent on the other rows of H_1, then this will leave the code unchanged. If it is independent, however, then only those codewords in the original code whose coordinates sum to zero will be in the new code. Any process that removes codewords from a code will be referred to as expurgation.

The only other process of modifying a code that we will be interested in is that of adding an overall parity check, i.e., to each codeword in the (n, k) code $\mathscr{C}$ over $GF(q)$ we add an $(n + 1)$th coordinate, which is the negative of the sum of the first n coordinates. This yields an $(n + 1, k)$ code, often called the extended code of $\mathscr{C}$, whose parity check matrix H' can be obtained from H by first adding a column of zeros to H and then a row of all ones. This gives an $(n - k + 1) \times (n + 1)$ matrix if H is defined over $GF(q)$ and is of rank $n - k + 1$. Many other operations of adding or deleting information or check bits may be carried out, but we prefer to avoid a discussion of these since it is very difficult to say much of a general nature on these operations.

Suppose now that we are given two codes, $\mathscr{C}_1$, an (n_1, k_1) code over $GF(q)$ with generator matrix G_1, and $\mathscr{C}_2$, an (n_2, k_2) code also over $GF(q)$ with generator matrix G_2. There are various ways of constructing a third code from these two given codes. For example, one could form a generator matrix $G = G_1 \oplus G_2$, the direct sum of G_1 and G_2, to give an $(n_1 + n_2, k_1 + k_2)$ code with minimum distance equal to the minimum of the minimum distances of codes $\mathscr{C}_1$ and $\mathscr{C}_2$. A more interesting construction is to form the Kronecker or direct product $G' = G_1 \otimes G_2$ of G_1 and G_2, which is readily seen to give an $(n_1 n_2, k_1 k_2)$ code $\mathscr{C} = \mathscr{C}_1 \otimes \mathscr{C}_2$ with minimum distance equal to the product of the minimum distances of codes $\mathscr{C}_1$ and $\mathscr{C}_2$. An equivalent formulation of a Kronecker product code $\mathscr{C} = \mathscr{C}_1 \otimes \mathscr{C}_2$ is to describe each codeword as an $n_1 \times n_2$ matrix such that each column is a codeword of $\mathscr{C}_1$ and each row is a codeword of $\mathscr{C}_2$. The set of all such matrices is the code $\mathscr{C}$.

The parity check matrix method of defining a linear code allows the basic problem of coding theory to be posed in a slightly different manner. Specifically, let H be an $r \times n$ matrix over $GF(q^s)$ and define the code $\mathscr{C}$ to be

$$\mathscr{C} = \{x \in V_n(q) \mid xH^{\mathrm{T}} = 0\}$$

If W is the null space of H over $GF(q^s)$, then $\mathscr{C} = W \cap V_n(q)$. For a given code distance d we want $|\mathscr{C}|$ to be as large as possible or, conversely, for a given $|\mathscr{C}|$ we want d to be as large as possible. In general, because of the method of specifying H and the different fields being used, it is very difficult to obtain exact expressions for either d or $|\mathscr{C}|$. This situation will become clear when we consider the class of BCH codes in Section 1.7. The problem of coding then may be stated as determining a method of specifying H [over $GF(q^s)$] for general parameters r, n, q, s such that the dimension of the null space of H in $V_n(q)$ and the code distance can be bounded below. The code distance corresponds to the minimum number of columns of H that are linearly dependent over $GF(q)$.

Until recently, by far the most successful solution to this problem was in the construction of BCH codes (Section 1.7). However, a recent discovery of Goppa (1971) has led to the following approach. We consider only the binary codes, although the approach is easily extended to arbitrary $GF(q)$. Let $L = \{\alpha_1, \ldots, \alpha_n\}$ be a set of elements from $GF(2^m)$ and for convenience we assume the elements are distinct. If $x = (a_1, \ldots, a_n) \in V_n(2)$, then we define the rational function over $GF(2^m)$

$$R_x(z) = \sum_{i=1}^{n} \frac{a_i}{z - \alpha_i}$$

Let $g(z)$ be a polynomial over $GF(2^m)$ and assume that it does not have roots in L.

1.5 Linear Codes

Definition The (L, g) Goppa code is the set

$$\mathscr{C} = \{x \in V_n(2) \mid R_x(z) \equiv 0 \pmod{g(z)}\}$$

We call $g(z)$ the generating polynomial of the code, although this term will be used in quite a different sense when we come to cyclic codes. Let the vector $x \in V_n(2)$ have nonzero elements only in positions $i_1, i_2, \ldots, i_k$ and, corresponding to x, define the polynomial

$$f(z) = \prod_{j=1}^{k} (z - \alpha_{i_j})$$

It follows that we have $R_x(z) = f'(z)/f(z)$, where the prime indicates the formal derivative operation, and all arithmetic is in the algebra $GF(2^m)[z]/(g(z))$. Clearly we have that $f'(z) \equiv 0 \bmod g(z)$. If $f(z) = \sum_i f_i z^i$, then $f'(z) = \sum_i i f_i z^{i-1}$ and $if_i = 0$ if i is even. Thus no odd powers of z are present in $f'(z)$, implying that it is a perfect square. We conclude that $f'(z) \equiv 0 \bmod g^2(z)$. Since the degree of $f(z)$ is the number of nonzero components of the corresponding code vector, the code distance d is such that $d \geq 2 \deg(g) + 1$.

To find the parity check matrix of this code, recall that $x = (a_1, a_2, \ldots, a_n) \in V_n(2)$ is a code vector iff

$$R_x(z) = \sum_{i=1}^{n} \frac{a_i}{z - \alpha_i} \equiv 0 \bmod g(z)$$

In $GF(2^m)[z]/(g(z))$ the inverse element of $z - \alpha_i$ is, by direct calculation,

$$(z - \alpha_i)^{-1} \equiv \frac{g(z) - g(\alpha_i)}{z - \alpha_i} g^{-1}(\alpha_i) \bmod g(z)$$

$$\equiv g(\alpha_i)^{-1} b_r z^{r-1} + g(\alpha_i)^{-1}(b_{r-1} + \alpha_i b_r) z^{r-2}$$
$$+ g(\alpha_i)^{-1}(b_{r-2} + \alpha_i b_{r-1} + \alpha_i^2 b_r) z^{r-3}$$
$$+ \cdots + g(\alpha_i)^{-1}(b_1 + \cdots + \alpha_i^{r-1} b_r) \tag{5.1}$$

where $g(z) = \sum_{i=0}^{r} b_i z^i$. Thus $x \in V_n(2)$ is a code vector iff $xH^T = 0$, where

$$H = ((z - \alpha_1)^{-1}, (z - \alpha_2)^{-1}, \ldots, (z - \alpha_n)^{-1})$$

where, again, inverses are in the algebra of polynomials mod $g(z)$. Replacing the inverse elements by the polynomial, in Eq. (5.1) and considering like powers of z, we have $x \in \mathscr{C}$ iff $xH^T = 0$, where

$$H = \begin{bmatrix} b_r g(\alpha_1)^{-1} & \cdots & b_r g(\alpha_n)^{-1} \\ (b_{r-1} + b_r \alpha_1) g(\alpha_1)^{-1} & \cdots & (b_{r-1} + b_r \alpha_n) g(\alpha_n)^{-1} \\ \vdots & & \vdots \\ (b_1 + b_2 \alpha_1 + \cdots + b_r \alpha_1^{r-1}) g(\alpha_1)^{-1} & \cdots & (b_1 + \cdots + b_r \alpha_n^{r-1}) g(\alpha_n)^{-1} \end{bmatrix}$$

The rows of this matrix span the same space as the rows of the matrix

$$H' = \begin{bmatrix} g^{-1}(\alpha_1) & \cdots & g^{-1}(\alpha_n) \\ \alpha_1 g^{-1}(\alpha_1) & \cdots & \alpha_n g^{-1}(\alpha_n) \\ \vdots & & \vdots \\ \alpha_1^{r-1} g^{-1}(\alpha_1) & \cdots & \alpha_n^{r-1} g^{-1}(\alpha_n) \end{bmatrix}$$

$$= \begin{bmatrix} 1 & \cdots & 1 \\ \alpha_1 & \cdots & \alpha_n \\ \vdots & & \vdots \\ \alpha_1^{r-1} & \cdots & \alpha_n^{r-1} \end{bmatrix} \begin{bmatrix} g^{-1}(\alpha_1) & \cdots & & 0 \\ & \ddots & & \\ \vdots & & \ddots & \vdots \\ 0 & & \cdots & g^{-1}(\alpha_n) \end{bmatrix} \quad (5.2)$$

Since $\alpha_i, g^{-1}(\alpha_i) \in GF(2^m)$, we can expand these elements into binary m-tuples and observe that the dimension of the null space of H, i.e., the dimension of the code, is $n - mr = n - m \deg g(z)$. Thus an (L, g) code has length $|L|$, dimension $\geq [n - m \deg g(z)]$, and distance $d \geq 2 \deg g(z) + 1$ if $g(z)$ has distinct roots. There will doubtless be much investigation of these codes.

The minimum distance of a linear code is the weight of the minimum weight codeword, and this is a critical parameter in the evaluation of a code. However, in many practical instances it is important to know the number of codewords of each weight, and this determination is labeled the weight enumeration problem. Although for certain classes of codes a complete weight enumeration is known, in general only very incomplete results are known. In this section we develop the results of MacWilliams and Pless relating the weight structure of a code to that of its dual. The results are important since, if, in an (n, k) code over $GF(q)$, k is greater than $n/2$, then it may be easier to determine the weight structure of the dual code, which contains fewer vectors.

We turn now to the formulas of MacWilliams and Pless. We define the weight enumerator of a code to be

$$A(z) = \sum_{i=0}^{n} A_i z^i$$

if the code contains A_i codewords of weight i. Following the notation of MacWilliams (1962), we will denote the weight enumerator of a given code $\mathscr{C}$ by $A(z)$ and that of its dual $\mathscr{C}^\perp$ by $B(z)$.

Theorem 5.2 (*MacWilliams*) Let $\mathscr{C}$ be a linear (n, k) code over $GF(q)$ with weight enumerator $A(z)$ and $\mathscr{C}^\perp$ the dual code with weight enumerator $B(z)$. Then

$$\sum_{i=0}^{n} A_i [1 + (q-1)z]^{n-i}(1-z)^i = q^k \sum_{i=0}^{n} B_i z^i$$

1.5 Linear Codes

or, stated another way,

$$[1 + (q-1)z]^n A\left(\frac{1-z}{1+(q-1)z}\right) = q^k B(z)$$

Proof Let V be any u-dimensional subspace and $V^\perp$ its dual. Then we claim that

$$|V \cap \mathscr{C}| = q^{k-(n-u)} |\mathscr{C}^\perp \cap V^\perp|$$

We first note that the dual of $V \cap \mathscr{C}$ is the smallest subspace containing both $V^\perp$ and $\mathscr{C}^\perp$, which we denote by $V^\perp \cup \mathscr{C}^\perp$. Since

$$\dim(V \cap \mathscr{C}) + \dim(V^\perp \cup \mathscr{C}^\perp) = n$$

we have

$$|V^\perp \cup \mathscr{C}^\perp| \cdot |V \cap \mathscr{C}| = q^n$$

Since, from the previous section,

$$\dim(V^\perp \cup \mathscr{C}^\perp) = \dim V^\perp + \dim \mathscr{C}^\perp - \dim(V^\perp \cap \mathscr{C}^\perp)$$

we have

$$|V^\perp \cup \mathscr{C}^\perp| = \frac{q^{n-u} q^{n-k}}{|V^\perp \cap \mathscr{C}^\perp|}$$

It follows that

$$|V \cap \mathscr{C}| = \frac{q^n}{|V^\perp \cup \mathscr{C}^\perp|} = \frac{q^n}{q^{n-u} q^{n-k}} |V^\perp \cap \mathscr{C}^\perp|$$

For a particular basis of $V_n(q)$ take the set of vectors v_i, $i = 1, \ldots, n$, where v_i has all components zero except the ith, which is the unit element. We choose for the subspace V the subspace generated by $\{v_i, i \in t\}$, where $t = \{i_1, i_2, \ldots, i_{n-u}\}$ is a collection of $n - u$ integers from the first n positive integers. To be more specific, we denote this subspace by V_t and it has dimension $n - u$. We denote by V_s the dual of V_t, and V_s is generated by the set $\{v_i, i \in s\}$, where s is the complement set of integers to t.

The number of code vectors in V_t is $|V_t \cap \mathscr{C}|$, and it is the number of code vectors orthogonal to V_s. Now suppose $x \in \mathscr{C}$ is a codeword of weight i; then it is orthogonal to precisely $n - i$ of the basis vectors v_i, $i = 1, \ldots, n$. We can choose u vectors from a set of $n - i$ such vectors in $\binom{n-i}{u}$ ways. For each choice of u vectors, the subscripts of which form a set s and the vectors generate V_s, the complement set of vectors generates V_t. Thus the number of code vectors orthogonal to a V_t as the set s varies over all sets of u subscripts must be

$$\sum_{i=0}^n A_i \binom{n-i}{u}$$

where, as usual, $\binom{n-i}{u} = 0$ if $u > n - i$. If $\sum_s$ denotes summation over all possible sets of subscripts of cardinality u, and $\sum_t$ a summation over all possible complement subsets of cardinality $n - u$, then the foregoing statement can be expressed as

$$\sum_t |\mathscr{C} \cap V_t| = \sum_{i=0}^{n} A_i \binom{n-i}{u}$$

The above argument is symmetric in that it can as well be applied to $\mathscr{C}^\perp$ and $V_t^\perp = V_s$ to give

$$\sum_s |\mathscr{C}^\perp \cap V_s| = \sum_{i=0}^{n} B_i \binom{n-i}{n-u}$$

However, for any fixed set s and t we have already seen that

$$|\mathscr{C} \cap V_t| = q^{-n+k+(n-u)} |V_t^\perp \cap \mathscr{C}^\perp| = q^{k-u} |V_t^\perp \cap \mathscr{C}^\perp| = q^{k-u} |V_s \cap \mathscr{C}^\perp|$$

where u is the cardinality of the set s and V_t is of dimension $n - u$. Summing over all sets of indices s of cardinality u and their complements, we have

$$\sum_t |\mathscr{C} \cap V_t| = q^{k-u} \sum_s |V_s \cap \mathscr{C}^\perp|$$

or, using previous equations,

$$\sum_{i=0}^{n} A_i \binom{n-i}{u} = q^{k-u} \sum_{i=0}^{n} B_i \binom{n-i}{n-u}$$

To obtain the desired identities, we multiply both sides of this equation by z^u, where z is indeterminate, and sum over allowable values of u, i.e., from zero to n, to give

$$\sum_{u=0}^{n} \left[\sum_{i=0}^{n} A_i \binom{n-i}{u} \right] z^u = \sum_{u=0}^{n} q^{k-u} \left[\sum_{i=0}^{n} B_i \binom{n-i}{n-u} \right] z^u$$

Interchanging the order of summation on both sides and using the familiar binomial expansion for $(1 + x)^a$, we obtain

$$\sum_{i=0}^{n} A_i (1 + z)^{n-i} = q^k \left(\frac{z}{q}\right)^n \sum_{i=0}^{n} B_i (1 + qz^{-1})^{n-i}$$

Cancelling a factor of $(1 + z)^n$ and setting $z_0 = 1/(1 + z)$ then gives

$$\sum_{i=0}^{n} A_i z_0^i = \frac{q^k}{q^n} [1 + (q-1)z_0]^n \sum_{i=0}^{n} B_i \left(\frac{1 - z_0}{1 + (q-1)z_0}\right)^i$$

or

$$A(z_0) = \frac{q^k}{q^n} [1 + (q-1)z_0]^n B\left(\frac{1 - z_0}{1 + (q-1)z_0}\right)$$

1.5 Linear Codes

The formula is symmetric in the sense that if we let

$$z_1 = \frac{1 - z_0}{1 + (q-1)z_0}$$

then this last equation can be written

$$[1 + (q-1)z_1]^n A\left(\frac{1 - z_1}{1 + (q-1)z_1}\right) = q^k B(z_1)$$

which is the result of the theorem. □

A very elegant proof of this theorem using character theory is given by van Lint (1971) and also in the original paper by MacWilliams (1962). Two generalizations of the MacWilliams identities have recently appeared. The first, due to MacWilliams *et al.* (1972) shows that if one defines the dual of an arbitrary (i.e., linear or nonlinear) code in a certain way, then the identities are still valid. Delsarte (1973a) investigates codes over Abelian groups. Again, with the proper definition of a dual code, the MacWilliams identities still hold. There are also interesting relationships between Krawtchouk polynomials and weight enumeration [see MacWilliams *et al.* (1972c) and Delsarte (1973a)], and some of these are mentioned in Exercise 22.

If the weight enumerator $B(z)$ of $\mathscr{C}^\perp$ is known, then the above theorem gives a set of $n + 1$ linear equations in the $n + 1$ unknowns A_i, $i = 0, \ldots, n$. In a practical situation it is possible that not all of the coefficients of $B(z)$ may be known and in this case the so-called power moment identities of Pless (1963) may be important.

Theorem 5.3 (*Pless*) Let $\mathscr{C}$ be a linear (n, k) code over $GF(q)$ with weight enumerator $A(z)$ and $\mathscr{C}^\perp$ the dual code with weight enumerator $B(z)$. Then

$$\sum_{j=0}^{n}(n-j)^r A_j = \sum_{j=0}^{n} B_j \left[\sum_{l=0}^{r} l!\, S(r, l) q^{k-l} \binom{n-j}{n-l}\right]$$

where $S(r, l)$ is a Stirling number of the second kind, which satisfies the equation

$$(n-j)^r = \sum_{l=0}^{r} l! \binom{n-j}{l} S(r, l)$$

Proof By direct calculation

$$\sum_{j=0}^{n}(n-j)^r A_j = \sum_{j=0}^{n}\left[\sum_{l=0}^{r} l!\binom{n-j}{l}S(r,l)\right]A_j$$

$$= \sum_{l=0}^{r} l!\, S(r, l)\left[\sum_{j=0}^{n}\binom{n-j}{l}A_j\right]$$

Using the result of MacWilliams for the expression in brackets, this gives

$$\sum_{j=0}^{n}(n-j)^r A_j = \sum_{l=0}^{r} l!\, S(r,l)\left[q^{k-l}\sum_{j=0}^{n}\binom{n-j}{n-l}B_j\right]$$

$$= \sum_{j=0}^{n} B_j\left[\sum_{l=0}^{r} l!\, S(r,l) q^{k-l}\binom{n-j}{n-l}\right]$$

which is the result of the theorem. □

As in the case of the MacWilliams identities, we can derive a second relationship, which is

$$\sum_{i=0}^{n} j^r A_j = \sum_{j=0}^{n}(-1)^j B_j\left[\sum_{l=0}^{r} l!\, S(r,l) q^{k-l}(q-1)^{l-j}\binom{n-j}{n-l}\right]$$

For $r = 0$ this last equation reduces to, for $q = 2$,

$$\sum_{j=0}^{n} A_j = 2^k$$

For $r = 1$ it yields

$$\sum_{j=0}^{n} j A_j = 2^{k-1} n - 2^{k-1} B_1$$

and for $r = 2$ it yields

$$\sum_{j=0}^{n} j^2 A_j = 2^{k-2} n(n+1) - 2!\, 2^{k-2} n B_1 + 2!\, 2^{k-2} B_2$$

and so on. In general the identities express the first r moments of $A_0, A_1, \ldots, A_n$ in terms of $B_0, B_1, \ldots, B_r$. Now suppose only r of the A_i are unknown and $B_0, B_1, \ldots, B_{r-1}$ are known. In matrix form the equations would read

$$\begin{bmatrix} 1 & 1 & 1 & \cdots & 1 \\ 0 & 1 & 2 & \cdots & n \\ 0 & 1 & 2^2 & \cdots & n^2 \\ \vdots & \vdots & \vdots & \cdots & \vdots \\ 0 & 1 & 2^{r-1} & \cdots & n^{r-1} \end{bmatrix} \begin{bmatrix} A_0 \\ A_1 \\ \vdots \\ A_n \end{bmatrix} = \begin{bmatrix} f_0 \\ f_1 \\ \vdots \\ f_{r-1} \end{bmatrix}$$

where f_i is a known function of $B_0, B_1, \ldots, B_i$. Transposing to the right-hand side all terms in the matrix product that are known leaves a matrix equation in the unknowns $A_{i_1}, \ldots, A_{i_r}$:

$$\begin{bmatrix} 1 & 1 & \cdots & 1 \\ i_1 & i_2 & \cdots & i_r \\ i_1^2 & i_2^2 & \cdots & i_r^2 \\ \vdots & \vdots & \cdots & \vdots \\ i_1^{r-1} & i_2^{r-1} & \cdots & i_r^{r-1} \end{bmatrix} \begin{bmatrix} A_{i_1} \\ A_{i_2} \\ \vdots \\ A_{i_r} \end{bmatrix} = \begin{bmatrix} \gamma_1 \\ \gamma_2 \\ \vdots \\ \gamma_r \end{bmatrix}$$

where the right-hand side is known. Since the matrix is invertible, we can solve for the unknowns.

The MacWilliams and the Pless identities have been used in many ways to resolve problems in weight enumeration. However, there is much yet to be learned on the subject.

1.6 Polynomials over Finite Fields

It was observed in Section 1.2 that finite extensions of finite fields provide some interesting examples of the general theory. The same is true for polynomials over finite fields the elementary properties of which we consider in this section.

Theorem 6.1 Let $f(x)$ be an irreducible polynomial of degree k over $GF(q)$. Then the splitting field for $f(x)$ is $GF(q^k)$.

Proof Let α be a root of $f(x)$ and consider the extension $GF(q)(\alpha)$. If $f(x) = \sum_{i=0}^{k} a_i x^i$, then

$$f(\alpha^{q^j}) = \sum_{i=0}^{k} a_i \alpha^{iq^j} = \left(\sum_{i=0}^{k} a_i \alpha^i \right)^{q^j} = [f(\alpha)]^{q^j} = 0$$

since it was previously observed that $(a+b)^p = a^p + b^p$ over a field of characteristic p. We conclude that α^{q^j} is also a root of $f(x)$ for any integer j. Suppose l is the smallest positive integer such that $\alpha^{q^l} = \alpha$. Consider the polynomial $g(x)$ defined as

$$g(x) = \prod_{i=0}^{l-1} (x - \alpha^{q^i})$$

Now

$$[g(x)]^q = \prod_{i=0}^{l-1} (x - \alpha^{q^i})^q = \prod_{i=0}^{l-1} (x^q - \alpha^{q^{i+1}}) = \prod_{i=0}^{l-1} (x^q - \alpha^{q^i}) = g(x^q)$$

and this implies that, if $g(x) = \sum_{i=0}^{l} b_i x^i$, then

$$[g(x)]^q = \sum_{i=0}^{l} b_i^q x^{iq} = g(x^q) = \sum_{i=0}^{l} b_i x^{iq}$$

whence $b_i^q = b_i$ and $g(x) \in GF(q)[x]$. Since $g(x) | f(x)$ in the splitting field for $f(x)$, then $g(x) | f(x)$ in $GF(q)[x]$ and hence $g(x) = f(x)$. It follows that $k = l$. Notice that the elements $1, \alpha, \alpha^2, \ldots, \alpha^{k-1}$ form a basis of $GF(q)(\alpha)$ over $GF(q)$ since, if a linear relationship over $GF(q)$ of these elements existed, α would be a root of a polynomial of degree less than k, which is not possible. Thus the splitting field of $f(x)$ is $GF(q^k)$. □

Theorem 6.2 Let $f(x)$ be an irreducible polynomial over $GF(q)$ of degree k. Then $f(x)|(x^{q^n} - x)$ iff $k|n$.

Proof Suppose $f(x)|(x^{q^n} - x)$ in $GF(q)[x]$. Then the splitting field for $f(x)$, $GF(q^k)$, must be contained in $GF(q^n)$ since $f(x)$ has its roots in $GF(q^n)$. From Theorem 3.5 we have that $k|n$. Conversely, suppose that $k|n$, which implies that $GF(q^k) \subset GF(q^n)$. Since both $f(x)$ and $x^{q^n} - x$ split in $GF(q^n)$, this implies $f(x)|(x^{q^n} - x)$ over $GF(q^n)$ and, by Theorem 2.2, over $GF(q)$. □

These last two theorems provide some information that we should note. From Theorem 6.2, since $x^{q^n} - x$ has only simple zeros, it follows that any irreducible polynomial over a finite field has simple zeros. Also, if α is a zero of an irreducible polynomial of degree k over $GF(q)$, then the other zeros are $\alpha^q, \alpha^{q^2}, \ldots, \alpha^{q^{k-1}}$. This fact will be important in the following.

Theorem 6.3 The polynomial $x^{q^n} - x$ equals the product of all distinct, monic, irreducible polynomials over $GF(q)$ whose degrees divide n.

Proof If $f(x)$ and $g(x)$ are two distinct, monic, irreducible polynomials whose degrees divide n, since distinct, irreducible polynomials are relatively prime, $f(x)g(x)|(x^{q^n} - x)$. The theorem then follows from Theorem 6.2, and the fact that $x^{q^n} - x$ has simple zeros and can be factored into distinct, irreducible, monic polynomials. □

We now digress slightly to consider in some detail the above factorization and to give some notation that will be useful later. Recall the following definition.

Definition The minimal polynomial of an element $\alpha \in GF(q^n)$ over $GF(q)$ is the monic polynomial that is irreducible in $GF(q)[x]$ and has α as a zero.

Notice from Theorem 6.2 that any element α in $GF(q^n)$ satisfies an irreducible polynomial over $GF(q)$ of degree at most n. Let $\alpha \in GF(q^n)$ be an arbitrary element and denote its minimal polynomial over $GF(q)$ by $m_\alpha(x)$. From the proof of Theorem 6.1, the roots of $m_\alpha(x)$ are $\alpha, \alpha^q, \alpha^{q^2}, \ldots, \alpha^{q^{k-1}}$, where deg $m_\alpha(x) = k$ and $k|n$. The polynomial $m_\alpha(x)$ is also the minimal polynomial of the other roots. The factorization of $x^{q^n} - x$ of Theorem 6.3 may then be considered as the product of all distinct minimal polynomials of elements of $GF(q^n)$ over $GF(q)$. Let z be a primitive element of $GF(q^n)$, i.e., nonzero elements of $GF(q^n)$ can be expressed by z^i, $i = 0, 1, \ldots, q^n - 2$. It is sometimes convenient to introduce the symbol ∞ to denote $z^\infty = 0$, and we shall use this occasionally. Now form sets of elements of $GF(q^n)$ such that each set is a collection of roots of a minimal polynomial, i.e., each set is of the form $\{z^i, z^{iq}, z^{iq^2}, \ldots, z^{iq^{k-1}}\}$, which corresponds to the minimal polynomial of z^i

1.6 Polynomials over Finite Fields

(or α^{iq^j}). Notice that for each set $k|n$. This process partitions elements into disjoint subsets. Elements of $GF(q)$ are in sets of cardinality one. All other elements are in sets with more than one element. This partition of elements of $GF(q^n)$ induces a partition of the integers $0, 1, \ldots, q^n - 2$ where each set of integers comprises the exponents of the elements in a set in the partition of $GF(q^n)$. Conversely, let i be an integer, $0 \leq i \leq q^n - 2$, and define the mapping

$$\sigma_{(q^n-1)}(i) = iq \bmod (q^n - 1)$$

This mapping partitions the integers into disjoint sets which we shall denote by s_i and, again, the cardinality of each such set is a divisor of n. Each set s_i is of the form $s_i = \{j, jq, jq^2, \ldots, jq^{k-1}\}$, where the integers are taken $\bmod(q^n - 1)$ and such a set corresponds to an irreducible polynomial $m_{\alpha^j}(x)$, the minimal polynomial of α^j, where α may be taken as any primitive element of $GF(q^n)$. We shall denote the collection of the sets s_i by $\Sigma_q(q^n - 1)$ and the number of sets in $\Sigma_q(q^n - 1)$ by $\sigma_q(q^n - 1)$. The element $\alpha^\infty = 0$ corresponds to the polynomial x and is neglected in these considerations. A set $s_i \in \Sigma_q(q^n - 1)$ will be called a q chain of $q^n - 1$.

A problem that is closely related to the foregoing, and one that we shall require for cyclic codes, is the factorization of $x^n - 1$ over $GF(q)$. Let $GF(q^m)$ be a splitting field for $x^n - 1$ and let η be a primitive nth root of unity in $GF(q^m)$. Then if $(n, q) = 1$, $x^n - 1$ factors into the product of distinct minimal polynomials of the elements η^i, $i = 0, 1, \ldots, n - 1$. Thus the zeros of the minimal polynomial of η^i are $\eta^{iq}, \eta^{iq^2}, \ldots, \eta^{iq^{k-1}}$, where k is the least positive integer such that $\eta^{iq^k} = \eta^i$ and the exponents are taken $\bmod n$. By precisely the same argument as before we can partition the elements η^i into sets, each set corresponding to an irreducible factor of $x^n - 1$. Equivalently, we can partition the integers $0, 1, \ldots, n - 1$ into sets s_i where each set is invariant under the mapping

$$\sigma_{(n)}(i) = iq \bmod n$$

To each set $s_i = \{j, jq, \ldots, jq^{k-1}\}$, integers mod n, there corresponds an irreducible polynomial with zeros $\eta^j, \eta^{jp}, \ldots, \eta^{jq^{k-1}}$, exponents mod n. The collection of these sets will be denoted by $\Sigma_q(n)$ and the cardinality of this set $\sigma_q(n)$. A set $s_i \in \Sigma_q(n)$ will be called a q chain of n or a q chain mod n. Finally, notice that if S is a union of sets $s_i \in \Sigma_q(n)$, then the polynomial

$$f(x) = \prod_{i \in S} (x - \eta^i)$$

is a polynomial over $GF(q)$ and is a factor of $x^n - 1$. Furthermore, any factor of $x^n - 1$ is formed in this manner.

Notice that if $(n, q) \neq 1$, then $x^n - 1$ can be expressed as a power of $x^{n_1} - 1$, $n_1 < n$, $n_1 | n$, and $(n_1, q) = 1$. Thus every irreducible factor of $x^{n_1} - 1$ has the same multiplicity in the factorization of $x^n - 1$.

Let α be an arbitrary nonzero element of $GF(q^n)$. If the order of α is m, then $m|(q^n-1)$ since $\alpha^{q^n-1} = 1$ for any $\alpha \in GF(q^n)$. As noted previously, the order of α^q is $m/(m,q)$. But $(m,q) = 1$ since a common divisor of m and q would be a common divisor of $q^n - 1$ and q, which are relatively prime. Thus α and α^q have the same order and all zeros of an irreducible polynomial have the same order. If $f(x)$ is an irreducible polynomial over $GF(q)$ whose zeros have order m, then $f(x)|(x^m - 1)$ since $x^m - 1$ has as zeros all elements of order m in its splitting field of order m. Also $f(x) \nmid (x^k - 1)$, $k < m$.

Definition Let $f(x)$ be an arbitrary polynomial in $GF(q)[x]$ not divisible by x. The exponent of $f(x)$ is the least positive integer e such that $f(x)|(x^e - 1)$.

For an irreducible polynomial the exponent is the order of its zeros. For an arbitrary polynomial $f(x)$ the exponent can be found in terms of the exponents of its irreducible factors, but we do not pursue the matter here [see Berlekamp (1968, Theorem 6.21)].

Definition A primitive polynomial of degree k over $GF(q)$ is a monic, irreducible polynomial whose zeros are primitive elements of $GF(q^k)$.

Alternatively, a primitive polynomial of degree k over $GF(q)$ has exponent $q^k - 1$. We have already seen that there are $\varphi(q^k - 1)$ primitive elements in $GF(q^k)$. If α is any primitive element of $GF(q^k)$, then, by definition, $\alpha^{q^k} = \alpha$ and $\alpha^i \neq \alpha$ for $i < q^k$. Thus the minimal polynomial of any primitive element is of degree k and there are precisely $\varphi(q^k - 1)/k$ primitive polynomials. It is clear that there may exist minimal polynomials of nonprimitive elements that are also of degree k over $GF(q)$. These correspond to elements α of order less than $q^k - 1$ but for which $\alpha, \alpha^q, \ldots, \alpha^{q^{k-1}}$ are distinct.

There are $\varphi(q^k - 1)$ primitive elements in $GF(q^k)$. If $d|(q^k - 1)$, how many elements of order d are there? Denote by $N_{q^k}(d)$ the number of elements of order exactly d in $GF(q^k)$. Then we have

$$\sum_{s|d} N_{q^k}(s) = d$$

To solve this equation, we use the Möbius inversion formula (see Appendix A) to give

$$N_{q^k}(d) = \sum_{s|d} \mu(s) \frac{d}{s}$$

where $\mu(s)$ is the Möbius function (also defined in Appendix A). In particular,

$$\varphi(q^k - 1) = N_{q^k}(q^k - 1) = \sum_{s|q^k-1} \mu(s) \frac{q^k - 1}{s}$$

1.6 Polynomials over Finite Fields

By a similar argument we can find the number of monic, irreducible polynomials of degree k over $GF(q)$, a quantity that we shall denote by $I_q(k)$.

Theorem 6.4 The number of monic, irreducible polynomials of degree k over $GF(q)$ is given by

$$I_q(k) = \frac{1}{k} \sum_{d|k} \mu\left(\frac{k}{d}\right) q^d$$

where the summation ranges over all divisors of k.

Proof The degree of the product of all distinct, monic, irreducible polynomials over $GF(q)$ whose degrees divide k is given by

$$\sum_{d|k} d I_q(d) = q^k$$

where the equality follows from Theorem 6.3. By the Möbius inversion formula, the result of the theorem follows. □

We should observe that $I_q(k) \geq 1$ for any prime power q and positive integer k. This is easily deduced from the fact that a primitive element always exists in $GF(q^k)$ and its minimal polynomial is of degree k. Alternatively we can argue from the above formula. Suppose $I_q(k) = 0$ for some q, k and let d' be the smallest value of d for which $\mu(k/d) \neq 0$. Factor out $q^{d'}$ and notice that we have expressed 1 as a sum of powers of q. Since by the construction this sum is divisible by q, while 1 is not, we have a contradiction.

This information suggests a method for constructing a representation of $GF(q^k)$ given an irreducible polynomial of degree k over $GF(q)$. Suppose $f(x) = \sum_{i=0}^{k} a_i x^i$ is such a polynomial and denote a zero of $f(x)$ by α. Then in a splitting field for $f(x)$ the elements $1, \alpha, \alpha^2, \ldots, \alpha^{k-1}$ would be independent. The set of all linear combinations of these elements over $GF(q)$ forms the elements of $GF(q^k)$. Addition follows from the addition in $GF(q)$ and since higher powers of α than $k - 1$ can be expressed as a linear combination of the basis elements, using the fact that $f(\alpha) = 0$, it can be shown that multiplication is also well defined. Since $I_q(k) \geq 1$, up to isomorphism any finite field can be constructed in this manner.

Example The ideas of the foregoing theory will be illustrated in the following example. Let α be a zero of the primitive polynomial $x^6 + x + 1$ over $GF(2)$. The minimal polynomials and their corresponding zeros and their orders are shown in Table I. It follows that

$$x^{2^6} - x = \prod_{i=1}^{14} f_i(x)$$

TABLE I

1	$\alpha, \alpha^2, \alpha^4, \alpha^8, \alpha^{16}, \alpha^{32}$	$f_1(x) = x^6 + x + 1$	Primitive
2	$\alpha^3, \alpha^6, \alpha^{12}, \alpha^{24}, \alpha^{48}, \alpha^{33}$	$f_2(x) = x^6 + x^4 + x^2 + x + 1$	Roots order 21
3	$\alpha^5, \alpha^{10}, \alpha^{20}, \alpha^{40}, \alpha^{17}, \alpha^{34}$	$f_3(x) = x^6 + x^5 + x^2 + x + 1$	Primitive
4	$\alpha^7, \alpha^{14}, \alpha^{28}, \alpha^{56}, \alpha^{49}, \alpha^{35}$	$f_4(x) = x^6 + x^3 + 1$	Roots order 9
5	$\alpha^9, \alpha^{18}, \alpha^{36}$	$f_5(x) = x^3 + x^2 + 1$	Roots order 7
6	$\alpha^{11}, \alpha^{22}, \alpha^{44}, \alpha^{25}, \alpha^{50}, \alpha^{37}$	$f_6(x) = x^6 + x^5 + x^3 + x^2 + 1$	Primitive
7	$\alpha^{13}, \alpha^{26}, \alpha^{52}, \alpha^{41}, \alpha^{19}, \alpha^{38}$	$f_7(x) = x^6 + x^4 + x^3 + x + 1$	Primitive
8	$\alpha^{15}, \alpha^{30}, \alpha^{60}, \alpha^{57}, \alpha^{51}, \alpha^{39}$	$f_8(x) = x^6 + x^5 + x^4 + x^2 + 1$	Roots order 21
9	α^{21}, α^{42}	$f_9(x) = x^2 + x + 1$	Roots order 3
10	$\alpha^{23}, \alpha^{46}, \alpha^{29}, \alpha^{58}, \alpha^{53}, \alpha^{43}$	$f_{10}(x) = x^6 + x^5 + x^4 + x + 1$	Primitive
11	$\alpha^{37}, \alpha^{54}, \alpha^{45}$	$f_{11}(x) = x^3 + x + 1$	Roots order 7
12	$\alpha^{31}, \alpha^{62}, \alpha^{61}, \alpha^{59}, \alpha^{55}, \alpha^{47}$	$f_{12}(x) = x^6 + x^5 + 1$	Primitive
13	$\alpha^{63} = 1$	$f_{13}(x) = x + 1$	
14	0	$f_{14}(x) = x$	

In this representation the subfields $GF(2) \subset GF(2^2) \subset GF(2^3) \subset GF(2^6)$ are given by

$$GF(2) = \{0, 1\}$$
$$GF(2^2) = \{0, 1, \alpha^{21}, \alpha^{42}\}$$
$$GF(2^3) = \{0, 1, \alpha^9, \alpha^{18}, \alpha^{27}, \alpha^{36}, \alpha^{45}, \alpha^{54}\}$$

The minimal polynomials and the corresponding zeros of $GF(2^6)$ over $GF(2^2)$ are given in Table II, where $\gamma = \alpha^{42}$ and $\varepsilon = \alpha^{21}$. These computations were carried out with the aid of Table III.

We introduce the interpolation problem for polynomials in order to consider the Van der Monde matrix and determinant. We want a polynomial $f(x) = \sum_{i=0}^{n} a_i x^i, a_i \in GF(q)$ that takes on the values $\beta_1, \beta_2, \ldots, \beta_n, \beta_i \in GF(q)$ at the distinct points $\alpha_1, \alpha_2, \ldots, \alpha_n$ of $GF(q)$. Denoting by $\mathbf{a}$ and $\boldsymbol{\beta}$ the row vectors of the a_i and β_i, respectively, a solution to the problem is a solution to the set of linear equations

$$\mathbf{a}V = \boldsymbol{\beta}$$

where V is the Van der Monde matrix of the elements $\alpha_1, \ldots, \alpha_n$ given by

$$V = \begin{bmatrix} 1 & 1 & \cdots & 1 \\ \alpha_1 & \alpha_2 & \cdots & \alpha_n \\ \alpha_1^2 & \alpha_2^2 & \cdots & \alpha_n^2 \\ \vdots & \vdots & & \vdots \\ \alpha_1^{n-1} & \alpha_2^{n-1} & \cdots & \alpha_n^{n-1} \end{bmatrix}$$

1.6 Polynomials over Finite Fields

TABLE II

1	$\alpha, \alpha^4, \alpha^{16}$	$g_1(x) = x^3 + x^2 + \gamma x + \varepsilon$
2	$\alpha^2, \alpha^8, \alpha^{32}$	$g_2(x) = x^3 + x^2 + \varepsilon x + \gamma$
3	$\alpha^3, \alpha^{12}, \alpha^{48}$	$g_3(x) = x^3 + \gamma x + 1$
4	$\alpha^6, \alpha^{24}, \alpha^{33}$	$g_4(x) = x^3 + \varepsilon x + 1$
5	$\alpha^5, \alpha^{20}, \alpha^{17}$	$g_5(x) = x^3 + \gamma x^2 + \varepsilon x + \gamma$
6	$\alpha^{10}, \alpha^{40}, \alpha^{34}$	$g_6(x) = x^3 + \varepsilon x^2 + \gamma x + \varepsilon$
7	$\alpha^7, \alpha^{28}, \alpha^{49}$	$g_7(x) = x^3 + \varepsilon$
8	$\alpha^{14}, \alpha^{56}, \alpha^{35}$	$g_8(x) = x^3 + \gamma$
9	$\alpha^9, \alpha^{36}, \alpha^{18}$	$g_9(x) = x^3 + x^2 + 1$
10	$\alpha^{11}, \alpha^{44}, \alpha^{50}$	$g_{10}(x) = x^3 + \gamma x^2 + \gamma x + \gamma$
11	$\alpha^{22}, \alpha^{25}, \alpha^{37}$	$g_{11}(x) = x^3 + \varepsilon x^2 + \varepsilon x + \varepsilon$
12	$\alpha^{13}, \alpha^{52}, \alpha^{19}$	$g_{12}(x) = x^3 + x^2 + x + \varepsilon$
13	$\alpha^{26}, \alpha^{41}, \alpha^{38}$	$g_{13}(x) = x^3 + x^2 + x + \gamma$
14	$\alpha^{15}, \alpha^{60}, \alpha^{51}$	$g_{14}(x) = x^3 + \gamma x^2 + 1$
15	$\alpha^{30}, \alpha^{57}, \alpha^{39}$	$g_{15}(x) = x^3 + \varepsilon x^2 + 1$
16	α^{21}	$g_{16}(x) = x + \varepsilon$
17	α^{42}	$g_{17}(x) = x + \gamma$
18	$\alpha^{23}, \alpha^{29}, \alpha^{53}$	$g_{18}(x) = x^3 + \varepsilon x^2 + x + \gamma$
19	$\alpha^{46}, \alpha^{58}, \alpha^{43}$	$g_{19}(x) = x^3 + \gamma x^2 + x + \varepsilon$
20	$\alpha^{27}, \alpha^{54}, \alpha^{45}$	$g_{20}(x) = x^3 + x + 1$
21	$\alpha^{31}, \alpha^{61}, \alpha^{55}$	$g_{21}(x) = x^3 + \gamma x^2 + \varepsilon x + \varepsilon$
22	$\alpha^{62}, \alpha^{59}, \alpha^{47}$	$g_{22}(x) = x^3 + \varepsilon x^2 + \gamma x + \gamma$
23	$\alpha^{63} = 1$	$g_{23}(x) = x + 1$
24	0	$g_{24}(x) = x$

TABLE III

$1 + \alpha = \alpha^6$	$1 + \alpha^{14} = \alpha^{52}$	$1 + \alpha^{27} = \alpha^{18}$	$1 + \alpha^{40} = \alpha^{55}$	$1 + \alpha^{53} = \alpha^{51}$	
$1 + \alpha^2 = \alpha^{12}$	$1 + \alpha^{15} = \alpha^{23}$	$1 + \alpha^{28} = \alpha^{41}$	$1 + \alpha^{41} = \alpha^{28}$	$1 + \alpha^{54} = \alpha^{36}$	
$1 + \alpha^3 = \alpha^{32}$	$1 + \alpha^{16} = \alpha^{33}$	$1 + \alpha^{29} = \alpha^{60}$	$1 + \alpha^{42} = \alpha^{21}$	$1 + \alpha^{55} = \alpha^{40}$	
$1 + \alpha^4 = \alpha^{24}$	$1 + \alpha^{17} = \alpha^{47}$	$1 + \alpha^{30} + \alpha^{46}$	$1 + \alpha^{43} = \alpha^{39}$	$1 + \alpha^{56} = \alpha^{19}$	
$1 + \alpha^5 = \alpha^{62}$	$1 + \alpha^{18} = \alpha^{27}$	$1 + \alpha^{31} = \alpha^{34}$	$1 + \alpha^{44} = \alpha^{37}$	$1 + \alpha^{57} = \alpha^{58}$	
$1 + \alpha^6 = \alpha$	$1 + \alpha^{19} = \alpha^{56}$	$1 + \alpha^{32} = \alpha^3$	$1 + \alpha^{45} = \alpha^9$	$1 + \alpha^{58} = \alpha^{57}$	
$1 + \alpha^7 = \alpha^{26}$	$1 + \alpha^{20} = \alpha^{59}$	$1 + \alpha^{33} = \alpha^{16}$	$1 + \alpha^{46} = \alpha^{30}$	$1 + \alpha^{59} = \alpha^{20}$	
$1 + \alpha^8 = \alpha^{48}$	$1 + \alpha^{21} = \alpha^{42}$	$1 + \alpha^{34} = \alpha^{31}$	$1 + \alpha^{47} = \alpha^{17}$	$1 + \alpha^{60} = \alpha^{29}$	
$1 + \alpha^9 = \alpha^{45}$	$1 + \alpha^{22} = \alpha^{50}$	$1 + \alpha^{35} = \alpha^{13}$	$1 + \alpha^{48} = \alpha^8$	$1 + \alpha^{61} = \alpha^{10}$	
$1 + \alpha^{10} = \alpha^{61}$	$1 + \alpha^{23} = \alpha^{15}$	$1 + \alpha^{36} = \alpha^{54}$	$1 + \alpha^{49} = \alpha^{38}$	$1 + \alpha^{62} = \alpha^5$	
$1 + \alpha^{11} = \alpha^{25}$	$1 + \alpha^{24} = \alpha^4$	$1 + \alpha^{37} = \alpha^{44}$	$1 + \alpha^{50} = \alpha^{22}$	$1 + \alpha^{63} = 0$	
$1 + \alpha^{12} = \alpha^2$	$1 + \alpha^{25} = \alpha^{11}$	$1 + \alpha^{38} = \alpha^{49}$	$1 + \alpha^{51} = \alpha^{53}$		
$1 + \alpha^{13} = \alpha^{35}$	$1 + \alpha^{26} = \alpha^7$	$1 + \alpha^{39} = \alpha^{43}$	$1 + \alpha^{52} = \alpha^{14}$		

The determinant of V is zero if $\alpha_i = \alpha_j$ for $i \neq j$ and hence $\Pi_{i>j}(\alpha_i - \alpha_j) | |V|$. Since the coefficient of the term $1\alpha_2 \alpha_3{}^2 \cdots \alpha_n^{n-1}$ is unity in both $|V|$ and $\Pi_{i>j}(\alpha_i - \alpha_j)$ and since they are polynomials of the same degrees in the α's, we must have that

$$|V| = \prod_{i>j}(\alpha_i - \alpha_j)$$

For certain cases an explicit form for V^{-1} may be exhibited if $|V| \neq 0$. Of course, $|V| \neq 0$ iff $\alpha_i \neq \alpha_j$; $i \neq j$, a fact that has been used previously in this chapter.

Any two polynomials of degree less than or equal to n whose values coincide at $n+1$ points must be equal since the difference of the polynomials, which is of degree at most n, has $n+1$ roots. Suppose then that $\beta_i = f(\alpha_i)$, $i = 0, 1, \ldots, n$, at the $n+1$ points $\alpha_0, \alpha_1, \ldots, \alpha_n$. The polynomial

$$f(x) = \sum \beta_i \left(\prod_{\substack{j=0 \\ j \neq i}}^{n} \frac{x - \alpha_j}{\alpha_i - \alpha_j} \right)$$

satisfies these conditions and since it is of degree n or less, it is the unique polynomial of this degree that does so. The above formula is the Lagrange interpolation formula.

Let the monic polynomial $f(x) = \sum_{i=0}^{m} a_{m-i} x^i$ over $GF(q)$ split in $GF(q^n)$ and let its roots be $\alpha_1, \ldots, \alpha_m$, i.e.,

$$f(x) = \sum_{i=0}^{m} a_{m-i} x^i = \prod_{i=1}^{m}(x - \alpha_i), \qquad a_0 = 1$$

Define the functions of the zeros

$$\beta_1 = \alpha_1 + \alpha_2 + \cdots + \alpha_m$$
$$\beta_2 = \alpha_1 \alpha_2 + \alpha_1 \alpha_3 + \cdots + \alpha_2 \alpha_3 + \alpha_2 \alpha_4 + \cdots + \alpha_{m-1}\alpha_m$$
$$\vdots$$
$$\beta_m = \alpha_1 \alpha_2 \cdots \alpha_m$$

where $\beta_i = (-1)^i a_i$. We can regard these functions as functions of m variables $\alpha_1, \alpha_2, \ldots, \alpha_m$. A function of m variables that is unchanged by a permutation of the variables is called a symmetric function. The functions β_i play a central role in the theory of such functions and are called the elementary symmetric functions. Notice that since $f(x)$ is a polynomial over $GF(q)$ the elementary symmetric functions are invariant under any automorphism of $G(q^n/q)$. If $f(x)$ is irreducible over $GF(q)$, then it splits in $GF(q^m)$ and $\alpha_i = \alpha_1^{q^{i-1}}$, $i = 1, \ldots, m$. Thus the function β_1 is the trace of any of the roots of $f(x)$ over $GF(q)$ and β_m is the norm function.

1.6 Polynomials over Finite Fields

Let P_i denote the sum of the ith powers of the roots of the arbitrary monic polynomial $f(x)$ of degree m over $GF(q)$. By direct computation the following recursive relationships can be derived:

$$0 = P_1 + a_1$$
$$0 = P_2 + a_1 P_1 + 2a_2$$
$$\vdots$$
$$0 = P_{m-1} + a_1 P_{m-2} + a_2 P_{m-3} + \cdots + (m-1)a_{m-1}$$

To these equations we can add the following:

$$0 = P_m + a_1 P_{m-1} + a_2 P_{m-2} + \cdots + ma_m$$
$$0 = P_{m+1} + a_1 P_m + a_2 P_{m-1} + \cdots + a_m P_1$$

and in general, for $l > m$,

$$0 = P_l + a_1 P_{l-1} + a_2 P_{l-2} + \cdots + a_m P_{l-m}$$

These equations are easily found by equating the quantities $\sum_{i=1}^{m} \alpha_i^k f(\alpha_i)$ to zero. Together with the previous equations they are called the Newton identities.

Apply these equations and the previous ones to the polynomial

$$f(x) = x^q - x$$

which, of course, splits over $GF(q)$ and has all the elements of $GF(q)$ as zeros. Since $a_i = 0$, $i \neq q-1$, and $a_{q-1} = -1$, we conclude that $P_i = 0$ for $0 < i < q - 1$ and $P_{q-1} = q - 1$. Since, in $GF(q)$, $\alpha^q = \alpha$, we can reduce exponents $(q - 1)$ for nonzero elements and state the following:

Theorem 6.5 Let θ_i, $i > 0$, denote the sum of the ith powers of the elements of $GF(q)$. Then

$$\theta_i = \begin{cases} q - 1 & \text{if } (q-1) | i \\ 0 & \text{otherwise} \end{cases}$$

There is a fascinating literature on the study of polynomials over a finite field but it would be inappropriate to review it here. The work tends to be on factorizing polynomials and determining irreducible polynomials. The books by Dickson (1958) and Albert (1956) contain much information on these problems. We have relegated to the exercises some of these interesting results. On the factorization of polynomials, the algorithm by Berlekamp is very efficient and is described in Chapter 6 of his book (Berlekamp, 1968). The problem has also been considered by Swan (1962) and McEliece (1969). The irreducibility of trinomials, polynomials of the form $x^n + x^k + 1$ has also received a great deal of attention. The problem is of importance in the generation of shift register sequences with specified properties. Swan (1962) gives some important results on the problem and much of the work is described by Berlekamp (1968). Zierler and Brillhart (1968) give a listing of all binary irreducible trinomials of degree <1000 as well as indicating which ones are primitive and the exponents of those that are not. Zierler (1970) lists all irreducible trinomials of the form $x^n + x + 1$, $n < 30{,}000$, together with their exponent. We conclude this section with a discussion of two problems, the irreducibility of $x^{k-1} + x^{k-2} + \cdots + x + 1$ and the cyclotomic factorization of $x^n - 1$. Since these problems illustrate much of the foregoing material, we present them as examples.

Example. Cyclotomic Polynomials In the factorization of $x^{q^m} - x$ over $GF(q)$ one approach is to sort out roots of unity according to their order. This approach has already been considered in another context. Consider the equation $x^n - 1 = 0$ in the field of complex numbers, which contains n roots of unity. There are $\varphi(n)$ primitive nth roots of unity.

Definition The polynomial

$$\Phi_n(x) = \prod_{(i,n)=1} (x - \beta^i)$$

where β is a primitive nth root of unity, is called the nth cyclotomic polynomial. It is of degree $\varphi(n)$.

If $n = kd$, then β^k is of order d and is a primitive dth root of unity. We can write the dth cyclotomic polynomial as

$$\Phi_d(x) = \prod_{(i,d)=1} (x - \beta^{ik})$$

Clearly each nth root of unity will be a primitive dth root of unity for precisely one d and this leads to a sorting of the n roots. The sorting leads to the factorization of $x^n - 1$:

$$x^n - 1 = \prod_{\substack{1 \leq d \leq n \\ d|n}} \Phi_d(x)$$

which is referred to as the cyclotomic factorization of $x^n - 1$. Applying the Möbius inversion formula to this equation gives

$$\Phi_n(x) = \prod_{d|n} (x^d - 1)^{\mu(n/d)} = \prod_{d|n} (x^{n/d} - 1)^{\mu(d)}.$$

The first three cyclotomic polynomials are

$$\Phi_1(x) = x - 1, \qquad \Phi_2(x) = x + 1, \qquad \Phi_3(x) = x^2 + x + 1$$

It is readily established, either by induction or by inspection of the formula for $\Phi_n(x)$, that all cyclotomic polynomials have integer coefficients. In fact, the first 104 cylotomic polynomials have coefficients that are either

1.6 Polynomials over Finite Fields

$+1$, 0, or -1. The 105th has a coefficient that is $+2$. It is also true that we can always find an n, for a given integer N, such that $\Phi_n(x)$ has a coefficient larger then N. The cyclotomic polynomials are irreducible over the rationals.

As an example of this situation that will also point out the relevance of factoring $x^n - 1$ over a finite field, consider the cyclotomic factorization of $x^{15} - 1$ over the complex numbers. We have that

$$x^{15} - 1 = \Phi_{15}(x)\Phi_5(x)\Phi_3(x)\Phi_1(x)$$

where

$$\Phi_1(x) = x - 1, \qquad \Phi_3(x) = x^2 + x + 1$$

$$\Phi_5(x) = \frac{(x^5 - 1)}{(x - 1)} = x^4 + x^3 + x^2 + x + 1$$

and

$$\Phi_{15}(x) = \frac{(x^{15} - 1)(x - 1)}{(x^5 - 1)(x^3 - 1)} = x^8 - x^7 + x^5 - x^4 + x^3 - x + 1$$

Although this factorization was over the complex numbers, it could equally well have been performed over any field containing primitive nth roots of unity. In particular, in $GF(p^n)$

$$x^{p^n - 1} - 1 = \prod_{d \mid p^n - 1} \Phi_d(x)$$

where, of course, the coefficients of $\Phi_d(x)$ are field integers. The $\Phi_d(x)$ are not, in general, irreducible over a subfield of $GF(p^n)$, but the foregoing is at least a partial factorization of $x^{p^n - 1} - 1$. Notice that $\Phi_d(x)$ has as roots all elements in $GF(p^n)$ of order d. An element of order d in an extension field of $GF(q)$ has a minimal polynomial of degree k over $GF(q)$ where k is the smallest integer such that $d \mid q^k - 1$. There are $\varphi(d)$ elements of order d and hence there are $\varphi(d)/k$ irreducible polynomials of degree k over $GF(q)$ of exponent d. Their product is equal to $\Phi_d(x)$, where the coefficients are taken as field integers.

From our previous factorization of $x^{15} - 1$ over $\mathbb{C}$, we consider the factorization over $GF(2)$. Since $15 \mid (2^4 - 1)$ and $\Phi_{15}(x)$ is a polynomial of degree eight, it must factor into two irreducible polynomials of degree four over $GF(2)$. In fact we have

$$\Phi_{15}(x) = x^8 + x^7 + x^5 + x^4 + x^3 + x + 1 \qquad [\text{over} \quad GF(2)]$$
$$= (x^4 + x + 1)(x^4 + x^3 + 1)$$

where the two polynomials in brackets are irreducible over $GF(2)$. The polynomials $\Phi_1(x)$ [$= x + 1$ over $GF(2)$], $\Phi_3(x)$, and $\Phi_5(x)$ are irreducible over $GF(2)$. Thus the problem of factorizing $x^n - 1$ over a finite field, which, as we will see later, has importance in coding theory, may be assisted by considering first the cyclotomic factorization over $\mathbb{C}$ and determining the number of irreducible polynomials of the same degree contained in the factorization of each cyclotomic polynomial.

Example. Irreducibility of Certain Polynomials Consider first the number of irreducible polynomials of degree m over $GF(q)$ of exponent e. Since the smallest extension field of $GF(q)$ containing the roots of an irreducible polynomial of degree m over $GF(q)$ is $GF(q^m)$, then $e|(q^m - 1)$ and $e \nmid (q^k - 1)$ $k < m$. Since $e|(q^m - 1)$, we have that $(x^e - 1) | (x^{q^{m-1}} - 1)$ and $GF(q^m)$ contains all the eth roots of unity in this case. Thus it contains $\varphi(e)$ primitive eth roots of unity, of which we suppose β is one. From the previous consideration the elements $\beta, \beta^q, \beta^{q^2}, \ldots, \beta^{q^{m-1}}$ are distinct, and thus the minimal polynomial of any primitive eth root of unity is of degree m. We conclude that there are precisely $\varphi(e)/m$ irreducible polynomials of exponent e and note that the only dependence of q is through the condition that $e|(q^m - 1)$ and $e \nmid (q^k - 1)$, $k < m$.

We now ask the question, when is

$$f(x) = \frac{x^e - 1}{x - 1} = x^{e-1} + x^{e-2} + \cdots + x^2 + x + 1$$

irreducible over $GF(q)$? We must have only a single irreducible polynomial of exponent e, which implies that $m = e - 1$ and $\varphi(e)/(e - 1) = 1$, which implies that e is a prime number. Furthermore, it must be true that $e|(q^{e-1} - 1)$ but $e \nmid (q^k - 1)$, $k < e - 1$, i.e., q is a primitive $(e - 1)$th root of unity mod e. Thus $f(x)$ is irreducible over $GF(q)$ iff these conditions are satisfied.

In a similar vein it is possible to show that the polynomial $\varepsilon^p x^p - \varepsilon x - \gamma$, $\varepsilon \neq 0$, ε, γ in $GF(q)$, $q = p^n$, is irreducible over $GF(q)$ iff the trace of γ in $GF(q)$ over $GF(p)$ is nonzero. First observe that $\varepsilon^p x^p - \varepsilon x - \gamma$ is reducible over $GF(q)$ iff it has a root in $GF(q)$. Notice that if $x^p - x = \gamma$, then $\gamma + \gamma^p + \cdots + \gamma^{p^{n-1}} = x^{p^n} - x$. Thus if x is any element of $GF(q)$, then $x^q - x = 0$ and it follows that tr $\gamma = 0$. Conversely, if tr $\gamma \neq 0$, then there can be no solution in $GF(q)$. There are many other constructions of irreducible polynomials and the reader is referred to the books by Dickson (1958, pp. 19–44) and Albert (1956, pp. 132–155). A few of the more interesting theorems have been included as exercises at the end of this chapter.

1.7 Cyclic Codes

The class of linear codes is unfortunately still too general for us to say much on the problem of constructing codes with good distance properties. We therefore add another restriction.

1.7 Cyclic Codes

Definition A code is cyclic if it is linear and if every cyclic shift of the coordinates of a codeword is a codeword.

Thus if $c = (\alpha_0, \alpha_1, \ldots, \alpha_{n-1})$ is a codeword, then $(\alpha_{n-1}, \alpha_0, \alpha_1, \ldots, \alpha_{n-2})$ is also a codeword. It is convenient to identify a codeword with a codeword polynomial. The codeword polynomial for the codeword c is

$$c(x) = \sum_{i=0}^{n-1} \alpha_i x^i$$

A cyclic shift of this codeword c is equivalent to multiplication by x and reducing exponents mod n. However, reducing exponents mod n is equivalent to reducing the polynomial mod $(x^n - 1)$.

Let $GF(q)[x]$ denote the set of all polynomials over $GF(q)$ in the variable x. This is a principal ideal domain and there is a natural isomorphism between the quotient ring $GF(q)[x]/(x^n - 1)$, where $(x^n - 1)$ is the ideal generated by the polynomial $x^n - 1$, and the ring of polynomials of degree less than n with multiplication defined mod $(x^n - 1)$. We shall denote the ring $GF(q)[x]/(x^n - 1)$ by A_n. It is also an algebra over $GF(q)$. The ideal consisting of all multiples of the element $a(x)$ will be denoted $(a(x))$. We mention again that a codeword, which is an n-tuple over $GF(q)$, is identified with an element of A_n. The algebra A_n is, of course, a commutative ring with identity and a principal ideal domain, so that every ideal has a single generator. In Chapter 4 we will consider the structure of A_n from the purely algebraic point of view. The algebra A_n is isomorphic to the group algebra $GF(q)C_n$, C_n the cyclic group of order n, and in this formulation it leads to an interesting generalization. The following theorem is of a fundamental nature.

Theorem 7.1 An (n, k) linear code $\mathscr{C}$ over $GF(q)$ is cyclic iff it is an ideal of A_n.

Proof Let $\mathscr{C}$ be an ideal in A_n. It is clearly a linear subspace of A_n as a vector space and it remains to check that it is a cyclic subspace. But since $\mathscr{C}$ is an ideal, it is closed under multiplication by x and hence cyclic.

Conversely, suppose $\mathscr{C}$ is a cyclic subspace. It is closed under addition and multiplication by x and hence closed under multiplication by any element of A_n. Thus $\mathscr{C}$ is an ideal. □

The polynomial $x^n - 1$ plays a key role in the theory of cyclic codes and we have considered its factorization in the previous section. In general, it is desirable that it should not have any factors of multiplicity greater than 1. From the previous section, a necessary and sufficient condition that $x^n - 1$ not have any factors of multiplicity greater than 1 over $GF(q)$ is that $(n, q) = 1$. We now investigate the ideals or codes A_n.

Theory 7.2 The unique monic polynomial $g(x)$ of minimal degree in any ideal A of A_n is a generator of A and divides $x^n - 1$. The dimension of A is $n - \deg g(x)$. Conversely, a divisor of $x^n - 1$ is a generator of an ideal in A_n.

Proof Note that by a generator $g(x)$ of an ideal A we mean that every element of A may be written $a(x)g(x)$ for some $a(x) \in A_n$. Let $g(x)$ be that unique monic polynomial of minimal degree in A and $c(x)$ any other element in the ideal. By the division algorithm

$$c(x) = q(x)g(x) + r(x), \quad \deg r < \deg g$$

Since A is an ideal, $(c(x) - q(x)g(x)) \in A$ and hence $r(x) \in A$, contradicting the fact that $g(x)$ was of minimal degree. Thus $r(x) = 0$ and $g(x)$ divides every element in the ideal.

We can apply exactly the same argument to the polynomial $x^n - 1$, i.e.,

$$x^n - 1 = q_1(x)g(x) + r_1(x), \quad \deg r_1 < \deg g$$

whence $-r_1(x) \in A$, which again gives a contradiction and implies that $g(x) | (x^n - 1)$.

Suppose $g(x)$ is of degree $n - k$. Then the elements $g(x), xg(x), x^2g(x), \ldots, x^{k-1}g(x)$ are linearly independent in A. Since any element in the ideal is of the form $a(x)g(x)$, $\deg a < k$, then these elements span the ideal. Thus A has dimension k. Notice that if $g(x) = \sum_{i=0}^{n-k} g_i x^i$, then the generator matrix of the code has the convenient form

$$G = \begin{bmatrix} g_0 & g_1 & g_2 & \cdots & g_{n-k-1} & 1 & 0 & \cdots & 0 & 0 \\ 0 & g_0 & g_1 & \cdots & g_{n-k-2} & g_{n-k-1} & 1 & \cdots & 0 & 0 \\ \vdots & & & & & & & & & \vdots \\ 0 & 0 & 0 & \cdots & 0 & g_0 & g_1 & \cdots & g_{n-k-1} & 1 \end{bmatrix}$$

Now, conversely, suppose that $g(x)$ is any monic polynomial that divides $x^n - 1$ and consider the ideal $(g(x)) = A_n g(x)$. Suppose $c(x) \in (g(x))$; then

$$c(x) = a(x)g(x) \bmod(x^n - 1)$$
$$= a(x)g(x) + b(x)(x^n - 1)$$

and since $g(x) | (x^n - 1)$, $g(x) | c(x)$. That $g(x)$ is the monic polynomial of minimum degree then follows readily, which completes the theorem. □

If $f(x)$ is any polynomial of degree t, then the reciprocal polynomial, which we shall denote by $f^*(x)$, is given by $f^*(x) = x^t f(1/x)$.

Theorem 7.3 Let $\mathscr{C}$ be a cyclic (n, k) code with generator polynomial $g(x)$. Then the dual code $\mathscr{C}^\perp$ is a cyclic $(n, n - k)$ code with generator polynomial $\alpha h^*(x)$, where $h(x) = (x^n - 1)/g(x)$, and α is chosen to make $\alpha h^*(x)$ monic.

1.7 Cyclic Codes

Proof Since $x^n - 1$ is the zero polynomial in $GF(q)C_n$, if $g(x) = \sum_{i=0}^{n-k} g_i x^i$ and $h(x) = \sum_{i=0}^{k} h_i x^i$, then

$$g(x)h(x) \equiv 0 = \sum_{i=0}^{n-1}\left(\sum_{j=0}^{n-1} g_j h_{i-j}\right) x^i$$

implies that

$$\sum_{j=0}^{n-1} g_j h_{i-j} = 0 \quad \text{for } i = 0, 1, \ldots, n-1$$

where $h_i = 0$ if $i < 0$ or $i > k$. However, this equation implies that the vector $(g_0, g_1, \ldots, g_{n-k}, 0, 0, \ldots, 0)$ is orthogonal to the vector $(h_k, h_{k-1}, \ldots, h_1, h_0, 0, \ldots, 0)$ and all cyclic shifts of it. Thus a parity check matrix for $\mathscr{C}$ can be written as

$$H = \begin{bmatrix} h_k & h_{k-1} & \cdots & h_1 & h_0 & 0 & \cdots & 0 & 0 \\ 0 & h_k & \cdots & & h_1 & h_0 & \cdots & 0 & 0 \\ \vdots & & & & & & & & \vdots \\ 0 & \cdots & 0 & 0 & h_k & h_{k-1} & \cdots & h_1 & h_0 \end{bmatrix}$$

which is of rank $n - k$ and hence this is the generator matrix of $\mathscr{C}^\perp$, a cyclic $(n, n-k)$ code with generator polynomial $h^*(x)$, up to a scalar multiple. $\square$

Let A_1 and A_2 be ideals in A_n. The intersection of A_1 and A_2, $A_1 \cap A_2$, is again an ideal. The union (set-theoretic) of A_1 and A_2 is not, in general, an ideal and so we define $A_1 \cup A_2$ (sometimes written as $A_1 + A_2$) as the smallest ideal in $GF(q)C_n$ containing A_1 and A_2. Finally, we define a product ideal $A_1 A_2$ to be the set of finite sums of the form $\sum_i a_{i_1} a_{i_2}$, $a_{i_1} \in A_1$, $a_{i_2} \in A_2$. It is readily checked that this is an ideal.

Theorem 7.4 Let A_1 and A_2 be two ideals in the cyclic group algebra $GF(q)C_n$ with generators $g_1(x)$ and $g_2(x)$, respectively. Then

(i) $A_1 \cup A_2$ is generated by $(g_1(x), g_2(x))$.
(ii) $A_1 \cap A_2$ is generated by $[g_1(x), g_2(x)]$.
(iii) $A_1 A_2$ is generated by $(g_1(x) g_2(x), x^n - 1)$.

Proof Every element of $A_1 \cup A_2$ is of the form $a_1(x)g_1(x) + a_2(x)g_2(x)$. Let $d(x) = (g_1(x), g_2(x))$; then clearly $d(x)$ divides each element of $A_1 \cup A_2$ and $A_1 \cup A_2 \subseteq (d(x))$. On the other hand, since there exist two polynomials $b_1(x)$ and $b_2(x)$ such that $b_1(x)g_1(x) + b_2(x)g_2(x) = d(x)$, then $d(x) \in A_1 \cup A_2$ and $(d(x)) \subseteq A_1 \cup A_2$ and $(d(x)) = A_1 \cup A_2$.

Every element of $A_1 \cap A_2$ is a multiple of $e(x) = [g_1(x), g_2(x)]$, the least common multiple of $g_1(x)$ and $g_2(x)$, since it is divisible by both $g_1(x)$ and $g_2(x)$. Since $e(x)$ is in both A_1 and A_2, $e(x) \in A_1 \cap A_2$ and $A_1 \cap A_2 = (e(x))$.

For two-sided ideals, $A_1 \cdot A_2 \subseteq A_1 \cap A_2$. But $A_1 A_2 = (g_1(x)g_2(x))$ and as a consequence of Theorem 7.2, the generator of $A_1 A_2$ is $(x^n - 1, g_1(x)g_2(x))$. This can also be seen as follows: Let

$$d_1(x) = a(x)g_1(x)g_2(x) + b(x)(x^n - 1) = (x^n - 1, g_1(x)g_2(x))$$

Then $d_1(x) = a(x)g_1(x)g_2(x) \bmod (x^n - 1)$ and $d_1(x) \in A_1 A_2$. Since $d_1(x)$ divides every element of the form $c(x)g_1(x)g_2(x)$, it follows that $A_1 A_2 = (d_1(x))$.

Notice that in general $A_1 A_2 \neq A_1 \cap A_2$. In the case when $(n, q) = 1$, however, $A_1 A_2 = A_1 \cap A_2$. This follows from the fact that in this case $x^n - 1$ has no repeated factors. Thus

$$(g_1(x)g_2(x), x^n - 1) = [g_1(x), g_2(x)]$$

since $g_1(x)$, $g_2(x)$, and $x^n - 1$ are all products of distinct irreducible factors of multiplicity one. □

Now suppose $g(x)$ is a generator of a cyclic (n, k) code $\mathscr{C}$, which is of exponent less than n, say e, so that $g(x)|(x^e - 1)$. Then, if $(x^e - 1) = a(x)g(x)$, the codeword $(x^e - 1)$ is in $\mathscr{C}$ and is a word of weight two. Thus the minimum distance of the code is at most two and the code can only detect one error and correct none. Clearly we are only interested in polynomial generators of exponent n for (n, k) cyclic codes. In a similar vein we note that if $(n, q) \neq 1$, then $x^n - 1 = (x^{n_1} - 1)^{k_1}$ for some n_1 and k_1. If $g(x)$ is a generator of a cyclic (n, k) code, then $g(x)$ may divide $x^{n_1} - 1$, leading to a code with minimum distance two. To eliminate these problems from consideration, we will often assume $(n, q) = 1$. It should be mentioned that the case $(n, q) \neq 1$ is not uninteresting and in fact some interesting and important codes arise in this manner. Assmus and Mattson (1969) describe one such class of optimal codes, which are of length p and have generator polynomials $(x - 1)^i$, $i = 1, \ldots, p - 1$, over $GF(p)$. We will consider this class of codes later. For the remainder of this chapter, however, we will assume that $(n, q) = 1$ unless explicitly mentioned otherwise. We will also assume that all codes do not have an all zero coordinate position.

A cyclic (n, k) code over $GF(q)$ corresponds to an ideal in A_n that is generated by $g(x)$, a divisor of $x^n - 1$. From the previous section there are $\sigma_q(n)$ irreducible factors of $x^n - 1$ over $GF(q)$ and hence there are $2^{\sigma_q(n)}$ cyclic codes of length n over $GF(q)$, $(n, q) = 1$. These include the trivial codes consisting of the null code and the whole algebra.

Also, some of these codes may be equivalent and perhaps we should consider this problem before proceeding. First note that if $\mathscr{C}$ is a cyclic (n, k) code generated by $g(x)$ with roots $\{\eta^i, i \in k\}$, where η is a primitive nth root of unity and k is a set of q chains mod n in $\sum_q(n)$, then $c(x)$ is a codeword if and only if $g(x)|c(x)$, which is equivalent to saying that $c(x)$ is a codeword if and only if $c(\eta^i) = 0$, $\forall i \in k$.

1.7 Cyclic Codes

Theorem 7.5 Let η and μ be any two primitive nth roots of unity in a suitable extension field of $GF(q)$. Then if k is a union of q chains mod n, the polynomials

$$g_\eta(x) = \prod_{i \in k}(x - \eta^i), \qquad g_\mu(x) = \prod_{i \in k}(x - \mu^i)$$

generate equivalent codes.

Proof Notice that there are $\varphi(n)$ primitive nth roots of unity and that if η is a given primitive root, the others are η^j, where $(j, n) = 1$. Let η and $\mu = \eta^j$ be two primitive roots and suppose $c(x) = \sum_{i=0}^{n-1} c_i x^i$ is a codeword in $(g_\eta(x))$. Then $\sum_{i=0}^{n-1} c_i(\eta^l)^i = 0$ for all $l \in k$. Define the permutation mapping on the integers $0, 1, \ldots, n-1$ by

$$\tau_j(i) = ij \bmod n$$

Then this polynomial can be written as

$$\sum_{i=0}^{n-1} c_{\tau_j(i)}(\eta^l)^{\tau_j(i)} = 0 = \sum_{i=0}^{n-1} c_{\tau_j(i)}((\eta^j)^l)^i$$

$$= \sum_{i=0}^{n-1} c_{\tau_j(i)}(\mu^l)^i, \qquad l \in k$$

which implies that $(c_{\tau_j(0)}, c_{\tau_j(1)}, \ldots, c_{\tau_j(n-1)})$ is a codeword in $(g_\mu(x))$ and establishes the equivalence of the codes $(g_\eta(x))$ and $(g_\mu(x))$. $\square$

Consider now the minimal ideals of A_n. Assume that over $GF(q)$, $x^n - 1$ factors into the irreducible factors $g_1(x)g_2(x) \cdots g_t(x)$, where $t = \sigma_q(n)$. Then the ideal $(g_i(x))$ for any i, $1 \leq i \leq t$, clearly corresponds to a maximal ideal since if A is an ideal such that $(g_i(x)) \subset A \subset A_n$, then A is generated by some polynomial $f(x)$ such that $f(x)|g_i(x)$ and $f(x)|(x^n - 1)$, which is impossible unless $f(x) = 1$ or $f(x) = g_i(x)$. Similarly, the polynomial $(x^n - 1)/g_i(x)$ generates a minimal ideal that we shall denote by M_i. These minimal ideals are disjoint and hence $M_i \cap M_j = 0$ and

$$M_i \cup M_j = M_i \oplus M_j = \{m_i + m_j, m_i \in M_i, M_j \in M_j\}$$

and this direct sum has a generator polynomial, from Theorem 7.4(i), of

$$\left(\frac{x^n - 1}{g_i(x)}, \frac{x^n - 1}{g_j(x)}\right) = \frac{x^n - 1}{g_i(x)g_j(x)}$$

Thus for any ideal $A = (g(x))$, where $g(x) = g_{i_1}(x)g_{i_2}(x) \cdots g_{i_s}(x)$, then

$$\frac{x^n - 1}{g_{j_1}(x) \cdots g_{j_r}(x)} = g_{i_1}(x)g_{i_2}(x) \cdots g_{i_s}(x)$$

and A is the direct sum of the minimal ideals $M_{j_1}, M_{j_2}, \ldots, M_{j_r}$.

For the remainder of the section we consider some well-known classes of cyclic codes, beginning with those of Hamming (1950), although in his original formulation the cyclic nature of the codes was not considered. The original codes of Hamming were restricted to the binary case and we shall treat these first before generalizing them. Define the $m \times (2^m - 1)$ parity check matrix H for these codes as the set of all nonzero, binary m-tuples. Since the addition of any column to another cannot be zero, and since it is easy to find three linearly dependent columns, this defines a code with distance three. The rank of H is clearly m and hence we have a $(2^m - 1, 2^m - m - 1, 3)$ code over $GF(2)$. In this formulation the ordering of the columns of H was irrelevant. A little thought will show that if we let α be a primitive element of $GF(2^m)$, then we can as well express H as the $1 \times (2^m - 1)$ matrix

$$H = [1 \; \alpha \; \alpha^2 \; \cdots \; \alpha^{2^m - 2}]$$

where it is understood that a code vector is one whose corresponding polynomial has α as a zero. In this formulation the code is cyclic. From Section 1.6, if $h(x)$ is the primitive polynomial with root α, then the generator polynomial of the code is $h(x)$.

There are many ways of generalizing Hamming codes but we shall only discuss two of them. Define the parity check matrix H of a code over $GF(q)$ as the $m \times [(q^m - 1)/(q - 1)]$ matrix with the property that no two of its columns are scalar multiples of one another. The matrix has rank m and by definition has distance three. The result is a $((q^m - 1)/(q - 1), (q^m - 1)/(q - 1) - m, 3)$ code over $GF(q)$. When $(n, q - 1) = 1$, $n = (q^m - 1)/(q - 1)$, a cyclic version of this code can be described as follows. Let α be a primitive element in $GF(q^m)$. Then $\beta = \alpha^{q-1}$ is a primitive nth root of unity, where

$$n = \frac{q^m - 1}{q - 1}$$

and α^n is a primitive element of $GF(q)$. Define a parity check matrix by

$$H = [1 \; \beta \; \beta^2 \; \cdots \; \beta^{n-1}]$$

and suppose two of these columns are linearly dependent over $GF(q)$, which implies

$$\beta^i = \alpha^{jn} \beta^k, \quad i \neq k, \quad i, k \leq n - 1, \quad j < q - 1$$

or

$$\alpha^{(q-1)(i-k)} = \alpha^{jn}$$

which implies that

$$(q - 1)(i - k) = jn$$

1.7 Cyclic Codes

Since $j < q - 1$, this equation will have no solutions if $(n, q - 1) = 1$, and in this case the powers of β will be linearly independent over $GF(q)$, as required. Notice that for $q = 2$ the parameters of this code reduce to those of the binary code and the arguments remain valid and give the same code as in the binary case.

That the above codes are perfect follows from the direct calculation

$$q^n = q^{(q^m-1)/(q-1)} = \left[\binom{n}{1}(q-1) + 1\right] q^{[(q^m-1)/(q-1)]-m}$$

which is readily verified.

By a straightforward argument, the weight enumerator polynomial for these codes can be obtained. As previously, we denote the number of words in the code of weight i by A_i and the weight enumerator polynomial by

$$A(z) = \sum_{i=0}^{n} A_i z^i$$

where n is the length of the code. We will consider linear combinations of columns of the parity check matrix H that correspond to codewords, i.e., we are interested in linear dependence relations among the columns. There are precisely A_j of these relationships with j terms. The number of linear combinations of $j - 1$ columns with the property that a scalar multiple of one more term may be added to give a linearly dependent relationship is jA_j since for each code n-tuple of weight j there are j ways to choose the column to be added. From the construction of the matrix H a linear sum of any $j - 1$ columns of H is a scalar multiple of some column of H. This column may or may not be in the original set of $j - 1$ columns. There are a total of

$$\binom{n}{j-1}(q-1)^{j-1}$$

linear combinations of $j - 1$ columns, of which

$$\binom{n}{j-1}(q-1)^{j-1} - A_{j-1}$$

yield nonzero answers. If our linear sum of $j - 1$ columns can be written as a linear sum of $j - 2$ columns which add to zero (which corresponds to a codeword of weight $j - 2$) plus a scalar multiple of one other column, then the number of ways this can be done is

$$[n - (j-2)](q-1)A_{j-2}$$

since there are $n - (j - 2)$ ways of choosing the $(j - 1)$th column. There is also the possibility that the $j - 1$ terms in the linear sum add to a scalar multiple of a column already included in the sum. If the scalar multiple were the negative of the coefficient of that column in the sum, it would cancel and we would be reduced to the above case. Thus we assume this does not happen and conclude that there are

$$(j-1)(q-2)A_{j-1}$$

ways in which this situation arises. Collecting the pieces, we have the recursion relationship

$$jA_j = \binom{n}{j-1}(q-1)^{j-1} - A_{j-1} - (j-1)(q-2)A_{j-1} \\ - [n(-j-2)](q-1)A_{j-2} \tag{7.1}$$

To obtain an explicit relationship, we multiply this equation by z^{j-1} and sum from $j = 0$ to $j = n + 2$ to obtain

$$A'(z) = [1 + (q-1)z]^n - A(z) - (q-2)zA'(z) \\ - n(q-1)zA(z) + (q-1)z^2 A'(z)$$

This is a first-order differential equation with the initial condition $A(0) = 1$. Its solution is

$$A(z) = \frac{1}{n(q-1)+1} \\ \times \{[1 + (q-1)z]^n + n(q-1)[1+(q-1)z]^{(n-1)/q}(1-z)^{[n(q-1)+1]/q}\}$$

which is the required weight enumerator for the $((q^m - 1)/(q-1), [(q^m-1)/(q-1)] - m, 3)$ Hamming codes over $GF(q)$.

We define a maximum length (n, k) code to be one whose null space is generated by a primitive polynomial of degree k over GF(q). It follows that $n = q^k - 1$. Suppose $a(x)$ is the primitive polynomial and $b(x) = (x^n - 1)/a(x)$; then $b(x)$ is the generator of an equivalent code. Consider a permutation of the codeword $b(x)$, say $x^l b(x)$, and suppose $x^l b(x) = b(x)$ mod $(x^n - 1)$. This implies that $a(x)|(x^l - 1)$, which, since $a(x)$ is primitive, implies $l \geq q^k - 1$, which in turn implies that $b(x)$ is distinct from its $q^k - 1$ permutations. Since the code only contains $q^k - 1$ nonzero codewords, we conclude that every nonzero codeword is a cyclic permutation of every other nonzero codeword. Since the code is constrained to be linear, any code with this property must have length $q^k - 1$ for some integer k. If the length of the code exceeded $q^k - 1$ and all permutations of a single codeword were distinct, then it would have more then q^k codewords, contradicting the fact that it is an (n, k) code. Thus the name maximal length is appropriate.

If the distance of the code is d, then it is clear that the weight enumerator polynomial is

$$A(z) = 1 + (q^k - 1)z^d$$

1.7 Cyclic Codes

To determine the code distance, we argue as follows. Denote the code space by V and consider the set of codewords that have a zero in some fixed coordinate position which, without loss of generality, we may take as the first. This set forms a subspace of V, say V_1, and consider the decomposition $V = V_1 \cup V_2 \cup \cdots \cup V_m$ of V into cosets of V_1, where the V_i are cosets. In any one coset the first coordinate position contains the same element. Since V is a subspace, all elements appear in the first position in the code and $m = q$. Since V_1 is also a subspace, it contains q^l vectors for some positive integer l and from the foregoing equation we must have that $q \cdot q^l = q^k$ or $l = k - 1$. It follows readily that each element of $GF(q)$ appears in the jth coordinate position exactly q^{k-1} times in any linear (n, k) code over $GF(q)$, unless the coordinate is identically zero, a case we have already excluded. To determine the distance d of the maximum length code, we simply count the number of nonzero elements appearing in all coordinate postions of each codeword. Since each nonzero codeword has weight d, this gives $(q^k - 1)d$. On the other hand, each coordinate position contains $(q-1)q^{k-1}$ nonzero elements and there are $q^k - 1$ coordinate positions, giving

$$(q^k - 1)d = (q - 1)q^{k-1}(q^k - 1)$$

or

$$d = (q - 1)q^{k-1}$$

The maximum length codes have the parameters $(q^k - 1, k, (q - 1)q^{k-1})$, and weight enumerator polynomial

$$A(z) = 1 + (q^k - 1)z^{(q-1)q^{k-1}}$$

Since a primitive polynomial of degree k exists for any positive integer k, maximum length codes with these parameters always exist. It can also be shown that these codes are optimum in the sense that no other $(q^k - 1, k)$ code can have a larger distance, which follows from the fact that the d attained with maximum length codes satisfies the Plotkin bound (see Section 1.11). For linear codes this bound states that an (n, k, d) code cannot exist unless

$$d \leq \frac{q^k(q - 1)}{(q^k - 1)q} n$$

and a maximum length code satisfies this bound with equality.

The relationship between maximum length and Hamming codes is clarified by the following discussion. Let G be a $k \times (q^k - 1)$ matrix over $GF(q)$ which is a generator matrix of a maximum length code. The columns of G comprise all the nonzero k-tuples over $GF(q)$ since, if two columns were identical, it would imply that two distinct permutations of the same word are identical, contrary to the assumption that the code is maximum length. Label the non-zero elements of $GF(q)$ by α^i, $i = 1, \ldots, q-1$, α a primitive element, and partition the matrix G into

$$G = [H \mid \alpha H \mid \alpha^2 H \mid \cdots \mid \alpha^{q-2} H]$$

where H is a $k \times l$ matrix over $GF(q)$, $l = (q^k - 1)/(q - 1)$, whose columns are linearly independent over $GF(q)$. Clearly H is a parity check matrix of a Hamming code with appropriate parameters. If H is considered as the generator matrix of a code $\mathscr{C}$, then we claim that every nonzero codeword in $\mathscr{C}$ has weight q^{k-1}. To see this, let $\mathbf{X}$ be a nonzero word in $\mathscr{C}$ and $\mathbf{y} = (\mathbf{X}, \alpha\mathbf{X}, \alpha^2\mathbf{X}, \ldots, \alpha^{q-2}\mathbf{X})$ the corresponding word in the maximum length code. If w is the weight of $\mathbf{X}$ then $(q-1)w = (q-1)q^{k-1}$, from which the result follows. Thus the weight enumerator of the code is

$$A(z) = \sum_{i=0}^{l} A_i z^i = 1 + (q^k - 1)z^{q^{k-1}}$$

Using the MacWilliams identity

$$B(z) = \sum_{i=0}^{l} B_i z^i = \frac{1}{q^k} \sum_{i=0}^{l} [1 + (q-1)z]^{l-i} (1-z)^i$$

it readily follows that

$$B(z) = \frac{1}{q^k} \{[1 + (q-1)z]^l + (q^k - 1)[1 + (q-1)z]^{l-q^{k-1}} (1-z)^{q^{k-1}}\}$$

from which the coefficients B_i can be obtained.

Certainly one of the most interesting, and general, classes of random error-correcting codes is given by the so-called BCH codes which were found independently by Bose and Ray-Chaudhuri (1960a,b) and Hocquenghem (1959). This original work was restricted to binary codes of length $2^m - 1$ for some integer m and was later extended by Gorenstein and Zierler (1961) to the nonbinary case, ie., $GF(q)$ and any length prime to q. We give the code construction for the general case of an (n, k) code over $GF(q)$. Let γ be an nth root of unity in $GF(q^m)$ and let $f(x) \in GF(q)[x]$ be the minimal polynomial of the set of elements

$$\gamma^{m_0}, \gamma^{m_0+1}, \gamma^{m_0+2}, \ldots, \gamma^{m_0+d-2}$$

where m_0 is an arbitary positive integer less than n. Since the order of γ^{m_0+i} divides n, n is a common multiple of the orders of these roots, and $f(x) \mid (x^n - 1)$. Hence we may take $f(x)$ to be the generator of an (n, k) code where

1.7 Cyclic Codes

k is the code dimension. We can express the parity check matrix of such a code as

$$H = \begin{bmatrix} 1 & \gamma^{m_0} & (\gamma^{m_0})^2 & \cdots & (\gamma^{m_0})^{n-1} \\ 1 & \gamma^{m_0+1} & (\gamma^{m_0+1})^2 & \cdots & (\gamma^{m_0+1})^{n-1} \\ \vdots & \vdots & \vdots & & \vdots \\ 1 & \gamma^{m_0+d-2} & (\gamma^{m_0+d-2})^2 & \cdots & (\gamma^{m_0+d-2})^{n-1} \end{bmatrix}$$

and the code is the null space of this matrix over $GF(q)$. We now show that any set of $d-1$ columns of H cannot be linearly dependent, a fact that will imply the code has distance at least d. Choose an arbitrary set of $d-1$ columns whose first entries are, say, $(\gamma^{m_0})^{i_1}$, $(\gamma^{m_0})^{i_2}$, ..., $(\gamma^{m_0})^{i_{d-1}}$. This set of $d-1$ columns forms a $(d-1) \times (d-1)$ determinant from which we can factor out $(\gamma^{m_0})^{i_j}$, $j = 1, \ldots, d-1$, to give

$$\gamma^{m_0 i_1 + m_0 i_2 + \cdots + m_0 i_{d-1}} \begin{vmatrix} 1 & 1 & \cdots & 1 \\ \gamma^{i_1} & \gamma^{i_2} & & \gamma^{i_{d-1}} \\ \gamma^{2i_1} & \gamma^{2i_2} & & \gamma^{2i_{d-1}} \\ \vdots & \vdots & & \vdots \\ \gamma^{(d-2)i_1} & \gamma^{(d-2)i_2} & \cdots & \gamma^{(d-2)i_{d-1}} \end{vmatrix}$$

Since the powers of γ in the second row are distinct and this is a Van der Monde matrix, its determinant is nonzero and hence any set of $d-1$ columns is linearly independent.

There are two slight problems with this construction. In general, it is very difficult to say much about the code dimension k and the distance d, the "designed distance," is only a lower bound on the actual distance of the code. At times this bound will be useful for codes that were not initially constructed as BCH codes, i.e., if we are given a cyclic code and can show that its generator polynomial has a sequence of roots of length $d-1$ that are successive powers of a given root, then the code has distance at least d. This bound is commonly referred to as the BCH bound regardless of the type of code under consideration. Stiffler (1971, p. 405) shows that the dimension k of a BCH code over $GF(q)$ with design distance d may be bounded below by the quantity $n - v$, where v is given for $m_0 = 0$ and $m_0 = 1$ by

$$v = \begin{cases} \left[\dfrac{q-1}{q}(d-1)\right]m + 1, & m_0 = 0 \\ \left[\dfrac{q-1}{q}d\right]m, & m_0 = 1 \end{cases}$$

A BCH code of length $q^m - 1$ for some positive integer m is termed a primitive BCH code. Otherwise, we call it a nonprimitive BCH code. This terminology will carry over to other codes as well. Binary BCH codes have received a considerable amount of attention and a great deal more is known about specific cases than the foregoing outline would indicate. The reader is referred to Peterson and Weldon (1972) and Berlekamp (1968).

Let us examine the relationship between these BCH codes and the (L, g) Goppa codes introduced in Section 1.5. From Equation (5.2) of that section the (L, g) code is defined as the null space of the matrix

$$H = \begin{bmatrix} 1 & \cdots & 1 \\ \alpha_1 & \cdots & \alpha_n \\ \vdots & & \vdots \\ \alpha_1^{r-1} & \cdots & \alpha_n^{r-1} \end{bmatrix} \begin{bmatrix} g^{-1}(\alpha_1) & & & 0 \\ & \cdot & & \\ & & \cdot & \\ 0 & & & g^{-1}(\alpha_n) \end{bmatrix}$$

If we choose the set $L = \{\alpha_1, \ldots, \alpha_n\}$ to be the set of all nonzero elements of $GF(2^m)$ and $g(z) = z^{2r}$, then

$$H = \begin{bmatrix} 1 & 1 & \cdots & 1 \\ 1 & \alpha & \cdots & \alpha^{n-1} \\ \vdots & \vdots & & \vdots \\ 1 & \alpha^{2r-1} & \cdots & \alpha^{(2r-1)(n-1)} \end{bmatrix} \begin{bmatrix} 1 & & & & 0 \\ & \alpha^{-2r} & & & \\ & & \alpha^{-4r} & & \\ & & & \alpha^{-6r} & \\ & & & & \cdot \\ 0 & & & & \alpha^{-2r(n-1)} \end{bmatrix}$$

$$= \begin{bmatrix} 1 & \alpha^{-2r} & \alpha^{-4r} & \cdots & \alpha^{-2r(n-1)} \\ 1 & & & & \\ \vdots & \vdots & \vdots & & \vdots \\ 1 & \alpha^{-1} & \alpha^{-2} & \cdots & \alpha^{-(n-1)} \end{bmatrix}$$

where $n = 2^m - 1$. Notice that although the roots of $g(z)$ are not contained in L, they are repeated and hence we have only the bound $d \geqslant 2r + 1$, the usual BCH bound. Thus, Goppa codes contain the class of primitive BCH codes. In order to compare the relative performance of BCH and Goppa codes, we list the roughly comparable double-error-correcting codes in Table IV.

A particular and interesting subclass of BCH codes are the Reed–Solomon or RS codes (Reed and Solomon, 1960), whose appearance actually antedated the BCH codes. If in the construction of BCH codes we choose $m = m_0 = 1$ then the generating polynomial is

$$f(x) = \prod_{i=1}^{d-1} (x - \gamma^i)$$

1.7 Cyclic Codes

TABLE IV

A COMPARISON OF (n, k) GOPPA AND BCH CODES

Redundancy	(L, g) code	BCH code
8	(14,6)	(15,7)
10	(30,20)	(31,21)
12	(62,50)	(63,51)
14	(126,112)	(127,113)
16	(254,238)	(255,239)

and we obtain an $(n, n - d + 1)$ code with minimum distance at least d. However, it is not difficult to show that for any (n, k, d) linear code it is always true that

$$d \leq n - k + 1$$

To establish this, we notice that an (n, k) code over $GF(q)$ with distance d will have an $(n - k) \times n$ parity check matrix H of rank $n - k$. It follows that every set of $n - k + 1$ columns of H will be a linearly dependent set, implying

$$d \leq n - k + 1$$

which is called the Singleton bound. The relationship between this bound and the Plotkin bound quoted for maximum length codes is made clear upon noting that a RS code over $GF(q)$ may not have length greater than $q - 1$ (we assumed length equal to $q - 1$ in our discussion). For a maximum length code over $GF(q)$ of length $q - 1$ we have $k = 1$. When $k = 1$ both the Plotkin and the Singleton bounds are identical. For $n \leq q - 1$, the Singleton bound is stronger than the Plotkin bound since it can always be achieved in this case.

Theorem 7.6 Let a (not necessarily linear) code have q^k codewords of block length n over an alphabet of q letters. If the minimum distance between any two codewords is d, then we have

$$d \leq n - k + 1$$

Proof Choosing any k coordinate positions, it is clear that we have at most q^k distinct k-tuples appearing in these positions among the q^k codewords. If any two of these k-tuples are the same, then $d \leq n - k$. If no two of these k-tuples agree in all k-places, then since there are q^k words, at least two must agree in $k - 1$ positions, which implies $d \leq n - k + 1$. □

Corollary If $d = n - k + 1$, then all q^k k-tuples appear in any k coordinate positions.

Any code satisfying the bound in Theorem 7.6 with equality will be called a maximum-distance-separable code. We avoid the term optimal, which some authors use, since its meaning tends to vary with the application.

We restrict our attention now to linear (n, k) codes over $GF(q)$ whose distance satisfies with equality the above bound. It turns out that we can completely determine the weight enumerator of this code. The following line of reasoning is due to Goethals (1969).

Theorem 7.7 Every k columns of a generator matrix of an (n, k) maximum-distance-separable code over $GF(q)$ are linearly independent.

Proof Suppose there exist k dependent columns of the $k \times n$ generator matrix G of rank k. This would imply that the k-tuples appearing in the rows corresponding to these k columns are also dependent. Thus there exists a nontrivial dependency relation among the k rows of G, yielding a nonzero codeword with zeros in the k-dependent columns. This is impossible since it would imply $d \leq n - k$. □

Corollary The dual of a maximum-distance-separable code is also a maximum-distance-separable code.

Proof If we let the matrix G of Theorem 7.7 be the parity check matrix of the dual code, it is clear that no codeword of the dual has weight less than $k + 1$ and that there trivially exist codewords of this weight. Thus the dual code is an (n, k', d') code, $k' = n - k$, $d' = k + 1$, and $d' = k + 1 = n - k' + 1 = n - (n - k) + 1$. □

Theorem 7.8 Let $\mathscr{C}$ be a maximum-distance-separable (n, k) code and $\mathscr{C}'$ be the subcode of $\mathscr{C}$ consisting of all those codewords with zeros in j fixed positions, $j \leq k - 1$, and with these j positions deleted. Then $\mathscr{C}'$ is a maximum-distance-separable $(n - j, k - j)$ code.

Proof Without loss of generality assume the j fixed positions to be the first j positions. Transforming the generator matrix into systematic form, it is evident that the last $n - j$ positions of the lower $k - j$ rows span an $(n - j, k - j)$ code with a distance of $n - k + 1$ inherited from the original code, i.e., a maximum-distance-separable code. □

With these facts we proceed to find the weight distribution of any linear (n, k) maximum-distance-separable code over $GF(q)$. Recall the MacWilliams weight enumeration identities:

$$\sum_{i=0}^{n-r} A_i \binom{n-i}{r} = q^{k-r} \sum_{j=0}^{r} \binom{n-j}{n-r} B_j$$

1.7 Cyclic Codes

where $A(z) = \sum_i A_i z^i$ and $B(z) = \sum_i B_i z^i$ are the weight enumerations of a linear code $\mathscr{C}$ and its dual $\mathscr{C}^\perp$, respectively. Since for a maximum-distance-separable code we have $A_0 = 1$ and $A_1 = A_2 = \cdots = A_{n-k} = 0$ and $B_0 = 1$ and $B_1 = B_2 = \cdots = B_k = 0$, then

$$\sum_{i=0}^{n-r} A_i \binom{n-i}{r} = q^{k-r}\binom{n}{r}, \qquad r = 0, 1, \ldots, k-1$$

Solving this set of equations, recursively or otherwise, yields the following result.

Theorem 7.9 The number of vectors of weight $n-i$ in a maximum-distance-separable linear (n, k) code over $GF(q)$ is given by

$$A_{n-i} = \sum_{j=i}^{k-1} (-1)^{j-i} \binom{j}{i}\binom{n}{j}(q^{k-j} - 1), \qquad i = 0, 1, \ldots, k-1$$

Thus the number of codewords of minimum (nonzero) weight in such a code is given by

$$A_{n-k+1} = \binom{n}{k-1}(q-1)$$

while the number of codewords of the next highest weight is given by

$$A_{n-k+2} = \binom{n}{k-2}(q^2 - 1) - \binom{k-1}{k-2}\binom{n}{k-1}(q-1)$$

$$= (q-1)\binom{n}{k-2}[(q+1) - (n-k+2)]$$

Since this quantity must be positive, we have that

$$q + 1 \geq n - k + 2 \qquad \text{or} \qquad q - 1 \geq n - k$$

Applying the same argument to the dual code gives the following theorem.

Theorem 7.10 If a linear (n, k) code over $GF(q)$ is maximum distance separable, $2 \leq k \leq n-2$, then

$$q - 1 \geq \max(k, n-k)$$

This theorem is, of course, not sufficient to yield a maximum-distance-separable code. For example, we could consider a given non-maximum-distance-separable code over $GF(q)$ as a code over an extension field chosen sufficiently large to satisfy the inequality.

For $q = 2$ and $k = 1$ the binary repetition code is maximum distance separable, while for $q = 2$ and $k = n - 1$ the set of all 2^{n-1} binary n-tuples with an overall parity check added (i.e., $d = 2$) is a maximum-distance-separable code. The Reed–Solomon codes of length $q - 1$ are also maximum distance separable. It can be shown in fact (Goethals, 1969) that if $k < n \leqslant q$, then there always exists a maximum-distance-separable code over $GF(q)$. The case when $k = 2$ has been completely solved (Singleton, 1964) in the sense that any maximum-distance-separable code with q^2 codewords over an alphabet of q letters is equivalent to a set of $n - 1$ pairwise orthogonal Latin squares. We will prove this in the next chapter and indicate how other interesting codes can be obtained from Latin squares. For $3 \leqslant k \leqslant n - 3$ the problem of existence of maximum-distance-separable codes is only partially solved. The following two classes of codes, both due to Assmus and Mattson, are examples of maximum-distance-separable codes that do not seem to fall into any of the particular classes of codes discussed so far.

Class 1   Consider the prime p and cyclic codes of length p over $GF(p)$. We should immediately remark that, by and large, cyclic codes whose length is not relatively prime to the field characteristic have not received much attention. There is one obvious reason for this situation, as shown previously. For example, let $2 \mid n$ and consider a cyclic code over $GF(2)$ generated by $g(x)$. Since $x^n + 1 = (x^{n/2} + 1)^2$, if $g(x) \mid (x^{n/2} + 1)$, then the code will have a word of weight two. If $g(x) \nmid (x^{n/2} + 1)$, it is possible to obtain interesting codes. A less obvious reason, as we will show in Chapter 4, is that when the characteristic of $GF(q)$ and code length n are not relatively prime then the algebra $GF(q)[x]/(x^n - 1)$ is nonsemisimple and, in general, has a considerably more complex structure than when $(n, q) = 1$. Returning to the codes at hand, we first observe that $x^p - 1 = (x - 1)^p$ over $GF(p)$ and since any ideal of $GF[x]/(x^p - 1)$ is generated by a divisor of $(x^p - 1)$, we have that all of the ideals, say A_i, have generator polynomials $(x - 1)^i$, $i = 0, \ldots, p$, and these account for all the ideals, including the two trivial ideals. Clearly, these ideals satisfy the strict inclusion relation

$$GF(p)[x]/(x^p - 1) = A_0 \supset A_1 \supset A_2 \supset \cdots \supset A_p = (0)$$

Since the inclusions are strict, the dimension of A_i is $p - i$. To determine the minimum weight of A_i, let $f(x)$ be a minimum weight code polynomial permuted so as to have a nonzero constant term. Since $(x - 1)^i \mid f(x)$, $f(x) = a(x)(x - 1)^i$, it follows that the formal derivative of $f(x)$, $f'(x) = a'(x)(x - 1)^i + a(x)i(x - 1)^{i-1}$, is a codeword polynomial of A_{i-1}. Its weight is, at most, one less than the minimum weight of A_i. Thus the minimum weight of A_i, d_i, is a strictly increasing function of i. Since $d_0 = 1$ and $d_{p-1} = p - 1$, we conclude that $d_i = i + 1$ and the codes are maximum distance separable.

Actually, Assmus and Mattson (1969), were able to prove a far stronger result contained in the following theorem:

1.7 Cyclic Codes

Theorem 7.11 Let p be any prime. Then for all but a finite number of primes p', each cyclic code of length p over $GF(p'^i)$ is maximum distance separable for all i.

Class 2 Let α be a primitive nth root of unity that we consider in some extension field of $GF(q)$, where q is assumed to be equal to $-1 \bmod n$. The polynomial $(x - \alpha)(x - \alpha^{-1}) = x^2 - (\alpha + \alpha^{-1})x + 1$ is irreducible over $GF(q)$ since, if $q + 1 = kn$, then

$$(\alpha + \alpha^{-1})^q = \alpha^q + \alpha^{-q} = \alpha^{-1+kn} + \alpha^{-(-1+kn)} = (\alpha + \alpha^{-1})$$

and hence $(\alpha + \alpha^{-1}) \in GF(q)$. When n is odd these irreducible polynomials, as α runs through half of the nth roots of unity, together with $x - 1$, form all the irreducible factors of $x^n - 1$. When n is even there is the additional factor $x + 1$.

Theorem 7.12 For $q \equiv -1 \bmod n$ there exists a cyclic (n, k) maximum-distance-separable code over $GF(q)$ for every dimension k if n is odd. If n is even, we can construct two such codes if k is odd.

Proof When n is odd we choose a generator polynomial with roots

$$\alpha^{-m}, \alpha^{-m+1}, \ldots, \alpha^{-1}, 1, \alpha, \ldots, \alpha^m$$

where $k = n - (2m + 1)$ and, as before, α is a primitive nth root of unity. From the BCH distance bound the resulting cyclic $(n, n - 2m - 1)$ code has distance $2m + 2$. For an odd dimension k we choose the generator polynomial roots to be the $2m$ consecutive powers

$$\alpha^{\frac{1}{2}(n-1)-(m-1)}, \alpha^{\frac{1}{2}(n-1)-(m-2)}, \ldots, \alpha^{\frac{1}{2}(n-1)}, \alpha^{\frac{1}{2}(n+1)}, \alpha^{\frac{1}{2}(n+1)+1},$$
$$\ldots, \alpha^{\frac{1}{2}(n+1)+(m-1)}$$

When n is even we can construct two different generating polynomials containing as roots $2m + 1$ consecutive powers of α, namely,

$$\alpha^{-m}, \ldots, \alpha^{-1}, 1, \alpha, \ldots, \alpha^m \quad \text{and} \quad -\alpha^{-m}, \ldots, -\alpha^{-1}, -1, -\alpha, \ldots, -\alpha^m$$

both of which yield $(n, n - 2m - 1)$ codes of distance $d = 2m + 2$. Notice that the dual of a code in this class is also in this class. □

We briefly discuss some elementary properties of primitive root (PR) codes discovered by Anderson (1968). As in his paper, we restrict our attention to codes over $GF(p)$, p a prime, and consider codes of length $p^m - 1$. Let $p_1(x), p_2(x), \ldots, p_l(x)$ be the primitive polynomials of degree m over $GF(p)$ and note that $l = \varphi(p^m - 1)/m$.

Definition A PR code of length $p^m - 1$ over $GF(p)$ is a code with a generator polynomial of the form

$$g(x) = (x^{p^m-1} - 1)/p_{i_1}(x)p_{i_2}(x) \cdots p_{i_r}(x)$$

where the $p_{i_j}(x)$ are distinct primitive polynomials of degree m over $GF(p)$.

Alternatively, a PR code may be described as one whose codewords contain all the nonprimitive elements of $GF(p^m)$ as roots. Since the degree of $g(x)$ is $p^m - 1 - rm$, a PR code has the parameters $(p^m - 1, rm)$. A simple bound on the minimum distance of the code is contained in the following theorem.

Theorem 7.13 A PR code of length $p^m - 1$ over $GF(p)$ with generator polynomial

$$g(x) = (x^{p^m-1} - 1)/p_{i_1}(x)p_{i_2}(x) \cdots p_{i_r}(x)$$

is the intersection of at most rm BCH codes, and at least one of these codes has a minimum distance at least $[(p^m - 1)/rm] + 1$.

Proof Let α be any primitive element of $GF(p^m)$. Then the roots of the generator polynomial $g(x)$ are all those powers of α that are not one of the rm roots of $p_{i_1}(x) \cdots p_{i_r}(x)$. This implies that in the sequence of powers of α that are roots of $g(x)$ there are at most rm subsequences of successive powers, and each subsequence of consecutive powers generates a BCH code. Thus the code generated by $g(x)$ is, from the previous section, an intersection of at most rm BCH codes. Among the at most rm subsequences of consecutive powers there must be at least one of length at least $(p^m - 1)/rm$ and hence the corresponding BCH code will have distance at least $[(p^m - 1)/rm] + 1$, which is thus also a lower bound on the minimum distance of the PR code. □

A more detailed discussion of these codes is given by Anderson (1968). We will encounter many more classes of codes in later sections.

1.8 Linear Transformations of Vector Spaces over Finite Fields

Perhaps one of the more important applications of finite fields in mathematics is the study of groups of linear transformations over finite fields. Many important group-theoretic notions arise in this manner. It can be shown that the group of automorphisms of a finite field, regarded as a vector space over its prime subfield, is isomorphic to the group of automorphisms of the finite field regarded as an additive Abelian p group. We consider certain subgroups of these automorphism groups in this section.

1.8 Linear Transformations of Vector Spaces over Finite Fields

As before, we denote the vector space of $GF(q^n)$ over $GF(q)$ by $V_n(q)$ and a k-dimensional subspace of it by $V_n^k(q)$. Before considering groups of automorphisms we deal with certain counting problems that will arise. The number of ways of choosing a basis for $V_n(q)$ is clearly equal to

$$(q^n - 1)(q^n - q)(q^n - q^2) \cdots (q^n - q^{n-1})$$

where the basis is ordered and there are $q^n - 1$ ways of choosing the first basis element, $q^n - q$ ways of choosing the second, since we delete all linear combinations of the first, etc. A more convenient way of writing this is

$$q^{(n-1)n/2} \prod_{i=1}^{n} (q^i - 1)$$

To find the number of distinct subspaces of dimension k of $V_n(q)$, we note that there are

$$(q^n - 1)(q^n - q) \cdots (q^n - q^{k-1})$$

ways of choosing k linearly independent vectors. Each such set of vectors generates a k-dimensional subspace $V_n^k(q)$ and each such subspace can be generated in

$$(q^k - 1)(q^k - q) \cdots (q^k - q^{k-1})$$

ways. Thus if we denote by $\begin{bmatrix} n \\ k \end{bmatrix}$ the number of distinct subspaces of dimension k in $V_n(q)$, then

$$\begin{bmatrix} n \\ k \end{bmatrix} = \frac{(q^n - 1)(q^n - q) \cdots (q^n - q^{k-1})}{(q^k - 1)(q^k - q) \cdots (q^k - q^{k-1})} = \prod_{i=0}^{k-1} \frac{(q^{n-i} - 1)}{(q^{k-i} - 1)}$$

The quantities $\begin{bmatrix} n \\ k \end{bmatrix}$ are called the Gaussian coefficients and satisfy the recursion relations

$$\begin{bmatrix} n \\ k \end{bmatrix} = \begin{bmatrix} n-1 \\ k-1 \end{bmatrix} + q^k \begin{bmatrix} n-1 \\ k \end{bmatrix} \quad \text{and} \quad \begin{bmatrix} n \\ k \end{bmatrix} = \frac{q^n - 1}{q^k - 1} \begin{bmatrix} n-1 \\ k-1 \end{bmatrix}$$

as can be checked by calculation. Notice the similarity of these relations to the binomial coefficient identities

$$\binom{n}{k} = \binom{n-1}{k-1} + \binom{n-1}{k} \quad \text{and} \quad \binom{n}{k} = \frac{n}{k} \binom{n-1}{k-1}$$

Indeed, they are identical if the limit as $q \to 1$ is considered.

As a final counting problem, we derive an expression for the number of l-dimensional subspaces of $V_n(q)$ containing a given k-dimensional subspace, for $n > l > k$. The number of ways of choosing $l - k$ additional independent vectors with which to extend the basis of the given $V_n^k(q)$ is simply

$$(q^n - q^k)(q^n - q^{k+1}) \cdots (q^n - q^{l-1})$$

These $l - k$ additional vectors generate a $V_n^{l-k}(q)$ that may be generated in

$$(q^{l-k} - 1)(q^{l-k} - q) \cdots (q^{l-k} - q^{l-k-1})$$

ways and hence the number of l-dimensional subspaces containing a given k-dimensional subspace is

$$\frac{(q^n - q^k)(q^n - q^{k+1}) \cdots (q^n - q^{l-1})}{(q^{l-k} - 1)(q^{l-k} - q) \cdots (q^{l-k} - q^{l-k-1})}$$

However, this quantity is equal to $\begin{bmatrix} n-k \\ l-k \end{bmatrix}$ since any l-dimensional subspace, say V_1, containing a given k-dimensional subspace V_2 is, by an extension of bases argument, the direct sum of V_2 and another subspace V_3, i.e., $V_1 = V_2 \oplus V_3$, where dim $V_3 = l - k$. The problem is then to find the number of distinct subspaces of dimension $l - k$ in a subspace of dimension $n - k$, i.e., $\begin{bmatrix} n-k \\ l-k \end{bmatrix}$.

Let A be an $n \times n$ nonsingular matrix over $GF(q)$ and let x, y, α be n-tuples over $GF(q)$. It is easily seen that the set of all transformations of the form

$$y = Ax + \alpha$$

which are called affine transformations, forms a group under composition. We shall call it the general linear nonhomogeneous group and denote it by $GLN(n, q)$. The number of nonsingular $n \times n$ matrices over $GF(q)$ is the number of ways that a basis of $V_n(q)$ can be taken to another basis. This is simply the number of bases of $V_n(q)$, which is

$$(q^n - 1)(q^n - q) \cdots (q^n - q^{n-1}) = q^{n(n-1)/2} \prod_{i=1}^{n} (q^i - 1)$$

The number of choices for α is q^n and so

$$|GLN(n, q)| = q^{n + [n(n-1)/2]} \prod_{i=1}^{n} (q^i - 1)$$

The set of all $n \times n$ nonsingular matrices forms a group, the general linear group, denoted by $GL(n, q)$. Its order is clearly

$$|GL(n, q)| = q^{n(n-1)/2} \prod_{i=1}^{n} (q^i - 1)$$

1.8 Linear Transformations of Vector Spaces over Finite Fields

The set of elements of $GL(n, q)$ with determinant unity forms a subgroup of $GL(n, q)$ called the special linear or unimodular group. It will be denoted by $SL(n, q)$ and it can be shown that it is generated by elements of the form

$$\begin{bmatrix} 1 & & & & & 0 & & \vdots & & \\ & 1 & & & & & & \vdots & & \\ & & 1 & & & & & \vdots & & \\ i \text{----} & & & \text{--------------} & & \text{----} & \lambda & \text{----} & & \text{----} \\ & & & & & & & \vdots & \ddots & \\ & & & & 0 & & & \vdots & & 1 \end{bmatrix}$$

(with column j indicated above)

obtained from the $n \times n$ identity matrix by adding a $\lambda \in GF(q)$ in the (i, j) position. In $GL(n, q)$ the number of elements of determinant unity is the same as the number of elements with determinant $\alpha \in GF(q)$, $\alpha \neq 0$. It follows that

$$|SL(n, q)| = \frac{1}{q-1} |GL(n, q)|$$

The center of $GL(n, q)$, $Z(GL(n, q))$, is the set of matrices that commute with all other matrices of $GL(n, q)$. It can be shown that

$$Z(GL(n, q)) = \{\lambda I \mid \lambda \in GF^*(q)\}$$

where I is the $n \times n$ identity matrix. The factor group $GL(n, q)/Z(GL(n, q))$ is called the projective general linear group and is denoted by $PGL(n, q)$. It follows that

$$|PGL(n, q)| = \frac{1}{q-1} |GL(n, q)|$$

The particular case $PGL(2, q)$ is often called the linear fractional group of q letters, while some authors use this term to denote $PGL(n, q)$. We shall consider $PGL(2, q)$ later.

The center of $SL(n, q)$ is the set of matrices

$$Z(SL(n, q)) = \{\alpha I \mid \alpha \in GF^*(q), \ \alpha^n = 1\}$$

If $n \mid (q-1)$, then we know that there are precisely n solutions to the equation $\alpha^n = 1$ in $GF(q)$. Also, an element $\alpha \in GF(q)$ satisfies this if and only if $\alpha^d = 1$, where $d = (n, q-1)$. This follows, for suppose that $d = an + b(q-1)$; then if $\alpha^n = 1$,

$$\alpha^d = \alpha^{an+b(q-1)} = (\alpha^n)^a (\alpha^{q-1})^b = 1$$

Conversely, if $\alpha^d = 1$, then $d|n$ and $\alpha^n = 1$ trivially. The factor group $SL(n, q)/Z(SL(n, q))$ is called the projective special linear group $PSL(n, q)$ and

$$|PSL(n, q)| = \frac{|SL(n, q)|}{d}, \qquad d = (n, q-1)$$

Many of the above-mentioned groups are of importance for their group-theoretic structure. We single out two types of groups for further study.

By a permutation group of order m and degree k is meant a set of permutations of k letters A that forms a group under composition. The group is said to be t-transitive if for every two ordered t-tuples of distinct letters, say $(a_1, a_2, \ldots, a_t)$ and $(b_1, b_2, \ldots, b_t)$, where $a_i, b_i \in A$, there exists an element σ of the group such that $\sigma(a_i) = b_i$, $i = 1, \ldots, t$.

Consider the special case of $GLN(n, q)$ obtained by setting $n = 1$, i.e., the set of mappings

$$y = \alpha x + \beta$$

where $\alpha, \beta \in GF(q)$ and $\alpha \neq 0$. The order of the group $GLN(1, q)$ is $q(q - 1)$. If for the set of letters A we use the elements of $GF(q)$, we see that the effect of such a mapping is to permute the letters of A. For if $y = \alpha\eta + \beta = \alpha\gamma + \beta$, then $\eta = \gamma$. Furthermore, this group is 2-transitive, or doubly transitive, on the letters of A. Suppose (α_1, α_2) and (β_1, β_2) are any 2-sets; then the equations

$$\beta_1 = \alpha\alpha_1 + \beta \qquad \text{and} \qquad \beta_2 = \alpha\alpha_2 + \beta$$

can be solved uniquely for α, β to give

$$\alpha = \frac{\beta_2 - \beta_2}{\alpha_2 - \alpha_1}, \qquad \beta = \frac{\alpha_1\beta_2 - \beta_1\alpha_2}{\alpha_1 - \alpha_2}$$

We conclude that $GLN(1, q)$ can be represented as a doubly transitive group of order $q(q - 1)$ and degree q. This group is often called the affine group.

Consider now the group of transformations of the form

$$y = \frac{\alpha_1 x + \beta_1}{\alpha_2 x + \beta_2}, \qquad \alpha_1\beta_2 - \beta_1\alpha_2 \neq 0$$

We call this the linear fractional group with one variable and it has a natural association with $PGL(2, q)$, which we do not pursue. The order of the group is $q(q^2 - 1)$ and we wish to show that it can be represented as a triply transitive permutation group on $q + 1$ letters. To the symbols of $GF(q)$ we adjoin the symbol ∞ to represent elements of the form $\beta/0$, $\beta \in GF(q)$, $\beta \neq 0$. That transformations of the above form permute the elements of $GF(q) \cup \{\infty\}$ is easily checked. The group is at least doubly transitive on $GF(q)$ since it contains the previous group as a subgroup. Since the group contains the element that maps

1.9 Code Invariance under Permutation Groups   63

0 to ∞ the group is at least 1-transitive on $GF(q) \cup \{\infty\}$. To show that it is 3-transitive on this set requres solving a set of equations for $\alpha_1, \beta_1, \alpha_2, \beta_2 \in GF(q)$ which takes the ordered triple (η_1, η_2, η_3) to $(\gamma_1, \gamma_2, \gamma_3)$, where η_i, $\gamma_i \in GF(q) \cup \{\infty\}$, i.e., solving the equations

$$\gamma_1 = \frac{\alpha_1 \eta_1 + \beta_1}{\alpha_2 \eta_1 + \beta_2}, \quad \gamma_2 = \frac{\alpha_1 \eta_2 + \beta_1}{\alpha_2 \eta_2 + \beta_2}, \quad \gamma_3 = \frac{\alpha_1 \eta_3 + \beta_1}{\alpha_2 \eta_3 + \beta_2}$$

That a solution exists follows from the fact that $\alpha_1 \beta_2 - \alpha_2 \beta_1 \neq 0$, the η_i are distinct, the γ_i are distinct, and that we have three equations in four unknowns.

There are many other problems of considerable interest on matrices over finite fields. Rather than extend the treatment here, we shall deal with those problems we require as we meet them.

1.9 Code Invariance under Permutation Groups

The relevance of the problem of finding as large a permutation group as possible on the coordinate positions of the code to the weight enumeration problem has already been indicated. Many of the results in this area tend to be quite specialized and tedious to prove. Instances of this have been observed in previous sections. We shall therefore confine our interest to a very elegant result of Kasami *et al.* (1968c).

It is clear that every cyclic code is invariant under the cyclic group of permutations of appropriate degree. Suppose $c(x)$ is a codeword polynomial of a cyclic (n, k) code over $GF(q)$. Then $c(x^q) \equiv c(x)^q$ (mod $x^n - 1$) and hence $c(x^q)$ (mod $x^n - 1$) is also a codeword. It follows that the code is invariant under the permutation group on n letters generated by the permutation

$$i \longrightarrow qi \bmod n$$

Also, if a linear code is invariant under some permutation group, then its dual code is also. The following theorem is both a simple and powerful result on this problem. If $\mathscr{C}$ is a code of length $q^m - 1$ over $GF(q)$ with generator polynomial $g(x)$, then we denote the extended code obtained by prefixing a digit to each word that is minus the sum of all the digits in the codeword by $\mathscr{C}_e$. Notice that if the unit element is a root of $g(x)$, then every codeword in the cyclic code is such that its digits sum to zero. Thus in this case the first position in each codeword of $\mathscr{C}_e$ is a zero. We exclude this case from consideration.

Assume that $\text{char}(GF(q)) = p$ and $q = p^l$ and suppose that

$$i = \sum_{t=0}^{ml-1} \delta_t p^t, \quad 0 \leq \delta_t \leq p-1, \quad \forall t$$

Then denote by $J(i)$ the set of integers with the representation

$$j = \sum_{t=0}^{ml-1} \sigma_t p^t, \qquad 0 \leq \sigma_t \leq \delta_t, \quad \forall t$$

Theorem 9.1 (Kasami et al. 1968c) The extended code $\mathscr{C}_e$ of length q^m over $GF(q)$ of a cyclic code $\mathscr{C}$ with generator polynomial $g(x)$, $g(1) \neq 0$, is invariant under the affine group of permutations if and only if for every α^i that is a root of $g(x)$ then for every $j \in J(i)$, $j \neq 0$, α^j is a root of $g(x)$, where α is a primitive root of $GF(q^m)$.

Proof Consider a codeword $c(x) \in \mathscr{C}_e$ of weight ω and let the coordinate positions of its nonzero components be $x_1, \ldots, x_\omega$ and the values of these nonzero components be $y_1, \ldots, y_\omega$. Then since $c(x) \in \mathscr{C}_e$ we must have

$$\delta_0 = \sum_{k=1}^{\omega} y_k = 0, \qquad S_i = \sum_{k=1}^{\omega} y_k x_k^i = 0$$

for each i for which α^i is a root of $g(x)$.

Define the affine transformation π_{ab} of $GF(q^m)$ by

$$\pi_{ab}: \beta \longrightarrow a\beta + b, \qquad a \neq 0, \quad a, b \in GF(q^m)$$

and denote by $c'(x)$ the result of the operation on the coordinate positions of $c(x)$ by π_{ab}. If $c'(x)$ is to be in $\mathscr{C}_e$, then defining

$$S_i' = \sum_{k=1}^{\omega} y_k (ax_k + b)^i$$

we must have $S_0' = S_i' = 0$ for each i for which α^i is a zero of $g(x)$. By a theorem of Lucas on the binomial coefficients (see Appendix B)

$$\binom{i}{j} = \prod_{t=0}^{ml-1} \binom{\delta_t}{\sigma_t} \mod p$$

where, as before, δ_t and σ_t are coefficients in the radix p expansions of i and j, respectively. This product is nonzero if and only if each term is nonzero, or, equivalently, if and only if $j \in J(i)$.

Returning to the equation of S_i', we see that

$$S_i' = \sum_{k=1}^{\omega} \sum_{j=0}^{i} y_k \binom{i}{j} a^j b^{i-j} x_k^j$$

$$= \sum_{k=1}^{\omega} \sum_{j \in J(i)} y_k \binom{i}{j} a^j b^{i-j} x_k^j$$

$$= \sum_{j \in J(i)} \binom{i}{j} a^j b^{i-j} S_j$$

1.9 Code Invariance under Permutation Groups

Suppose now that we choose $a = 1$ and consider

$$S_i' = \sum_{j \in J(i)} \binom{i}{j} b^{i-j} S_j$$

as a polynomial in b. If the code $\mathscr{C}_e$ is to be invariant under the doubly transitive affine group, then we must have $S_i' = 0$. Since, by definition of $J(i)$, $\binom{i}{j} \neq 0$, this implies that

$$S_j = 0, \qquad \forall j \in J(i)$$

and hence that if α^i is a root of $g(x)$ and $\mathscr{C}_e$ is invariant under the doubly transitive affine group, then α^j is also a root of $g(x)$ for each $j \in J(i)$. Conversely, if, whenever α^i is a root of $g(x)$ implies that α^j is also a root for any $j \in J(i)$, then $S_j = 0 \;\forall j \in J(i)$ and it follows that $S_i' = 0$ for every (a, b), implying that $\mathscr{C}_e$ is invariant under the permutation group, which completes the theorem. □

Implicit in the proof of this theorem is the fact that any code of length q^m that is invariant under the doubly transitive permutation group is actually an extended cyclic code, i.e., the code obtained by deleting the first coordinate position is cyclic. An inspection of the conditions on the roots used in defining the cyclic primitive BCH codes over $GF(q)$ reveals that the extended version of this code is invariant under the doubly transitive affine group.

Some other results of a general nature on this problem were obtained by Delsarte *et al.* (1970) and Delsarte (1970a, b).

In an interesting and significant paper Delsarte (1970a) discusses the invariance problem to some depth and defines classes of codes using these concepts. We prove one of his results here. Let η be a primitive bth root of unity in $GF(q^r)$, where $b|(q^r - 1)$. Let $x \in V_m(q^r)$ and denote the set of vectors

$$\{x, \eta x, \eta^2 x, \ldots, \eta^{b-1} x\}$$

by (x). Clearly this is an equivalence relation that divides the nonzero elements of $V_m(q^r)$ into $[(q^{rm} - 1)/b] = n$ classes, which are denoted by V_b. The general linear homogeneous group $GL(m, q^r)$ is the set of $m \times m$ nonsingular matrices over $GF(q^r)$ that acts as a permutation group of degree $q^{rm} - 1$ on the nonzero elements of $V_m(q^r)$. It fixes the all-zeros vector. However, it also acts as a permutation group of degree n on the classes of V_b and with this interpretation we denote the group $G_b(m, q^r)$.

Theorem 9.2 Let $\mathscr{C}$ be a linear code of length n over $GF(q)$. If $\mathscr{C}$ is invariant under $G_b(m, q^r)$, then it is equivalent to a cyclic code.

Proof Let σ be a primitive element of $GF(q^{rm})$ and $\boldsymbol{\alpha} = \{\alpha_1, \alpha_2, \ldots, \alpha_m\}$ a basis of $GF(q^{rm})$ over $GF(q^r)$. Since $\sigma\boldsymbol{\alpha} = \{\sigma\alpha_1, \sigma\alpha_2, \ldots, \sigma\alpha_m\}$ is also a basis there must exist a matrix, say M, such that its transpose M^T transforms one basis to the other

$$M^T\alpha = \sigma\alpha$$

where α is considered as a column vector. If ε is a vector such that $(\alpha, \varepsilon) = 1$, then it follows that

$$\sigma^i = \alpha^T M^i \varepsilon, \quad i = 0, 1, \ldots, q^{rm} - 2$$

which implies that $M^i \varepsilon$, $0 \leq i \leq q^{rm} - 2$, runs through all the elements of $V_m(q^r)$. Now σ^n is a primitive bth root of unity in $GF(q^r)$ and thus $M^n = \sigma^n I = \eta I$. Thus the vectors $M^i \varepsilon$, $0 \leq i \leq n - 1$, belong to different classes in V_b. If we relabel the n coordinate positions with these vectors, then since $M \in G_b(m, q^r)$, the new code will be cyclic and equivalent to the original code. □

We consider, without proof, a related result of Delsarte (1970a). Let $GLN(m, q^r)$ be the general linear nonhomogeneous group whose elements are transformations on $V_m(q^r)$ of the form

$$V_m(q^r) \longrightarrow V_m(q^r)$$
$$x \longmapsto Ax + b$$

where A is a nonsingular matrix. This group acts as a permutation group of degree q^{rm} on $V_m(q^r)$. $G_1(m, q^r)$ is the subgroup of $G(m, q^r)$ that fixes the all-zero vector of $V_m(q^r)$. Delsarte (1970a, Theorem 4) shows that if a nontrivial linear code of length q^{rm} over $GF(q)$ is invariant under $GLN(m, q^r)$, then it is equivalent to an extended cyclic code. Using a polynomial approach to codes, he is able to characterize rather completely the codes of appropriate length which are invariant under $G_b(m, q^r)$ and $G(m, q^r)$.

1.10 The Polynomial Approach to Coding

In this section we consider an alternative approach to coding, due to Mattson and Solomon (1961). This approach, introduced first for the single-variable case and then for the multivariable case, will yield the important class of generalized Reed–Muller (GRM) codes and the more general class of polynomial codes. While these are interesting in their own right, they will also lead to the finite geometry codes to be discussed in the next chapter.

Let $\mathscr{C}$ be a cyclic (n, k) code over $GF(q)$ generated by the polynomial $g(x)$ which has the roots $\{\beta^i, i \in k\}$, where β is a primitive nth root of unity and k is a set of q chains mod n. We assume $(n, q) = 1$ unless stated otherwise. The code $\mathscr{C}$ can be described by the set of codeword polynomials

$$\mathscr{C} = \left\{ f(x) \in GF(q)[x]/(x^n - 1) \,\Big|\, g(x) | f(x) \right\}$$

1.10 The Polynomial Approach to Coding

or equivalently,

$$\mathscr{C} = \left\{ f(x) \in GF(q)[x]/(x^n - 1) \,\middle|\, f(\beta^i) = 0, \quad i \in k \right\}$$

Let $c(x) = \sum_i c_i x^i \in \mathscr{C}$, where as usual we identify a codeword and the polynomial. To each such codeword we associate the polynomial

$$F_c(x) = \frac{1}{n} \sum_{j=1}^{n} f_j x^{n-j}$$

where

$$f_j = c(\beta^j) = \sum_{i=0}^{n-1} c_i (\beta^j)^i, \quad j = 1, \ldots, n$$

By a simple calculation we observe that

$$F_c(\beta^l) = \frac{1}{n} \sum_{j=1}^{n} f_j (\beta^l)^{n-j} = \frac{1}{n} \sum_{j=1}^{n} \left[\sum_{i=0}^{n-1} c_i (\beta^j)^i \right] \beta^{l(n-j)}$$

$$= \frac{1}{n} \sum_{i=0}^{n-1} c_i \left(\sum_{j=1}^{n} \beta^{j(i-l)} \right) = c_l$$

since the sum of all nth roots of unity is zero, for any n, and if $i = l$, then the term in the last set of parentheses is n. Thus, we could characterize the code $\mathscr{C}$ as

$$\mathscr{C} = \{(F_c(1), F_c(\beta), F_c(\beta^2), \ldots, F_c(\beta^{n-1})), \quad c(x) \in \mathscr{C}\}$$

which is not a significant improvement over our original characterization.

To improve upon this characterization, consider first some properties of the polynomials $F_c(x)$, the Mattson–Solomon polynomials. If the set k contains the integers $1, 2, \ldots, d-1$, then

$$c(\beta^i) = 0, \quad i = 1, 2, \ldots, d-1$$

which implies that the coefficients $f_i = 0$, $i = 1, 2, \ldots, d-1$. Since these are the coefficients of the terms $x^{n-1}, x^{n-2}, \ldots, x^{n-d+1}$ in $F_c(x)$, we conclude that $\deg F_c \leq n - d$ for any $c(x) \in \mathscr{C}$. Now the weight of a codeword $c(x)$ is simply n minus the number of nth roots of unity that are roots of F_c. Since a polynomial of degree $n - d$ cannot have more than $n - d$ such roots, we conclude the weight of any codeword polynomial is at least $n - (n - d) = d$. This, of course, is just the BCH bound on the code distance.

The situation can be viewed in a slightly different manner. We view the set of polynomials in this paragraph as elements of $GF(q)[x]/(x^n - 1)$. As before, we have

$$\mathscr{C} = \{c(x) \,|\, c(\beta^i) = 0, \quad i \in k\}$$

or, equivalently,

$$\mathscr{C} = \{(F(1), F(\beta), \ldots, F(\beta^{n-1})) | F(x) = \sum f_i x^{n-i} \text{ and } f_j = 0, \ j \in k\}$$
$$= \{(F(1), F(\beta), \ldots, F(\beta^{n-1})) | F(x) = \sum_{i_j \in \bar{k}} f_{i_j} x^{n-i_j}\}$$

where $\bar{k}$ is the complement of k in $\{1, 2, \ldots, n\}$. Now one basis of $GF(q)[x]/(x^n - 1)$ is $(x^{n-1}, x^{n-2}, \ldots, x, 1)$ and with respect to this basis the polynomial $F(x)$ can be expressed as $(f_1, \ldots, f_n)$ where $f_j = 0$ if $j \in k$. Rather than this basis, choose the polynomials

$$G_i(x) = \prod_{\substack{j=1 \\ j \neq i}}^{n} \frac{x - \beta^j}{\beta^i - \beta^j}, \qquad i = 1, \ldots, n$$

Clearly, we have $G_i(\beta^j) = \delta_{ij}$ and these polynomials form a basis of $GF(q)[x]/(x^n - 1)$. It is easily verified that

$$F(x) = \sum_{i=0}^{n-1} f_i x^{n-i} = \sum_{i=1}^{n} F(\beta^i) G_i(x)$$

since $F(x)$ is a polynomial of degree n or less. Thus we can characterize the code $\mathscr{C}$ as the set of polynomials

$$\{F(x) = \sum f_i x^{n-i}, \ f_j = 0, \ j \in k\}$$

expressed with respect to the basis $G_i(x)$, $i = 1, \ldots, n$. Although we will make no direct use of this fact, it is interesting to keep in mind.

The idea of associating a polynomial with each codeword in an (n, k) linear code, which when evaluated on the nth roots of unity yields the codeword coordinates, can be extended to the multivariate case. For the remainder of this section we will be concerned with the relationship between these two approaches. The generalized Reed–Muller (GRM) codes (Kasami *et al.* 1968a) are a particularly interesting class of codes in this respect and we will investigate them quite extensively. The polynomial codes also introduced by Kasami *et al.* (1968b) are also defined by a multivariable approach. We give some of their basic properties later in this section but our real interest in them will be in the next chapter where we will show that certain duals of polynomial codes are finite geometry codes.

Denote by $\bar{X}$ the set of variables $X_1, X_2, \ldots, X_m$ and by $P(\bar{X})$ a polynomial in these variables with coefficients from $GF(q)$. The degree of a monomial

$$X_1^{i_1} X_2^{i_2} \cdots X_m^{i_m}$$

is defined as $\sum_{j=1}^{m} i_j$ and the degree of a linear combination over $GF(q)$ of such monomials is the degree of the monomial of largest degree. Since we are interested in the case where X_i assumes values in $GF(q)$, all exponents of X_i

1.10 The Polynomial Approach to Coding

are taken mod $(q-1)$, i.e., $X_i^q = X_i$. We say such a polynomial is in reduced form and hence assume this to be the case unless stated otherwise.

If a polynomial assumes a value of zero for every $\bar{\alpha} = (\alpha_1, \ldots, \alpha_m) \in V_m(q)$, i.e., $X_i = \alpha_i \in GF(q)$, then it is identically zero. A simple proof of this by induction on m is easily constructed.

Recalling that

$$\prod_{\alpha \in GF(q)} (X - \alpha) = X^q - X$$

we see that

$$\prod_{\substack{\alpha \in GF(q) \\ \alpha \neq \omega}} (X - \alpha) = (X - \omega)^{q-1} - 1$$

By setting X equal to ω in this expression, it follows that the product of all nonzero elements of a finite field is equal to -1. This is an extension of Wilson's theorem, which states that $(p-1)! \equiv -1 \bmod p$. We now define the indicator function $F_{\bar{\alpha}}(\bar{X})$ by

$$F_{\bar{\alpha}}(\bar{X}) = \prod_{i=1}^{m} [1 - (X_i - \alpha_i)^{q-1}]$$

so that $F_{\bar{\alpha}}(\bar{X}) = 0$ if $\bar{X} \neq \bar{\alpha}$ and 1 if $\bar{X} = \bar{\alpha}$. Clearly, for a polynomial $P(\bar{X})$ we have

$$P(\bar{X}) = \sum_{\bar{\alpha} \in V_m(q)} P(\bar{\alpha}) F_{\bar{\alpha}}(\bar{X})$$

since X_i is a q-ary variable and P is assumed in reduced form. If we denote by $P(m, q)$ the set of all (reduced form) polynomials over $GF(q)$, then it is clear that $P(m, q)$ is a vector space of dimension q^m over $GF(q)$. An obvious basis is given by the set of monomials

$$X_1^{i_1} X_2^{i_2} \cdots X_m^{i_m}, \quad 0 \leq i_k \leq q-1, \quad k = 1, \ldots, m$$

Denote by $P_v(m, q)$ the subspace of $P(m, q)$ generated by the monomials of degree v or less.

Following Delsarte *et al.* (1970), as we will for much of the work on GRM codes, we prove the following two lemmas.

Lemma 10.1 If $\deg P(\bar{X}) < m(q-1)$, then

$$\sum_{\bar{\alpha} \in V_m(q)} P(\bar{\alpha}) = 0$$

Proof We have seen that

$$P(\bar{X}) = \sum_{\bar{\alpha} \in V_m(q)} P(\bar{\alpha}) F_{\bar{\alpha}}(\bar{X})$$

Since the coefficient of the monomial of degree $m(q-1)$ in the expansion

$$P_{\bar{\alpha}}(\bar{X}) = \prod_{i=1}^{m}[1-(X_i-\alpha_i)^{q-1}] = \prod_{i=1}^{m}\left[1 - \sum_{j=0}^{q-1} X_i^j(-\alpha_i)^{q-1-j}\binom{q-1}{j}\right]$$

is $(-1)^m$, and since the coefficient of this monomial in $P(\bar{X})$ is zero by assumption, we conclude that

$$\sum_{\bar{\alpha}\in V_m(q)} P(\bar{\alpha}) = 0$$

Alternatively, we could have shown that since

$$\prod_{\alpha\in GF(q)}(X-\alpha) = X^q - X$$

then, by examining the coefficient of X^{q-1} we have that

$$\sum_{\alpha\in GF(q)} \alpha = 0$$

From this it follows that

$$\sum_{\bar{\alpha}\in V_m(q)} \alpha_1^{i_1}\alpha_2^{i_2}\cdots\alpha_m^{i_m} = 0, \qquad \sum_{j=1}^{m} i_j < m(q-1)$$

and finally that

$$\sum_{\bar{\alpha}\in V_m(q)} P(\bar{\alpha}) = 0 \qquad \deg P < m(q-1) \qquad \square$$

The proof of the following lemma is, in a sense, crucial to the understanding of much of the remainder of this section. It will eventually enable us to relate the important notion of the q weight of an integer to the zeros of a cyclic code. Recall that by the q weight of an integer h we will mean the quantity $\omega_q(h)$, where

$$h = \sum_{i=0}^{\infty} h_i q^i \quad \text{and} \quad \omega_q(h) = \sum_{i=0}^{\infty} h_i$$

Lemma 10.2 If $\omega_q(h) \leq \mu$ and $P(\bar{X}) \in P_\nu(m, q)$, $\mu + \nu < m(q-1)$, then

$$\sum_{\bar{\alpha}\in V_m(q)} (\bar{\alpha}, \bar{\beta})^h P(\bar{\alpha}) = 0 \quad \text{in} \quad GF(q^r)$$

where $(\cdot, \cdot)$ indicates the usual inner product and $\bar{\beta}$ is some fixed element of $V_m(q^r)$.

Proof

$$\sum_{\bar{\alpha}\in V_m(q)} (\bar{\alpha}, \bar{\beta})^h P(\bar{\alpha}) = \sum_{\bar{\alpha}\in V_m(q)} \left(\sum_i \alpha_i\beta_i\right)^{\sum_k h_k q^k} P(\bar{\alpha})$$

$$= \sum_{\bar{\alpha}\in V_m(q)} \prod_k \left(\sum_i \alpha_i\beta_i^{q^k}\right)^{h_k} P(\bar{\alpha})$$

1.10 The Polynomial Approach to Coding

In the inner summation, raised to the power h_k for fixed k, each term is of the form $C(\bar{\beta})\alpha_1^{i_1}\alpha_2^{i_2}\cdots\alpha_m^{i_m}$, $\sum_j i_j = h_k$. Each term in the product of the sum of such terms is again a term of the form $C'(\bar{\beta})\alpha_1^{j_1}\alpha_2^{j_2}\cdots\alpha_m^{j_m}$, where $\sum_l j_l \leq \sum_i h_i = \omega_q(h)$. The summation is then of the form

$$\sum_{\bar{\alpha}\in V_m(q)} [\sum C'(\bar{\beta})\alpha_1^{j_1}\alpha_2^{j_2}\cdots\alpha_m^{j_m}]P(\bar{\alpha})$$

$$= \sum C'(\bar{\beta})\left[\sum_{\bar{\alpha}\in V_m(q)} \alpha_1^{j_1}\alpha_2^{j_2}\cdots\alpha_m^{j_m}P(\bar{\alpha})\right]$$

where the nonindexed outer summation is over the terms obtained in the inner product. Since $\alpha_1^{j_1}\alpha_2^{j_2}\cdots\alpha_m^{j_m}P(\bar{\alpha})$ is a polynomial in $\bar{\alpha}$ of degree at most $\mu + \nu < m(q-1)$, the summation over $\bar{\alpha}$ is, by the previous lemma, zero. □

We now give the first of three definitions of GRM codes.

Definition 1 The νth-order GRM code, denoted by $C_\nu(m, q)$, is the set of vectors of length q^m over $GF(q)$

$$C_\nu(m, q) = \{(P(\bar{0}), P(\bar{\alpha}_1), \ldots, P(\bar{\alpha}_{q^m-1})),\quad P(\bar{X}) \in P_\nu(m, q)\}$$

where $\bar{0}, \bar{\alpha}_1, \ldots, \bar{\alpha}_{q^m-1}$ is an arbitrary ordering of the elements of $V_m(q)$.

As in the single-variable case, if we identify the polynomial

$$P(\bar{X}) = \sum_{(i_1\cdots i_m)} C_{i_1 i_2 \cdots i_m} X_1^{i_1} X_2^{i_2} \cdots X_m^{i_m}$$

with the q^m-tuple over $GF(q)$ with the coefficients $C_{i_1 i_2 \cdots i_m}$ in some ordering in the coordinate vectors (i.e., w.r.t. the basis of monomials), then the code $C_\nu(m, q)$ is simply the image of $P_\nu(m, q)$ with respect to the basis $F_{\bar{\alpha}}(\bar{X})$, i.e.,

$$P(\bar{X}) = \sum_{\bar{\alpha}\in V_m(q)} P(\bar{\alpha})F_{\bar{\alpha}}(\bar{X})$$

and with respect to this new basis the corresponding coordinate vector contains the elements $P(\bar{\alpha})$, $\bar{\alpha} \in V_m(q)$, again in some ordering.

This code will be shown later to be equivalent to an extension of a cyclic code. Clearly the dimension of $C_\nu(m, q)$ over $GF(q)$ is the same as the dimension of $P_\nu(m, q)$ over $GF(q)$, as a subspace of $P(m, q)$. To determine this dimension, choose the set of monomials

$$\left\{X_1^{i_1} X_2^{i_2} \cdots X_m^{i_m}, \sum_{k=1}^m i_k \leq \nu\right\}$$

as a basis of $P_\nu(m, q)$. The number of polynomials in such a basis is just the number of ways we can place ν or fewer objects in m cells where no cell is to contain more than $q-1$ objects. The number of ways of placing t objects in m cells, no cell to contain more than $q-1$ objects, is (Riordan, 1958)

$$N(t, m, q) = \sum_{k=0}^{m} (-1)^k \binom{m}{k} \binom{t - kq + m - 1}{t - kq}$$

Thus the dimension of the vth-order GRM code $C_v(m, q)$ is

$$\sum_{t=0}^{v} N(t, m, q) = \sum_{t=0}^{v} \sum_{k=0}^{m} (-1)^k \binom{m}{k} \binom{t - kq + m - 1}{t - kq}$$

To find the dual code of $C_v(m, q)$, we let $\mu = m(q-1) - v - 1$. The product of a polynomial $P_1(\bar{X}) \in P_v(m, q)$ and $P_2(\bar{X}) \in P_\mu(m, q)$ is of degree less than $m(q-1)$. Thus, by the first lemma in this section,

$$\sum_{\bar{\alpha} \in V_m(q)} P_1(\bar{\alpha}) P_2(\bar{\alpha}) = 0$$

which implies that the corresponding codewords are orthogonal. Thus if $\bar{x} \in C_v(m, q)$ and $\bar{y} \in C_\mu(m, q)$, $(\bar{x}, \bar{y}) = 0$.

Theorem 10.1 $C_\mu(m, q)$ is the dual code of $C_v(m, q)$.

Proof From the above comments we need only verify that the dimensions of the two codes add to q^m. Let r be the number of monomials of degree less than or equal to v. By replacing the exponent i of each variable by $(q-1) - i$, for each monomial of degree $\leq v$ we obtain a monomial of degree $\geq \mu + 1$. Thus the number of monomials of degree $\geq \mu + 1$ is again r. The number of monomials of degree $\leq \mu$ is thus $q^m - r$. Thus the dimensions of $C_\mu(m, q)$ and $C_v(m, q)$ add to q^m. □

For some fixed ordering of the elements of $V_m(q)$ we defined

$$C_v(m, q) = \{(P(\bar{0}), P(\bar{\alpha}_1), \ldots, P(\bar{\alpha}_{q^m - 1})), \quad P(\bar{x}) \in P_v(m, q)\}$$

We identify the ith coordinate position of the codeword with $\bar{\alpha}_i$. We prove the following:

Theorem 10.2 The action of $GLN(m, q)$ on the coordinate positions of $C_v(m, q)$ leaves the code invariant.

Proof Consider the permutation which takes $\bar{\alpha}_i$ to $\bar{\alpha}_i'$, where

$$\bar{\alpha}_i' = \bar{\alpha}_i A + \bar{\beta}, \quad \bar{\alpha}_i, \bar{\beta} \in V_m(q)$$

We have to show that the permuted codeword in which $P(\bar{\alpha}_i)$ appears in co-ordinate position $\bar{\alpha}_i'$, or equivalently, $P((\bar{\alpha}_i' - \bar{\beta})A^{-1})$ appears in position

1.10 The Polynomial Approach to Coding

$\bar{\alpha}_i'$, is also a codeword. Consider the polynomial $P'(\bar{x}) = P((\bar{x} - \bar{\beta})A^{-1})$ which is an element of $P_v(m, q)$. The corresponding codeword is

$$(P'(\bar{0}), P'(\bar{\alpha}_1), \ldots, P'(\bar{\alpha}_{q^m - 1}))$$
$$= (P(-\bar{\beta}A^{-1}), P((\bar{\alpha}_1' - \bar{\beta})A^{-1}), \ldots, P((\bar{\alpha}_{q^m-1}' - \bar{\beta})A^{-1}))$$

which is thus also a codeword, i.e., $P'(\bar{\alpha}_i') = P(\bar{\alpha}_i)$, as required. □

Since

$$\sum_{\bar{\alpha} \in V_m(q)} P(\bar{\alpha}) = 0, \qquad \deg P < m(q - 1)$$

we may view

$$P(\bar{0}) = -\sum_{\substack{\bar{\alpha} \in V_m(q) \\ \bar{\alpha} \neq \bar{0}}} P(\bar{\alpha})$$

as an overall parity check on the code. Also, if $A \in GL(m, q)$, then A acting on the coordinate positions fixes the first position. Now choose an element M of $GL(m, q)$ that is of order $q^m - 1$. This is always possible since, for example, the companion matrix of the minimal polynomial of a primitive element of $GF(q^m)$ over $GF(q)$ is such a matrix. If $\bar{\gamma} \in V_m(q)$ is a fixed nonzero element, then the set $\{\bar{\gamma}M^i, 0 \leq i \leq q^m - 2\}$ exhausts the nonzero elements of $V_m(q)$. It is clear then that the shortened $C_v(m, q)$ code

$$\{(P(\bar{\gamma}), P(\bar{\gamma}M), P(\bar{\gamma}M^2), \ldots, P(\bar{\gamma}M^{q^m - 2}), \quad P(\bar{x}) \in P_v(m, q)\}$$

is a cyclic code. We refer to this code as the cyclic GRM code and denote it by $HC_v(m, q)$, following Delsarte et al. (1970).

Theorem 10.3 The code $HC_v(m, q)$ of length $q^m - 1$ is invariant under the action of $GL(m, q)$ on its coordinate positions.

In order to describe $HC_v(m, q)$ as a cyclic code, it is sufficient to determine the roots of its generator polynomial. As before, let α be a primitive element of $GF(q^m)$, $m(x)$ its minimal polynomial over $GF(q)$, and M its companion matrix. Let $\bar{e} = (1, 0, 0, \ldots, 0) \in V_m(q)$ and define $\bar{\alpha} = (1, \alpha, \alpha^2, \ldots, \alpha^{m-1})$, where it is recognized that this set of elements forms a basis of $GF(q^m)$ over $GF(q)$. Clearly $(\bar{e}M^i, \bar{\alpha}) = \alpha^i$, $i = 0, 1, \ldots, m - 1$, and from this it is readily established that the mapping

$$V_m(q) \longrightarrow GF(q^m)$$
$$\bar{e}M^i \longmapsto (\bar{e}M^i, \bar{\alpha})$$

is a vector space [over $GF(q)$] isomorphism and $(\bar{e}M^i, \bar{\alpha}) = \alpha^i$, $i = 0, 1, \ldots, q^m - 2$, e.g.,

$$(\bar{e}M^m, \bar{\alpha}) = (\bar{e}M^{m-1}, \bar{\alpha}M^T) = (\bar{e}M^{m-1}, \alpha\bar{\alpha}) = \alpha \cdot \alpha^{m-1} = \alpha^m$$

The mapping
$$\mathcal{N}: GF(q^m) \longrightarrow \langle M \rangle$$
$$\alpha \longmapsto M$$
is a field isomorphism, where $\langle M \rangle$ is the multiplicative subgroup of $GL(m, q)$ with the zero matrix added and where addition and multiplication are as in a matrix ring.

Now let
$$\bar{P} = (P(\bar{e}), P(\bar{e}M), \ldots, P(\bar{e}M^{q^m-2})), \quad P(\bar{X}) \in P_v(m, q)$$
be a codeword of $HC_v(m, q)$ and define the corresponding codeword polynomial by
$$P(z) = \sum_{i=0}^{q^m-2} P(\bar{e}M^i) z^i$$

Then α^j is a root of the generator polynomial of $HC_v(m, q)$ iff $P(\alpha^j) = 0$ for all $P(\bar{x}) \in P_v(m, q)$, i.e.,
$$P(\alpha^j) = \sum_{i=0}^{q^m-2} P(\bar{e}M^i)(\alpha^j)^i$$
$$= \sum_{i=0}^{q^m-2} P(\bar{e}M^i)(\bar{e}M^i, \bar{\alpha})^j$$
$$= \sum_{\substack{\bar{\beta} \in V_m(q) \\ \bar{\beta} \neq \bar{0}}} P(\bar{\beta})(\bar{\beta}, \bar{\alpha})^j \tag{10.1}$$

and, from a previous lemma, this last quantity is zero if $\omega_q(j) \leq \mu = m(q-1) - v - 1$. We have proved:

Theorem 10.4 The element α^j is a zero of the generator polynomial of $HC_v(m, q)$ iff $0 < \omega_q(j) \leq m(q-1) - v - 1$.

The "only if" part of this theorem results from considering the code dimension.

Corollary The element α^j is a zero of the generator polynomial of the dual code of $HC_v(m, q)$ iff $0 \leq \omega_q(j) \leq m(q-1) - \mu - 1$.

Notice in the theorem that $\omega_q(j) > 0$. This follows since the summation in (10.1) is over $\bar{\beta} \in V_m(q)$, $\bar{\beta} \neq \bar{0}$, and a previous lemma stated that
$$\sum_{\bar{\beta} \in V_m(q)} P(\bar{\beta}) = 0$$
Since, in general, $P(\bar{0}) \neq 0$, we require the condition $\omega_q(j) > 0$. The corollary is easily derived from the characterization of the dual of the code $C_v(m, q)$.

1.10 The Polynomial Approach to Coding

If i is an integer such that $0 < \omega_q(i) \leq m(q-1) - v - 1$, then every integer $j \in J(i)$ (See Section 1.9) also has a q weight less than $m(q-1) - v - 1$ and from Theorem 9.1 we have:

Theorem 10.5 The code $C_v(m, q)$ is invariant under the doubly transitive affine group of permutations.

From Theorem 10.4 we have our second definition of GRM codes:

Definition 2 The cyclic code $HC_v(m, q)$ has α^j as a zero iff $0 < \omega_q(j) \leq \mu = m(q-1) - v - 1$. The code $C_v(m, q)$ is the code $HC_v(m, q)$ extended.

For our third definition we return to the original approach of Kasami et al. (1968a, b). This approach is a more natural generalization of the binary Reed–Muller codes than the previous work but we will only outline it here. Again let α be a primitive element of $GF(q^m)$ and

$$\alpha^j = \sum_{i=0}^{m-1} a_{ij} \alpha^i, \quad 0 \leq j \leq q^m - 2, \quad a_{ij} \in GF(q)$$

Define the matrix G as

$$G = \begin{bmatrix} 1 & 1 & 1 & \cdots & 1 \\ a_{00} & a_{01} & a_{02} & \cdots & a_{0, n-1} \\ a_{10} & a_{11} & a_{12} & \cdots & a_{1, n-1} \\ \vdots & \vdots & \vdots & & \vdots \\ a_{m-1, 0} & a_{m-1, 1} & a_{m-1, 2} & \cdots & a_{m-1, n-1} \end{bmatrix}, \quad n = q^m - 1$$

Label the rows successively as $v_I, v_0, v_1, \ldots, v_{m-1}$. As with the binary Reed–Muller codes, define a componentwise multiplication on $V_n(q)$. Define the matrix G_v to be the matrix G with all additional rows of the form

$$v_0^{i_0} v_1^{i_1} \cdots v_{m-1}^{i_{m-1}}, \quad \sum_{j=0}^{m-1} i_j \leq v$$

added.

Definition 3 The rows of the matrix G_v generate the cyclic code $HC_v(m, q)$.

The equivalence between Definitions 2 and 3 is established in Theorem 1 (Kasami et al., 1968a) and we will not demonstrate it here. With minor modifications this is the same as the multivariable polynomial approach.

Theorem 10.6 The code $HC_v(m, q)$ is a subcode of the BCH code with design distance $(R+1)q^Q - 1$, where $\mu + 1 = Q(q-1) + R$, $0 \leq R < q - 1$.

Proof Recalling that α^j is a zero of $HC_v(m, q)$ iff $0 < \omega_q(j) \leq m(q-1) - v - 1 = \mu$, we let h be the smallest integer such that $\omega_q(h) = \mu + 1$. Then we can write

$$h = Rq^Q + \sum_{i=0}^{Q-1} (q-1)q^i = (R+1)q^Q - 1$$

From the construction of h, every integer j, $0 \leq j < h$, is such that $\omega_q(j) \leq \mu$. Thus the elements $\alpha^1, \alpha^2, \ldots, \alpha^{(R+1)q^Q - 2}$ are all zeros of the code and $HC_v(m, q)$ is a subcode of the BCH code of design distance

$$d = (R+1)q^Q - 1. \quad \square$$

Actually, $HC_v(m, q)$ has exactly this minimum distance, as is shown (Kasami *et al.*, 1968a) by constructing a vector of minimum weight. Notice that for $m = 1$ the code $HC_v(m, q)$ is just the Reed–Solomon code of length $q - 1$ and distance $d = v + 1$.

To see the relationship between the single-variable and multivariable approaches, suppose that $L = \{\alpha^{-j}, j \in k\}$ are the zeros of a cyclic code $\mathscr{C}$. Then an equivalent characterization of $\mathscr{C}$ is

$$\mathscr{C} = \left\{ (F(1), F(\beta), \ldots, F(\beta^{n-1})) \,\Big|\, F(x) = \sum_{i=0}^{n-1} f_i x^i, \; f_i = 0 \; \text{if} \; \alpha^{-i} \in L \right\}$$

(Note: It is convenient in this instance to write $F_c(x)$ as $\sum_{i=0}^{n-1} f_i x^i$ rather than $\sum_{i=1}^{n} f_i x^{n-i}$ as before—hence we write the code zeros as α^{-i}.) We need to generalize the situation slightly. Let $F[z]$ be the set of polynomials

$$F[z] = \left\{ F(z) \in GF(q^m)[z]/(z^{q^m - 1} - 1) \,\Big|\, F(\alpha^i) \in GF(q), \; 0 \leq i \leq q^m - 2 \right\}$$

where α is a primitive element of $GF(q^m)$. Actually, $F[z] \cong GF(q)[z]/(z^{q^m-1} - 1)$. Define the subspace of $F[z]$, $F_v[z]$, to be the set

$$F_v[z] = \left\{ F(z) \in F[z] \,\Big|\, F(z) = \sum_{i=0}^{q^m - 2} f_i z^i, \; f_i = 0 \; \text{if} \; \omega_q(i) > v \right\}$$

For the code $HC_v(m, q)$ we had the theorem saying that α^i is a code zero iff $\omega_q(i) \leq m(q-1) - v - 1$. Equivalently, since $\omega_q(-i) = m(q-1) - \omega_q(i)$, we can say that α^{-i} is a zero of $HC_v(m, q)$ iff $\omega_q(i) \geq v + 1$. It follows then that the single-variable Mattson–Solomon characterization of $HC_v(m, q)$ would be

$$HC_v(m, q) = \{(F(1), F(\alpha), F(\alpha^2), \ldots, F(\alpha^{q^m - 2})) \,|\, F(z) \in F_v[z]\}$$

The precise relationship between this single-variable approach and the multivariable approach is explored in detail by Delsarte *et al.* (1970). We briefly

1.10 The Polynomial Approach to Coding

indicate the relationship here. Clearly the sets of polynomials $P_v(m, q)$ and $F_v[z]$ must somehow be related.

Choose the elements $\lambda_1, \lambda_2, \ldots, \lambda_m \in GF(q^m)$ as a dual basis to $1, \alpha, \alpha^2, \ldots, \alpha^{m-1}$, with α a primitive element of $GF(q^m)$, i.e.,

$$T(\lambda_i \alpha^j) = \delta_{ij}$$

where $T(x) = \sum_{i=0}^{m-1} x^{q^i}$ is the trace function of $GF(q^m)$ over $GF(q)$. It is then established that the mapping

$$\eta: P(m, q) \longrightarrow F[z]$$
$$P(\bar{x}) \longmapsto F(z) = P(T(\lambda_1 z), T(\lambda_2 z), \ldots, T(\lambda_m z))$$

is onto and that

$$F(\alpha^i) = P(\bar{\alpha}_i)$$

where $\bar{\alpha}_i = \bar{e}M^i$, using the previous notation. The key property of the mapping is that if $P(\bar{x}) \in P_v(m, q)$ and

$$F(z) = P(T(\lambda_1 z), T(\lambda_2 z), \ldots, T(\lambda_m z)) = \sum_{i=0}^{q-2} f_i z^i$$

then $f_i = 0$ if $\omega_q(i) > v$. This establishes the connection between the two polynomial characterizations of $HC_v(m, q)$

$$HC_v(m, q) = \{(F(1), F(\alpha), \ldots, F(\alpha^{q^m-2})) \,|\, F(z) \in F_v[z]\}$$
$$= \{(P(\bar{y}), P(\bar{y}M), \ldots, P(\bar{y}M^{q^m-2})) \,|\, P(\bar{x}) \in P_v(m, q)\}$$

Consider now the nonprimitive GRM codes, i.e., those whose length is not of the form $q^m - 1$ for some integer m. The construction of these codes is much like that of the nonprimitive BCH codes, although slightly more complicated. We begin by considering the generator matrix of the primitive vth-order GRM code over $GF(q)$ of length $q^m - 1$ and recall that if α is a primitive root of $GF(q^m)$, then α^h is a root of the generator $g_d(x)$ of the dual of the vth-order GRM code iff $\omega_q(h) \leq v$. Thus all code vectors in the dual of the vth-order GRM code are in the null space over $GF(q)$ of the matrix

$$G_v = \begin{bmatrix} 1 & 1 & 1 & \cdots & 1 \\ 1 & \alpha^{h_1} & \alpha^{2h_1} & \cdots & \alpha^{(q^m-2)h_1} \\ 1 & \alpha^{h_2} & \alpha^{2h_2} & \cdots & \alpha^{(q^m-2)h_2} \\ \vdots & \vdots & \vdots & & \vdots \\ 1 & \alpha^{h_i} & \alpha^{2h_i} & \cdots & \alpha^{(q^m-2)h_i} \end{bmatrix}$$

and hence we may take this as a representation of the generator matrix of the vth-order GRM code, where $\{h = 0, h_1, h_2, \ldots, h_i\}$ represent all integers less than $q^m - 1$ whose weight is v or less. If we delete the first row of G_v, then the result will be the parity check matrix of the $[m(q-1) - v - 1]$th-order GRM code, as discussed previously.

Now let β be a primitive nth root of unity of $GF(q^m)$, where $b = (q^m - 1)/n$, and consider those rows of G_v for which $b | h_j$ and denote, for these rows $t_j = h_j/b$. Notice that

$$(\alpha^{h_j})^{nl} = (\alpha^{bn})^{t_j l} = (\alpha^{q^m - 1})^{t_j l} = 1, \qquad l = 1, \ldots, e - 1$$

and hence each of the rows being considered consists of e replications of the first n bits. If there are $n - k$ rows such that $b | h_j$ then define the parity check matrix H to be the $(n - k) \times n$ matrix consisting of the first n bit positions of these $n - k$ rows. Clearly $n - k$ is the number of integers less than $q^m - 1$ that are divisible by b and whose weight is v or less. We can express the matrix H as

$$H = \begin{bmatrix} 1 & \beta^{t_1} & \beta^{2t_1} & \cdots & \beta^{(n-1)t_1} \\ 1 & \beta^{t_2} & \beta^{2t_2} & \cdots & \beta^{(n-1)t_2} \\ \vdots & \vdots & \vdots & & \vdots \\ 1 & \beta^{t_{n-k}} & \beta^{2t_{n-k}} & \cdots & \beta^{(n-1)t_{n-k}} \end{bmatrix}$$

and we define the vth-order nonprimitive GRM code as the null space of this matrix. In other words, the dual of the $((q^m - 1)/b, k)$ cyclic code whose generator polynomial contains α^h as a root iff $b | h$ and $\omega_q(h) \leq v$ will be called a vth-order nonprimitive GRM code.

When b is of the form $b = q^s - 1$ for some positive integer s, then the vth-order GRM code is closely related to the projective geometry codes which will be discussed in the next chapter.

The primitive GRM codes are related to the Euclidean geometry codes, also considered in the next chapter.

We now discuss a rather remarkable class of codes, polynomial codes, which were discovered by Kasami et. al. (1968b). Many of the previous classes of codes occur as a subclass of polynomial codes. Historically, they appear as a natural culmination of the work of the GRM codes and the finite geometry codes. While the basic work on finite geometry codes preceded that of polynomial codes, it is convenient to discuss the geometry codes as a subclass of polynomial codes. We shall examine both the Euclidean and projective geometry codes in some detail in the next chapter.

We begin with some preliminary terminology needed to define polynomial codes. Let m and s be two positive integers and α a primitive element of $GF(q^{ms})$. Choosing $1, \alpha, \alpha^2, \ldots, \alpha^{m-1}$ as a basis for $GF(q^{ms})$ over $GF(q^s)$, to each power of α we associate the m-tuple over $GF(q^s)$ by

$$\alpha^j = \sum_{i=1}^{m} a_{ij} \alpha^{i-1}, \qquad 0 \leq j \leq q^{ms} - 2, \quad a_{ij} \in GF(q^s)$$

1.10 The Polynomial Approach to Coding

and shall designate such an m-tuple as the coordinate vector of α^j over $GF(q^s)$. Let the positive integer b be a divisor of $q^s - 1$ and define

$$z = \frac{q^s - 1}{b}, \qquad n = \frac{q^{ms} - 1}{b}$$

where n will be the code length. Let $X_1, X_2, \ldots, X_m$ be m variables over $GF(q^s)$ and define $\overline{X}$ as $(X_1, X_2, \ldots, X_m)$. Let $Q_m(\mu, b)$ be the set of all polynomials in the variables $X_1, \ldots, X_m$ of the form

$$F(\overline{X}) = \sum_{i, v_i} c_{v_1 v_2 \cdots v_m} X_1^{v_1} X_2^{v_2} \cdots X_m^{v_m}$$

with the properties that:

(i) The coefficients $c_{v_1 v_2 \ldots v_m} \in GF(q^s)$.
(ii) $0 \leq v_i \leq q^s - 1$, $1 \leq i \leq m$.
(iii) $\sum_{i=1}^{m} v_i = jb$ with $0 \leq j \leq \mu$.
(iv) $f(a_{1l}, a_{2l}, \ldots, a_{ml}) \in GF(q)$, $0 \leq l < n$, where $(a_{1l}, a_{2l}, \ldots, a_{ml})$ is the coordinate vector of α_l.

Now for each such polynomial $f(\overline{X})$ in $Q_m(\mu, b)$ we define an associated n-tuple over $GF(q)$

$$v(f) = (v_0, v_1, \ldots, v_{n-1})$$

whose components are given by

$$v_j = f(a_{1j}, a_{2j}, \ldots, a_{mj}) \in GF(q), \qquad 0 \leq j \leq n - 1$$

Definition An (n, m, s, μ, q) polynomial code over $GF(q)$ of length n is defined as a set of vectors

$$\{v(f) \mid f(\overline{X}) \in Q_m(\mu, b)\}$$

It is clear that the code is linear since the sum of two polynomials in $Q_m(\mu, b)$ is also in $Q_m(\mu, b)$. The following theorem, however, is surprising.

Theorem 10.7 (Kasami et al. 1968b) An (n, m, s, μ, q) polynomial code over $GF(q)$ is cyclic and α^h is a root of its generator polynomial if and only if $b \mid h$ and

$$\min_{0 \leq l < s} \omega_{q^s}(hq^l) = jb$$

with $0 < j < mz - \mu$.

For $s = 1$ this condition is similar to that specifying the roots of GRM codes. Also, conditions specifying roots of the generator polynomial of the cyclic Euclidean and projective geometry codes, which historically appeared before the above theorem, turn out to be disguised versions of this theorem, as we will see in the next chapter. We shall not prove this important theorem, since the notation required is formidable. We refer the reader to the original paper (Kasami et al. 1968b, Theorem 6, p. 809). Notice that if α^h is to be a root of the generator polynomial $g(x)$, then, since the code is cyclic and of length n, the order of α^h must divide n. Since the nth roots of unity are $1, \alpha^b, \alpha^{2b}, \ldots, \alpha^{(n-1)b}$, then h is a multiple of b. If h is such that α^h is a root of $g(x)$ then it necessarily follows from the condition in Theorem 10.7, that α^{hq} is also a root and we conclude that the codimension of the code is the number of integers h, $0 < h \leqslant q^{ms} - 2$, such that $b \mid h$ and

$$\min_{0 \leqslant l < s} \omega_{q^s}(hq^l) = jb, \qquad 0 < j < mz - \mu$$

Theorem 10.8 The dual code of an (n, m, s, μ, q) polynomial code has α^h as a root of its generator polynomial $g_d(x)$ iff $b \mid h$ and

$$\max_{0 \leqslant l < s} \omega_{q^s}(hq^l) = jb, \qquad 0 \leqslant j \leqslant \mu$$

Proof The proof is actually an immediate consequence of the previous theorem and the arguments are similar to those used for the GRM codes. However, we shall elaborate a little further. If $g(x)g^*(x) = x^n - 1$, where $g(x)$ is the generator polynomial of the (n, m, s, μ, q) polynomial code of dimension k, say, then $g_d(x) = x^k g^*(1/x)$. From Theorem 10.7, α^h is a root of $g(x)$ iff

$$\min_{0 \leqslant l < s} \omega_{q^s}(hq^l) = jb, \qquad 0 < j < mz - \mu$$

If α^r is an nth root of unity, then $b \mid r$, and if $b \mid r$, then $b \mid \omega_{q^s}(r)$ since, if

$$r = \sum_{i=0}^{m-1} \delta_i q^{is} = \sum_{i=0}^{m-1} \delta_i + \sum_{i=0}^{m-1} \delta_i(q^{is} - 1)$$

and, since $b \mid (q^{is} - 1)$, this implies $b \mid \sum_{i=0}^{m-1} \delta_i = \omega_{q^s}(r)$. From the previous theorem we conclude that α^r is a root of $g^*(x)$ iff

$$\min_{0 \leqslant l < s} \omega_{q^s}(rq^l) = jb, \qquad mz - \mu \leqslant j \leqslant mz$$

and in this case α^{-r} will be a root of $g_d(x)$. Letting $r' = -r \equiv q^{ms} - 1 - r$, we have that

$$\omega_{q^s}(r'q^l) = \omega_{q^s}((q^{ms} - 1 - r)q^l) = m(q^s - 1) - \omega_{q^s}(rq^l)$$

and hence α^r is a root of $g_d(x)$ iff

$$\max_{0 \leqslant l < s} \omega_{q^s}(r'q^l) = m(q^s - 1) - jb, \qquad mz - \mu \leqslant j \leqslant mz$$
$$= (mz - j)b$$
$$= j'b, \qquad 0 \leqslant j' \leqslant \mu$$

as required. □

1.10 The Polynomial Approach to Coding

A BCH type of bound on the distance of an (n, m, s, μ, q) polynomial code is readily established by the following reasoning. Let h_0 be the smallest positive integer that is divisible by b such that

$$\min_{0 \leq l < s} \omega_{q^s}(h_0 q^l) = (mz - \mu)b$$

Since the smallest integer with q^s weight $(mz - \mu)b$ is

$$h_0' = (q^s - 1) + (q^s - 1)q^s + \cdots + (q^s - 1)q^{(Q-1)s} + Rq^{Qs}$$

we have that $h_0 \geq h_0'$, where Q and R are the quotient and remainder obtained by dividing $(mz - \mu)b$ by $q^s - 1$ or

$$(mz - \mu)b = Q(q^s - 1) + R, \qquad 0 \leq R < q^s - 1$$

Now consider any positive integer $h < h_0'$ that is divisible by b. Clearly, the q^s weight of any such integer is less than $(mz - \mu)b$ and hence

$$\min_{0 \leq l < s} \omega_{q^s}(hq^l) = jb, \qquad 0 < j < mz - \mu$$

and thus, for any such integer h, α^h is a root of the generator polynomial. This implies that the elements

$$\alpha^b, \alpha^{2b}, \ldots, \alpha^{h_0' - b}$$

are zeros of the generator polynomial.

Since $\beta = \alpha^b$ is a primitive nth root of unity, we have that

$$\beta, \beta^2, \ldots, \beta^{(h_0' - b)/b}$$

are roots of $g(x)$ and hence the minimum distance of the code is at least

$$\frac{h_0'}{b} = \frac{(q^s - 1) + (q^s - 1)q^s + \cdots + (q^s - 1)q^{(Q-1)s} + Rq^{Qs}}{b}$$

$$= \frac{q^{Qs} - 1 + Rq^{Qs}}{b} = \frac{(R + 1)q^{Qs} - 1}{b}$$

Theorem 10.9 An (n, m, s, μ, q) polynomial code has minimum distance at least

$$\frac{(R + 1)q^{Qs} - 1}{b}$$

where Q and R are the quotient and remainder obtained by dividing $(mz - \mu)b$ by $q^s - 1$.

In certain instances the lower bound on the minimum distance given in the previous theorem can be shown to be exact. In particular, if either (i) $R = 0$, or (ii) $s = 1$, or (iii) there exists a BCH code with design distance R/b and of length $(q^s - 1)/b$ whose minimum distance is exactly R/b, then the bound

$$d_{\min} \geq \frac{(R+1)q^{Qs} - 1}{b}$$

is exact. It may also be that the polynomial code is a subcode of the BCH code with this design distance and thus in these instances the design distance of the BCH code is exact.

The relationship between BCH and polynomial codes is further examined by Chen and Lin (1969), who also show that any primitive polynomial code (i.e., $b = 1$) is a subcode of the m-fold direct product of an extended primitive BCH code with itself.

To investigate subclasses of polynomial codes, notice that $b \mid \omega_q^s(h)$ iff $b \mid h$. This follows from the fact that if

$$h = \sum_{i=1}^{m} \delta_i q^{(i-1)s}$$

then

$$h - \omega_{q^s}(h) = \sum_{i=1}^{m} \delta_i q^{(i-1)s} - \sum_{i=1}^{m} \delta_i = \sum_{i=2}^{m} \delta_i [q^{(i-1)s} - 1]$$

and b divides the right-hand side of this equation. It follows that $b \mid [h - \omega_{q^s}(h)]$ and hence if it divides either h or $\omega_{q^s}(h)$, it must also divide the other. Now consider the case when $m = 1$ and $\mu n = -d$. The condition that α^h be a zero of the code then reduces to

$$\min_{0 \leq l < s} \omega_{q^s}(hq^l) = jb, \qquad 0 < j < d$$

If we set $\beta = \alpha^b$, then, since $\omega_{q^s}(b) = b$ and $b \mid \omega_{q^s}(bq^l)$, we have that

$$\min_{0 \leq l < s} \omega_{q^s}(bq^l) = b$$

and α^b is a zero of the code. Similarly $\beta^2, \beta^3, \ldots, \beta^{d-1}$ are zeros of the code of length $n = (q^s - 1)/b$ over $GF(q)$ and β is a primitive nth root of unity. Thus the code is a BCH code with design distance d. If, in addition to the foregoing conditions, $s = 1$, we obtain a Reed–Solomon code of length $(q-1)/b$ over $GF(q)$.

If we set $s = 1$, $b = 1$, then α^h is a zero of the code iff

$$\omega_q(h) = j, \qquad 0 < j < m(q-1) - \mu$$

The code is of length $q^m - 1$ over $GF(q)$ and these conditions describe the GRM code $HC_\mu(m, q)$.

The problem of determining the number of information symbols of polynomial codes has been considered by Chen and Lin (1969), who obtain exact expressions for certain subclasses. Lin (1972b) obtains a combinatorial expression for this quantity.

The GRM codes were the immediate predecessors of the polynomial codes and the two classes are, of course, closely related. To define this relationship, we consider the idea of a subfield subcode. Let $\mathscr{C}(q^s)$ be a cyclic code over $GF(q^s)$ with generator polynomial $g(x)$ and let $\mathscr{C}(q)$ be the set of codewords in $\mathscr{C}(q^s)$ with coordinates from $GF(q)$. Since $\mathscr{C}(q)$ is cyclic, denote its generator polynomial by $g_s(x)$. Suppose that α is a root of $g_s(x)$, so that α^{q^i}, $i = 1, 2, \ldots$, are also roots of $g(x)$. Suppose that α^{q^l} is also a root of $g(x)$ for some l. Since $g(x)|g_s(x)$, α^{q^l} is also a root of $g_s(x)$, as are $(\alpha^{q^l})^{q^j}$. In particular, $(\alpha^{q^l})^{q^{s-l}} = \alpha$ is a root of $g_s(x)$. Conversely, suppose α is a root of $g_s(x)$. If α^{q^i} is not a root of $g(x)$ for any i, then we can factor $g_s(x) = g_{1s}(x)g_{2s}(x)$, where $g_{1s}(x)$ is the minimum polynomial of α over $GF(q)$. Clearly $g_{2s}(x) \in GF(q)[x]$ and $g(x)|g_{2s}(x)$, implying that $g_s(x)$ is not the generator polynomial of $\mathscr{C}(q)$, contrary to assumption. Thus α^{q^i} must be a root of $g(x)$ for some i. Thus α is a root of $g_s(x)$ iff α^{q^i} is a root of $g(x)$ for some i.

This fact makes clear the relationship between GRM and polynomial codes. Let α be a primitive root of $GF(q^{ms})$. Then α^h is a root of the μth-order GRM code of length $(q^{ms} - 1)/b$ over $GF(q^s)$ iff $b|\omega_{q^s}(h)$ and

$$0 < \omega_{q^s}(h) < m(q^s - 1) - \mu b$$

for some m, μ, and b. On the other hand, from Theorem 10.7, $\alpha^{h'}$ is the root of an (n, m, s, μ, q) polynomial code over $GF(q)$ iff $b|h'$, and

$$\min_{0 \leq l < s} \omega_{q^s}(h'q^l) = jb, \qquad 0 < j < \frac{m(q^s - 1) - \mu}{b}$$

Note that if $b|(q^s - 1)$, then $b|h'$ iff $b|\omega_{q^s}(h')$. Thus, if $\alpha^{h'}$ is a root of the polynomial code, then for some l, $\alpha^{h'q^l}$ will be a root of the GRM code over $GF(q^s)$ and conversely. Thus the generator of the polynomial code, an element of $GF(q)[x]$, divides the generator of the GRM code, an element of $GF(q^s)[x]$, and the polynomial code is a subfield subcode of the GRM code.

1.11 Bounds on Code Dictionaries

Bounds that have been mentioned or used in this chapter, or that will be used in succeeding chapters, are proven here. The section is then by no means a complete discussion of the results and techniques of bounding. To begin on an optimistic note, we first prove the Gilbert (1952) lower bound on linear codes, which assures the existence of interesting codes. It was discovered independently by Varshamov (1957) and is usually referred to as the Varshamov–Gilbert bound.

In order to demonstrate the existence of a linear (n, k) code with distance d over $GF(q)$, it is only necessary to show the existence of an $(n - k) \times n$ matrix H such that no set of $d - 1$ columns is linearly independent. In finding such a matrix, we can choose the first column as any of the $q^{n-k} - 1$ nonzero $(n - k)$-tuples over $GF(q)$. The second column can be chosen as any nonzero $(n - k)$-tuple over $GF(q)$ which is not a scalar multiple of the first column. In general, suppose $n - 1$ columns have been chosen so that any $d - 1$ of them are linearly independent. There are

$$\sum_{i=0}^{d-2} \binom{n-1}{i} (q - 1)^i$$

vectors generated by nonzero linear combinations of $d - 2$ or fewer of the chosen $n - 1$ columns. At worst, these will be distinct, implying that if

$$q^{n-k} > \sum_{i=0}^{d-2} \binom{n-1}{i} (q - 1)^i$$

then it will be possible to choose an nth column that is independent of any $d - 2$ of the first $n - 1$ columns. We have shown:

Theorem 11.1 It is possible to construct an (n, k) linear code over $GF(q)$ with minimum distance d if

$$q^{n-k} > \sum_{i=0}^{d-2} \binom{n-1}{i} (q - 1)^i \tag{11.1}$$

The bound in the following theorem, called the Plotkin bound, will be of interest in the next chapter, where classes of codes that meet the bound with equality are constructed using balanced incomplete block designs.

Theorem 11.2 If a block code $\mathscr{C}$ of length n, size N, and minimum distance d exists, then

$$d \leqslant \frac{nN(q - 1)}{(N - 1)q} \tag{11.2}$$

Proof Form the $N \times n$ matrix D whose rows are the codewords and consider the sum

$$S = \sum_{(u, v \in \mathscr{C})} d(u, v)$$

1.11 Bounds on Code Dictionaries

Since, if $u \neq v$, we have $d(u, v) \geq d$, we can lower bound S by $S \geq N(N-1)d$. To obtain an upper bound on S, we consider the first column of D. Label the elements of the alphabet by the integers $0, 1, \ldots q-1$ and let t_i be the number of times i appears in the first column. For a fixed $u, v \in \mathscr{C}$, if their first coordinates are the same, they contribute nothing to S, while if they are different, they contribute one. If a given word u has first coordinate i, then we have $\sum_{j \neq i} t_j = N - t_i$ since $\sum_{i=0}^{q-1} t_i = N$. Since there are t_i such words, they contribute $t_i(N - t_i)$ to S. Finally, we have

$$\sum_{i=0}^{q-1} t_i(N - t_i) = R \tag{11.3}$$

is the contribution of the first coordinate position to S. An elementary induction argument yields that the expression in Equation (11.3) is maximized by choosing $t_i = N/q$. Since t_i is an integer, we conclude that

$$R = \sum_{i=0}^{q-1} t_i(N - t_i) = N^2 - \sum_{i=0}^{q-1} t_i^2 \leq N^2 - q\left(\frac{N}{q}\right)^2 = N^2 \frac{q-1}{q}$$

Since the same applies for any column, we have

$$N(N-1)d \leq S \leq nN^2 \frac{q-1}{q}$$

which implies that

$$d \leq \frac{nN(q-1)}{(N-1)q} \tag{11.4}$$

and the bound is satisfied iff each element appears N/q times in each column of D, which requires that $q | N$. □

Let $A(n, d)$ be the maximum size of a binary code of length n and minimum distance d. Hamming (1950) showed that for a binary alphabet

$$A(n, 1) = 2^n, \quad A(n, 2) = 2^{n-1}, \quad A(n+1, 2k) = A(n, 2k-1)$$

as well as finding the Hamming sphere packing bound for odd distance codes

$$A(n, 2k-1) \leq \frac{2^n}{\sum_{i=0}^{k-1} \binom{n}{i}}$$

Although Hamming was concerned with linear dictionaries, the results are still valid. The fact that $A(n+1, 2k) = A(n, 2k-1)$ is easily seen as follows. Suppose we have an $A(n, 2k-1)$ code. Adding a parity check position increases both the length and distance by one, giving that $A(n+1, 2k) \geq A(n, 2k-1)$. Conversely, suppose that we are given an $A(n+1, 2k)$ code. If we delete the first coordinate position, we obtain a code of length n and distance at least $2k - 1$, giving $A(n, 2k - 1) \geq A(n + 1, 2k)$. Thus we have $A(n + 1, 2k) = A(n, 2k - 1)$. From the Plotkin bound of Eq. (11.4) we have that for $q = 2$

$$2d(N - 1) \leq nN \quad \text{or} \quad (2d - n)N \leq 2d$$

Thus, if $2d > n$, we conclude that

$$A(n, d) \leq 2d/(2d - n)$$

Applying this formula to the various parameters yields:

(i) $A(n, n) = 2$.
(ii) $A(4m - 1, 2m) \leq 4m$.
(iii) $A(4m - 2, 2m) \leq 2m$.

Lemma $A(n, d) \leq 2A(n - 1, d)$.

Proof Let $\mathscr{C}$ be a binary code with $A(n, d)$ words. Suppose, without loss of generality, that at least half the codewords begin with zero. If, from this set of codewords, we delete the first coordinate position, we obtain a code of length $n - 1$, distance at least d, with at least $A(n, d)/2$ words, i.e., $A(n - 1, d) \geq A(n, d)/2$. □

Corollary Since $A(4m - 1, 2m) \leq 4m$, we have that $A(4m, 2m) \leq 8m$. If $A(4m, 2m) = 8m$, then $A(4m - 1, 2m) = 4m$ and $A(4m - 2, 2m) = 2m$.

Plotkin (1960) showed that if $4m - 1$ is a prime, then a code with $8m$ words can be constructed, showing that $A(4m, 2m) = 8m$, in this case. We will discuss this situation further in the next chapter in relating codes that achieve this bound to Hadamard matrices.

Johnson (1962) provides an interesting extension of the Hamming sphere packing bound. Again we are interested in bounding $A(n, d)$ for binary codes. We introduce the function $R(n, d, e)$ to denote the maximum number of binary n-tuples of weight exactly d such that any two of them are distance at least d apart, where $d = 2e + 1$. Without loss of generality we assume that the zero n-tuple is in the code. Let $S_i(x)$ be the sphere of radius i about x. The spheres of radius e about each codeword of our code do not intersect and this argument yields the Hamming bound. To refine this bound, we consider the points in spheres of radius $e + 1$ around codepoints.

There are $\binom{n}{e+1}$ points in $V_n(2)$ of weight $e + 1$, and the maximum number of nearest neighbors of the origin in a code of distance $d = 2e + 1$ is $R(n, d, e)$. Any codeword x of weight d has $\binom{d}{e}$ points of $V_n(2)$ in $S_e(x)$ of weight $e + 1$.

1.11 Bounds on Code Dictionaries

Thus $\binom{n}{e+1} - \binom{d}{e}R(n, d, e)$ points of $e + 1$ are a distance at least $e + 1$ from any codeword. However, each such point can be a distance $e + 1$ from no more than $\lfloor n/(e + 1) \rfloor$ codewords and these codewords would have to be of weight $2(e + 1)$. We obtain our bound by arguing as follows. The spheres of radius e about codewords are nonintersecting. A point of $V_n(2)$ that is a distance $e + 1$ from some codeword could be a distance $e + 1$ from at most $\lfloor n/(e + 1) \rfloor$ codewords. Thus the number of points of $V_n(2)$ that are at a distance $e + 1$ from some codeword is upper-bounded by the quantity

$$\frac{A(n, d)\binom{n}{e+1} - \binom{d}{e}R(n, d, e)}{\lfloor n/(e + 1) \rfloor}$$

where we divide by $\lfloor n/(e + 1) \rfloor$ to ensure that each such point is counted only once. We have shown:

Theorem 11.3

$$A(n, d)\left\{\sum_{i=0}^{e}\binom{n}{i} + \frac{\binom{n}{e+1} - \binom{d}{e}R(n, d, e)}{\lfloor n/(e+1) \rfloor}\right\} \leq 2^n \quad (11.5)$$

It follows directly from this that the number of minimum weight vectors in a perfect binary code must be $\binom{n}{e+1}/\binom{d}{e}$. This is also demonstrated in Chapter 3, where the relationship between t designs and codes is considered. Codes that meet the bound in Theorem 11.3 will be constructed in Chapter 3.

The quantity $R(n, d, e)$ is crucial to the bound (11.5) and the following theorem gives two interesting upper bounds to it.

Theorem 11.4

(i) $R(n, d, e) \leq \left\lfloor \frac{n}{d} \left\lfloor \frac{n-1}{d-1} \cdots \left\lfloor \frac{n-e}{d-e} \right\rfloor \cdots \right\rfloor \right\rfloor$

(ii) If $d^2 > ne$, then $R(n, d, e) \leq \left\lfloor \frac{n(d-e)}{d^2 - ne} \right\rfloor$

Proof (i) Consider a matrix with $R(n, d, e)$ rows and n columns where the rows are of weight d and are a distance at least d apart. The maximum number of ones in any column is $R(n - 1, d - 1, e - 1)$ since if we consider those rows of the matrix with ones in, say, the first column, then these rows with this column deleted have length $n - 1$, weight $d - 1$, and distance at least d or, equivalently, any two rows have an inner product of at most $e - 1$. The maximum number of ones in the original matrix is $n \cdot R(n - 1, d - 1,$

$e-1$) by counting the maximum number of ones in each column. On the other hand, each row is of weight d and we conclude that

$$dR(n, d, e) \leqslant nR(n-1, d-1, e-1)$$

or

$$R(n, d, e) \leqslant \left\lfloor \frac{n}{d} R(n-1, d-1, e-1) \right\rfloor$$

since $R(n, d, e)$ is an integer. Repeated application of this yields part (i).

(ii) The calculation for this part is almost identical to that for the Plotkin bound. Let λ_{ij} be the inner product of the ith and jth rows of the $R = R(n, d, e) \times n$ matrix described in part (i). Clearly the quantity

$$S = \sum_{\substack{i,j=1 \\ i \neq j}}^{R} \lambda_{ij}$$

may be upper-bounded by $R(R-1)e$ since the inner product of any two distinct rows is at most e. Doing the same calculation on a column basis and letting the ith column have $k_i \leqslant R(n-1, d-1, e-1)$ ones in it, we have

$$S = \sum_{i=1}^{n} k_i(k_i - 1)$$

Since $Rd = \sum_{i=1}^{n} k_i$, then

$$S = \sum_{i=1}^{n} k_i^2 - Rd$$

and, as in the Plotkin bound, the first term is minimized when $k_i = Rd/n$. Comparing the two quantities yields

$$R(R-1)e \geqslant \frac{R^2 d^2}{n} - Rd$$

or

$$R \leqslant \left\lfloor \frac{n(d-e)}{d^2 - ne} \right\rfloor$$

where the integral value signs arise since R is an integer. □

It has been observed that for a linear (n, k) code over $GF(q)$ of distance d it is true that

$$d \leqslant n - k + 1$$

We now prove this bound for nonlinear codes. Let $\mathscr{C}$ be a block code of length n, size N, and distance d. Arranging the codewords in a matrix, observe that in the first column at least one of the alphabet symbols appears at least $\lceil N/q \rceil$ times. If $\mathscr{C}'$ is such a set of $\lceil N/q \rceil$ words with the first digit deleted, then $\mathscr{C}'$ is code of block length $n - 1$, size at least $\lceil N/q \rceil$, and distance at least d. Repeating the argument t times would yield a code of block length $n - t$ with size

$$\lceil \cdots \lceil \lceil N/q \rceil / q \rceil \cdots \rceil \geqslant \lceil N/q^t \rceil$$

and distance d, providing $\lceil N/q^t \rceil \geqslant q$, or equivalently, $t \leqslant \lceil \log_q N \rceil - 1$. Since $d \leqslant n - t$, we conclude that

$$d \leqslant n - \lceil \log_q N \rceil + 1$$

which is the required bound.

1.12 Comments

The aim of the chapter was to provide a reasonably complete account of the theory of finite fields and of polynomials and vector spaces over finite fields, although in some respects the treatment was necessarily shallow. The background material assumed of the reader can be found in any good introductory book on algebra. Particularly good books are those of Artin (1959), Herstein (1964), and Kaplansky (1969b). Several books contain accounts of finite field theory from different points of view. The books on coding theory by Peterson (1961) and Berlekamp (1968) contain chapters devoted to finite fields and polynomials. The classic reference is perhaps the book by Dickson (1958), which is entirely devoted to finite fields and groups of linear transformations defined over them. The books by Albert (1956) and Carmichael (1956) also contain excellent sections on the theory. Van der Waerden (1953) gives a nice account of symmetric functions, among other things.

The initial part of Section 1.4 concerned with the more geometric aspects of vector spaces over finite fields was influenced by the excellent monograph of Kaplansky (1969a). The paper by Conway (1968a) was particularly helpful in the discussion of dual bases of vector spaces as well as for its material on computation over a finite field. Other information on this problem is contained in the paper by Alanen and Knuth (1964). Gaussian coefficients are investigated by Goldman and Rota (1969). Discussions of linear groups can be found in many places, but we followed mainly Artin (1957) and Carmichael (1956).

The sources of the coding material in this chapter are quoted in the text. We mention only the paper of Delsarte *et al.* (1970) for their polynomial approach to coding and to GRM codes, which was followed in Section 1.10.

Exercises

1 Let R be a commutative ring with identity. Show that $R[x]$ is a principal ideal domain iff R is a field.

2 Following Theorem 1.1 it was stated that the splitting field of any polynomial $f(x) \in K[x]$, $\deg f = n$, is at most degree $n!$ over K. Demonstrate that this upper bound can be achieved by showing that the splitting field of $x^3 - 2$ over the rational numbers $\mathbb{Q}$ is of degree six over $\mathbb{Q}$.

3 Show that if the field K has characteristic p not equal to zero, then if $f(x) \in K[x]$ is such that $f'(x) = 0$, then $f(x) = h(x^p)$ for some $h(x) \in K[x]$.

4 Prove that the distinct isomorphisms of the field K_1 onto the field K_2 are linearly independent.

5 (Albert, 1956, p. 133) Let $f(x)$ be an irreducible polynomial of degree m over $GF(q)$ and of exponent $(q^m - 1)/d$. Let t be a prime divisor of e and $t \nmid d$. Then $f(x^t)$ is irreducible over $GF(q)$ and has exponent et.

6 (Dickson, 1958, p. 34) Let $f(x)$ be an irreducible polynomial of degree m over $GF(q)$, $q = p^n$. Show that $f(x^p - x)$ is an irreducible polynomial over $GF(q)$ of degree mp if the trace of α in $GF(q)$ over $GF(p)$ is nonzero, where α is the coefficient of x^{m-1} in $f(x)$.

7 (Dickson, 1958, p. 32) Show that an irreducible polynomial of degree m over $GF(q)$, $q = p^d$, is also irreducible over $GF(q^n)$ if $(n, m) = 1$. Further, show that an irreducible polynomial of degree l over $GF(q)$ decomposes into k irreducible factors of degree l/k over $GF(q^s)$, $k = (l, s)$.

8 (Dickson, 1958, p. 42) Show that $x^p - x - \alpha$ is a primitive irreducible polynomial over $GF(p)$ iff (i) α is a primitive root of $GF(p)$ and (ii) a root of the equation $y^p \equiv y + 1 \pmod{p}$ has order $(p^p - 1)/(p - 1)$.

9 (Albert, 1956, p. 132) Let $f(x)$ be an irreducible polynomial of degree n over $GF(q)$. Let T be the automorphism $x \to x^p$ of $GF(q)$, i.e., $T \in G(q/p)$. Define the polynomial $g(y)$, where $f(T)y = yg(y)$. Then show that $g(y)$, which has degree $q^n - 1$, is irreducible over $GF(q)$ iff $f(x)$ is primitive.

10 (Berlekamp, 1968, p. 243) To find the solutions of a quadratic equation over a finite field, we first note that the equation $x^2 + ax + b$, $a, b \in GF(2^m)$, has a solution over $GF(2^m)$ iff $y^2 + y + \alpha$, $\alpha = a^{-2}b$, has a solution. Show (as in Problem 8) that this equation has a solution iff $\text{tr}(\alpha) = 0$. Let β be a root of an irreducible polynomial of degree m over $GF(2)$ and let β^k be such that

Exercises

tr $\beta^k = 1$. Then it is always possible to determine elements y_i, $i = 0, 1, \ldots, m - 1$, such that

$$y_i^2 + y_i = \begin{cases} \beta^i & \text{if } \operatorname{tr} \beta^i = 0 \\ \beta^i + \beta^k & \text{if } \operatorname{tr} \beta^i = 1 \end{cases}$$

If $\alpha = \sum_{i=0}^{m-1} a_i \beta^i$, $a_i \in GF(2)$, then show that the two solutions of the quadratic equation $y^2 + y + \alpha$ are given by $\sum_{i=0}^{m-1} a_i y_i$ and $1 + \sum_{i=0}^{m-1} a_i y_i$.

11 Let V_1 and V_2 be subspaces of $V_n(q)$. Show that $\dim(V_1 \cap V_2) + \dim(V_1 + V_2) = \dim V_1 + \dim V_2$.

12 (Kaplansky, 1969b, p. 47) Show that in a finite field any element can be written as a sum of two squares.

13 Let $f(x) = \sum_{i=0}^{n} f_i x^i$ be an irreducible polynomial over $GF(2)$ of exponent e and let $f^{\alpha}(x) = \sum_{i=0}^{n} f_i x^{2^i - 1}$. Show that all irreducible factors of $f^{\alpha}(x)$ are of exponent e. Show also that $f(x)$ is primitive iff $f^{\alpha}(x)$ is irreducible.

14 (Trinomials) Let $T_{n,k}(x) = x^n + x^k + 1$, $0 < k < n$.

(i) (Kaplansky, 1969b) For $n \geqslant 3$ show that $T_{2n,1}(x)$ is reducible over $GF(2)$. Show also that the degree of every irreducible factor divides $2n$ but not n.

(ii) (Mills and Zierler, 1969) Show that the degree of every irreducible factor of $T_{2n+1, 2^{n-1}-1}(x)$ divides either $2(n - 1)$ or $3(n - 1)$ but not $n - 1$.

(iii) (Zierler and Brillhart, 1968) If $n \equiv 1$ or 2 and $k \equiv 2$ or $1 \pmod{3}$, then $T_{n,k}(x)$ is divisible by $x^2 + x + 1$ and conversely.

15 (Swan, 1962) Let $f(x)$ be a monic polynomial over $GF(q)$, where q is the power of an odd prime. The discriminant of $f(x)$ is defined to be $D(f) = \delta(f)^2$, where

$$\delta(f) = \prod_{i<j} (\alpha_i - \alpha_j)$$

where $\alpha_1, \ldots, \alpha_n$ are the roots of $f(x)$, including repeated roots in some extension field of $GF(q)$. If $f(x)$ has no repeated roots and has r irreducible factors over $GF(q)$, then show that $r \equiv n \bmod 2$ iff $D(f)$ is a square in $GF(q)$.

16 (Ireland and Rosen, 1972) If $q \equiv 1 \bmod n$, show that the equation $x^n = \alpha \neq 0$, $\alpha \in GF(q)$, has either no solutions or n solutions in $GF(q)$. Show that it has n solutions in $GF(q^n)$.

17 (Ireland and Rosen, 1972) If $(q, n) \equiv 1$, show that the smallest extension field of $GF(q)$ in which $x^n - 1$ splits is $GF(q^a)$, where a is the smallest integer such that $q^a \equiv 1 \bmod n$.

18 Let $\mathscr{C}_1$ be an (n_1, k_1) code with generator matrix G_1 over $GF(q)$ and $\mathscr{C}_2$ an (n_2, k_2) code with generator matrix G_2 over $GF(q)$. Let $\mathscr{C}$ be the code with generator matrix $G_1 \otimes G_2$ (Kronecker product). Show that $\mathscr{C}$ is an $(n_1 n_2, k_1 k_2)$ code with distance $d_1 d_2$.

19 Let α be a root of the irreducible polynomial $x^5 + x^2 + 1$ over $GF(2)$. Find the generator polynomial of the BCH code of length 31 and design distance five.

20 (Relationship between Maximum Length and Hamming Codes) Let A be a $q^k \times n$ matrix over $GF(q)$ whose rows are the codewords of an (n, k) code over $GF(q)$. Show that each (nonzero) column of A contains each field element exactly q^{k-1} times. Let G be a generator matrix of dimension $k \times (q^k - 1)$ whose columns consist of all nonzero k-tuples over $GF(q)$. Show that in the corresponding code all codewords have weight $(q-1)q^{k-1}$. If G' is the generator matrix whose columns contain one representative from each of the $(q^k - 1)/(q - 1)$ classes, each class containing all nonzero scalar multiples of a given nonzero k-tuple, then all codewords have weight q^{k-1}.

21 In the Mattson–Solomon polynomial description of a code we associated a polynomial $F_\mathbf{c}(x)$ with each codeword $\mathbf{c}$ such that $F_\mathbf{c}(\beta^i) = c_i$. Show that, in general, for a polynomial $f(x) = \sum f_i x_i \in GF(q^m)[x]$ that if β is a primitive nth root of unity and $(n, q) = 1$, then

$$F_\mathbf{c}(\beta^i) \in GF(q) \quad \text{iff} \quad f_{q^i} = (f_i)^q \quad (q^i \text{ taken mod } n)$$

22 (Delsarte, 1973b) The generating function of Krawtchouk polynomials is given by

$$(z - y)^i (z + \lambda y)^{n-i} = \sum_{k=0}^{n} P_k(i) y^k z^{n-k}$$

and the Krawtchouk polynomial of degree k is

$$P_k(x) = \sum_{j=0}^{k} (-1)^j \lambda^{k-j} \binom{x}{j} \binom{n-x}{k-j}$$

where $\binom{x}{j} = x(x-1) \cdots (x - j + 1)/j!$. Show that these polynomials satisfy the relationships

(i) $\binom{n}{i} P_s(i) = \binom{n}{s} P_i(s)$, with i, s positive integers.

(ii) $\sum_{i=0}^{n} v_i P_r(i) P_s(i) = q^n v_r \delta_{rs}$, where $v_i = \binom{n}{i} \lambda^i$ and $\lambda = q - 1$.

To an $(n + 1)$-tuple $A = (A_0, A_1, \ldots, A_n)$ associate the bivariate homogeneous polynomial

$$A(y, z) = \sum_{i=0}^{n} A_i y^i z^{n-i}$$

The Krawtchouk or K-transform of this $(n+1)$-tuple is given by

$$A'(y, z) = A(z - y, z + \lambda y) = \sum_{i=0}^{n} A_i' y^i z^{n-i}$$

If the $(n+1)$-tuple A is the weight distribution for a code and $B = (B_0, B_1, \ldots, B_n)$ the weight distribution of its dual, then show that $A_i' = q^k B_i$. If A is an arbitrary $(n+1)$-tuple, A' its K-transform, $d(A)$ the smallest i, $1 \leq i \leq n$, such that $A_i \neq 0$, and $s(A)$ the number of nonzero components in A, then show that

$$s(A') \geq [d(A) - 1]/2$$

23 (Van Lint, 1971) Let β be a primitive nth root of unity in $GF(q)$, where $q = p^k$, p a prime, and k is the multiplicative order of p mod n. Show that

$$\mathscr{C} = \{(\text{tr}(\xi), \text{tr}(\xi\beta), \ldots, \text{tr}(\xi\beta^{n-1})) \mid \xi \in GF(q)\}$$

is a minimal ideal in $GF(q)[x]/(x^n - 1)$ of dimension k.

24 (Mattson and Solomon, 1961) Let $f(x) = \sum_{i=0}^{k} b_i x^{k-i}$, $b_0 = 1$, be a polynomial over $GF(2)$ such that $f(x) | (x^n + 1)$. Define the set

$$\mathscr{C} = \{(a_0, a_1, \ldots, a_{n-1}) \Big| \sum_{j=0}^{k} a_{i+k-j} b_j = 0; \quad i = 0, 1, \ldots, n - k + 1\}$$

Show that $\mathscr{C}$ is a linear cyclic (n, k) code over $GF(q)$ with generator polynomial $g(x) = (x^n + 1)/f(x)$. [In this case the codeword $(a_0, \ldots, a_{n-1})$ corresponds to the polynomial $\sum_{i=0}^{n-1} a_i x^{n-1-i}$.]

25 Determine the conditions under which a binary cyclic (n, k) code is left invariant under complementation.

26 Let $\mathscr{C}$ be a cyclic binary $(127, 112)$ code. Show that every codeword has even weight.

27 How many distinct binary cyclic codes of length 255 are there? What is the minimum number of zeros necessary to specify a $(255, 239)$ code? How many such codes are there?

28 Let $\mathscr{C}$ be a code of block length n which is invariant under a transitive group of permutations. If A_i is the number of codewords of weight i, show that $n | i A_i$.

29 Let $g(x) \in GF(2)[x]$ be self-reciprocal, i.e., if $\deg g = k$, then $g^*(x) = x^k g(1/x)$ and $g(x) = g^*(x)$. Show that $g(x)$ has at least five consecutive roots.

30 (Kasami et al., 1968c) If a code of length q^m is invariant under the doubly transitive affine group of permutations, then show that the code obtained by

deleting the first (coordinate position zero) digit is cyclic. Show that the extended primitive BCH and GRM codes are invariant under the doubly transitive affine group of permutations.

31 (Kasami *et al.*, 1968c, Corollary 4) Let $\mathscr{C}$ be a cyclic code of length $q^m - 1$ whose extension is invariant under the doubly transitive affine group of permutations, and that has minimum distance less than or equal to $q^s - 1$. Then show that its dual is a subcode of a primitive BCH code with minimum distance greater than or equal to $q^{m-s} + 1$.

32 (Sloane and Whitehead, 1970) Let $\mathbf{X}, \mathbf{Y} \in V_n(2)$ and define addition by the equation

$$\mathbf{X} \oplus' \mathbf{Y} = (\mathbf{X} + \mathbf{Y}, \mathbf{Y})$$

where the operation on the right is one of concatenation. Show that if $\mathscr{C}_1$ is an (n, M_1, d_1) code and $\mathscr{C}_2$ is an $(n, M_2, [\frac{1}{2}(d_1 + 1)])$, then

$$\mathscr{C}_1 \oplus \mathscr{C}_2 = \{\mathbf{X} \oplus' \mathbf{Y} \,|\, \mathbf{X} \in \mathscr{C}_1, \mathbf{Y} \in \mathscr{C}_2\}$$

is an $(2n, M_1 M_2, d_1)$ code.

33 (Griesmer, 1960) Let $\mathscr{C}$ be a linear (n, k) code of distance d over $GF(q)$. Derive a linear code of length $n - d$, dimension $k - 1$, and distance $[d/q]$.

2 Combinatorial Constructions and Coding

2.1 Introduction

Not surprisingly, certain combinatorial structures have played a significant role in coding theory. In this chapter we present a reasonably self-contained account of these structures, their interrelationships, and their relationships to codes. It is interesting to note the part that finite fields play here. There are many deep and intriguing questions concerning some of these configurations that have not yet been answered, mainly to do with existence and enumeration. We avoid these interesting problems here and instead present the basic configurations along with their elementary properties and one or two methods for constructing them.

2.2 Finite Geometries: Their Collineation Groups and Codes

The study of geometries is the study of a set of axioms concerning abstract quantities that we call points, lines, and flats. The axioms are usually framed to satisfy some geometric notions and there is a considerable interplay between the algebraic and geometric properties of such systems. In this section we confine ourselves to the definition of finite projective and Euclidean geometries and an examination of their elementary properties. For the material on projective geometries we will follow quite closely the first few sections of the paper by Veblen and Bussey (1906).

Definition A finite projective geometry consists of a finite set of elements called points and a family of subsets called lines, which satisfy the following axioms:

(i) The set contains a finite number (>2) of points. Each line contains at least three points.

(ii) If A and B are distinct points, there is one and only one line that contains A and B.

(iii) If A, B, and C are noncollinear points and if a line L contains a point D of the line AB and a point E of the line BC but does not contain A, B, or C, then the line L contains a point F of the line CA.

By a plane we mean a collection of points in the projective geometry that are collinear with a point A and any point on the line BC, where A, B, and C are noncollinear points. We define a k space of the geometry inductively by saying that a point is a 0 space and that if the points $A_1, A_2, \ldots, A_{k+1}$ are not all in the same $k-1$ space, then the set of points collinear with A_{k+1} and points in $\langle A_1, \ldots, A_k \rangle$ forms a k space denoted by $\langle A_1, \ldots, A_{k+1} \rangle$. With these axioms it is then possible to prove that in a k space an l space and an m space have at least an r space in common if $l + m - k = r$. To ensure the existence of dimensionality of the k space we add the two axioms to the foregoing definition:

(iv) If l is a positive integer less than k, then not all of the points considered are in the same l space.

(v) If (iv) is satisfied, then there exists in the set of points considered no $k+1$ space.

Finally, we mention that any proposition that we can prove about points and hyperplanes using these five axioms is also true if the words point and hyperplane are interchanged. This is a consequence of the principle of duality, which follows from these axioms. From duality we conclude that the number of points on any line in a projective plane is a constant that equals the number of lines intersecting in a given point. If $n+1$ is the number of points on a line in a projective plane, i.e., a 2-space, then from the definition of a plane there exists a point A not on a given line L such that the set of points on the set of lines through A and each point of L accounts for all the points of the plane. It follows that there are $n^2 + n + 1$ points in such a finite projective plane and we call n the order of the plane. In general, a projective geometry that is a k space contains $n^k + n^{k-1} + \cdots + n + 1$ points, where $n+1$ is the number of points on any line.

Example One of the simplest projective geometries is the Fano geometry, which is a projective plane of order two. For this geometry we have the helpful diagram shown in Fig. 2. The points of the geometry are A, B, C, D, E, F, and G and the lines are the sets

$$(A, B, C), \quad (C, D, E), \quad (A, F, E), \quad (B, G, E)$$
$$(C, G, F), \quad (A, G, D), \quad (B, F, D)$$

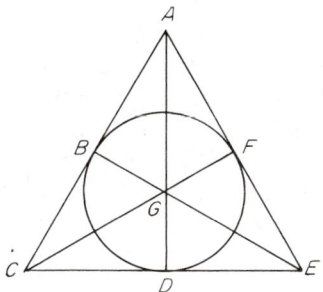

FIG. 2

and the plane, as stated, is of order two. Recall from the axioms that every line must intersect every other line and hence the notion of parallelism does not exist for such geometries.

Fortunately, there exists a construction of projective geometries of any dimension using the elements of a finite field $GF(q)$. Consider the vector space of dimension $n + 1$ over $GF(q)$, $V_{n+1}(q)$, and denote its points by the $(n + 1)$-tuples $(\alpha_1, \alpha_2, \ldots, \alpha_{n+1})$, $\alpha_i \in GF(q)$. For the points of the projective geometry, we take a nonzero point in $V_{n+1}(q)$ and identify all nonzero scalar multiples of it with a single point in the projective geometry. This construction gives

$$\frac{q^{n+1} - 1}{q - 1} = q^n + q^{n-1} + \cdots + q + 1$$

points. For the line containing the two distinct points $P_1 = (\alpha_1, \alpha_2, \ldots, \alpha_{n+1})$ and $P_2 = (\beta_1, \beta_2, \ldots, \beta_{n+1})$ we take the set of points

$$\eta P_1 + \gamma P_2 = (\eta \alpha_1 + \gamma \beta_1, \eta \alpha_2 + \gamma \beta_2, \ldots, \eta \alpha_{n+1} + \gamma \beta_{n+1})$$

where η and γ are not both equal to zero. There are $q^2 - 1$ possible choices for the pair (η, γ), but since we identify scalar multiples, there are $(q^2 - 1)/(q - 1) = q + 1$ points on a line and hence each point has $q + 1$ lines intersecting at it. We leave as an exercise the verification that the set of points, lines, and flats, defined inductively, does indeed give a projective geometry. We denote the geometry constructed in this manner by $PG(n, q)$.

Example If in the preceding example of the Fano geometry we coordinatize the points by assigning a triple over $GF(2)$ to each point in the following manner:

$$A = (0, 0, 1), \quad B = (0, 1, 0), \quad C = (0, 1, 1), \quad D = (1, 0, 1)$$
$$E = (1, 1, 0), \quad F = (1, 1, 1), \quad G = (1, 0, 0)$$

then we can identify the Fano geometry with $PG(2, 2)$.

The interesting property of the above construction technique is that, up to isomorphism (i.e., a permutation of points and lines), all finite projective geometries of dimension three or greater can be obtained in this manner. We sketch the proof of this statement as given by Veblen and Bussey (1906).

When the number of points in the projective geometry is finite and it is of dimension three or more, then the following statement is true:

Let A, B, and C be three collinear points and A', B', and C' be three other collinear points not on the same line. Then if the pairs of lines AB' and $A'B$, BC' and $B'C$, CA' and $C'A$ intersect, the three points of intersection are collinear.

The situation is illustrated in Fig. 3. Such a configuration is often referred to as a Pappus configuration, and if the statement holds for all allowable points in a geometry, we say the geometry has the Pappus property.

We first sketch the proof that all finite projective geometries of dimension greater than two have the Pappus property. From the first four axioms on the projective geometry it is possible to show the truth of Desargues' theorem when the dimension is greater than two and the number of points is finite, i.e., if two triangles A, B, C and A', B', C' are in the same plane and if the lines

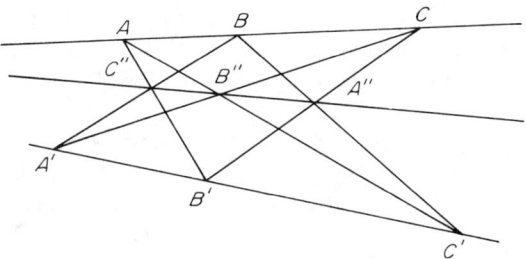

Fig. 3

AA', BB', CC' are concurrent, then the lines AB and $A'B'$, BC and $B'C'$, CA and $C'A'$ intersect in points that are collinear. This statement is illustrated in Fig. 4. Extending the lines AA', BB' and CC', we see that they meet in a point O and the statement is often called the Desargues theorem on perspective triangles. As stated, all finite projective geometries of dimension greater than

2.2 Finite Geometries: Their Collineation Groups and Codes

two have this property and are called Desarguesian. There exist non-Desarguesian planes. However, if the theorem of Desargues is valid in a finite projective plane, then we may identify it as a subgeometry of $PG(n, q)$ for some q, and it follows that the plane is a $PG(2, q)$. An important theorem of finite geometry states that the points in any finite geometry in which the theorem of

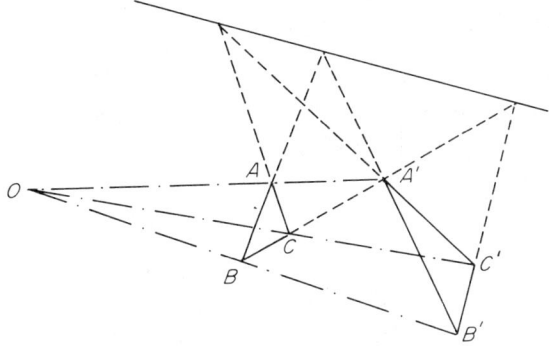

Fig. 4

Desargues is valid can be coordinatized by an $(n + 1)$-tuple over a finite division ring D (which is thus a finite field), where all scalar multiples of a point are identified and lines are defined as previously. Thus any finite projective geometry in which the theorem of Desargues holds is isomorphic, as a geometry, to a $PG(n, q)$ for some n and q. Finally, if the theorem of Pappus is valid in a finite projective geometry, then so is the theorem of Desargues and hence, again, the geometry is $PG(n, q)$ for some n and q.

As stated, in the case of finite projective planes the situation is not quite so simple and consequently this is the case where most of the current interest in finite geometries lies. Indeed a major unsolved problem is the enumeration of all such planes. The order of all known non-Desarguesian planes, however, is the power of a prime. It is known that no projective plane of order six exists. The next lowest order for which the question arises is ten and the answer for this order is as yet unknown. However, a significant result of this problem is the famous Bruck–Ryser theorem, which states that if n is a positive integer, $n \equiv 1$ or $2 \pmod 4$, and p is a prime, $p \equiv 3 \pmod 4$ such that p divides the square free part of n, then there cannot exist a finite projective plane of order n. The theorem tells us nothing about $n = 10$, but excludes, for instance, $n = 14$.

It has become conventional to call an m space in $PG(n, q)$, $m < n$, an m flat, presumably to clearly distinguish the projective geometry spaces from the subspaces of the underlying vector space. It should also be noted that, because of the properties of the underlying vector space, any m flat in $PG(n, q)$ is a $PG(m, q)$.

For future reference we determine some useful information on m flats in $PG(n, q)$. The number of such flats is the number of ways of choosing $m + 1$ independent points in $PG(n, q)$ divided by the number of ways of choosing $m + 1$ independent points in $PG(m, q)$. The number of bases of an m flat is

$$(1 + q + q^2 + \cdots + q^m)(q + \cdots + q^m) \cdots (q^{m-1} + q^m)q^m$$

where the number of ways of choosing the first point is $(1 + q + q^2 + \cdots + q^m)$, the number of ways of choosing the second point $(q + q^2 + \cdots + q^m)$, etc. By similar reasoning the number of ways of choosing $m + 1$ independent points in $PG(n, q)$ is

$$(1 + q + \cdots + q^n)(q + \cdots + q^n) \cdots (q^m + \cdots + q^n)$$

and the number of distinct $PG(m, q)$ in $PG(n, q)$ is

$$\frac{(1 + q + \cdots + q^n)(q + \cdots + q^n) \cdots (q^m + \cdots + q^n)}{(1 + q + \cdots + q^m)(q + \cdots + q^m) \cdots (q^{m-1} + q^m)q^m} = \prod_{t=0}^{m} \frac{q^{n+1-t} - 1}{q^{m+1-t} - 1} \tag{2.1}$$

We shall also require the number of $PG(m, q)$ containing a given $PG(l, q)$ in $PG(n, q)$ where $l < m < n$. The number of ways of adding an independent point to $PG(l, q)$ to give a $PG(l + 1, q)$ is $(q^{l+1} + \cdots + q^n)$. We can add $m - l$ independent points in

$$(q^{l+1} + \cdots + q^n)(q^{l+2} + \cdots + q^n) \cdots (q^m + \cdots + q^n)$$

ways. For any given $PG(m, q)$ the number of ways of choosing the extra $m - l$ independent points to add to the given $PG(l, q)$ is

$$(q^{l+1} + \cdots + q^m)(q^{l+2} + \cdots + q^m) \cdots (q^{m-1} + q^m)q^m$$

and the number of $PG(m, q)$ containing a given $PG(l, q)$ is

$$\frac{(q^{l+1} + \cdots + q^n)(q^{l+2} + \cdots + q^n) \cdots (q^m + \cdots + q^n)}{(q^{l+1} + \cdots + q^m)(q^{l+2} + \cdots + q^m) \cdots (q^{m-1} + q^m)q^m}$$

If $(\alpha_1, \alpha_2, \ldots, \alpha_{n+1})$ represents a point in $PG(n, q)$, then the nonzero solutions to the equation

$$\sum_{i=1}^{n+1} \beta_i \alpha_i = 0$$

form an $n - 1$ flat. Similarly, l independent equations of this form define an $n - l$ flat, where by independence we mean the l points are independent in the underlying vector space.

2.2 Finite Geometries: Their Collineation Groups and Codes

The elements of $GF(q^{n+1})$ can be used to represent the points of $PG(n, q)$ in the following manner. Let ξ be a primitive element of $GF(q^{n+1})$ with minimal polynomial over $GF(q)$

$$m_\xi(x) = \sum_{i=0}^{n+1} m_i x^i, \quad m_i \in GF(q), \quad m_{n+1} = 1$$

Every nonzero element of $GF(q^{n+1})$ can be written uniquely as a power of ξ, and using the minimum polynomial, we can identify uniquely every such power with an $(n+1)$-tuple over $GF(q)$. Note that $\gamma = \xi^{q-1}$ is a primitive vth root of unity, where $v = (q^{n+1} - 1)/(q - 1)$ and $\beta = \xi^v$ is a primitive element in $GF(q)$. Thus, in $PG(n, q)$ two points ξ^i and ξ^j are identified iff $i \equiv j \bmod v$, i.e., iff $\xi^i = \beta^k \xi^j = \xi^{kv+j}$. Thus the first v powers of ξ can be taken as the points in $PG(n, q)$. To determine which of these points are in an m flat of $PG(n, q)$, $m < n$, let $\xi^{s_1}, \xi^{s_2}, \ldots, \xi^{s_{m+1}}$ be a set of $m + 1$ independent points in $PG(n, q)$. Then an m flat is defined as the set of points

$$\alpha_i = \sum_{j=1}^{m+1} \varepsilon_{ij} \xi^{s_j} \tag{2.2}$$

where the ε_{ij} are chosen such that no two vectors $(\varepsilon_{i1}, \varepsilon_{i2}, \ldots, \varepsilon_{i(m+1)})$ are linear multiples over $GF(q)$. There are $(q^{m+1} - 1)/(q - 1)$ such points. Now each such α_i can be expressed in the form

$$\alpha_i = \beta^{k_i} \xi^{l_i}, \quad l_i < v = \frac{q^{n+1} - 1}{q - 1} \tag{2.3}$$

and the $(q^{m+1} - 1)/(q - 1)$ points ξ^{l_i} obtained in this maner form an m flat in $PG(n, q)$. That the points so described are distinct follows easily.

An alternative but equivalent method of describing r flats in $PG(n, q)$ is to consider the set of points $\{\sum_{i=0}^{r} \beta_i \alpha_i\}$, where $\{\alpha_i\}$ is a set of $r + 1$ independent points of $GF(q^{n+1})$ over $GF(q)$ and $\beta_i \in GF(q)$, and the β_i are never all identically zero. The r flat contains $(q^{r+1} - 1)/(q - 1)$ points since scalar multiples over $GF(q)$ are identified. There is an obvious one-to-one correspondence between subspaces of dimension $r + 1$ in a vector space of dimension $n + 1$, $V_{n+1}(q)$, and r flats in $PG(n, q)$.

If from a $PG(n, q)$ we delete the points of an $n - 1$ flat, we obtain a structure called a finite Euclidean geometry, which we denote by $EG(n, q)$. That all $EG(n, q)$ have the same structure follows from the fact that any $n - 1$ flat in $PG(n, q)$ is isomorphic to any other $n - 1$ flat. As an example, consider the points of a $PG(n, q)$, which we may divide into two types, those of the form $(1, \alpha_2, \ldots, \alpha_{n+1})$ and those of the form $(0, \alpha_2, \ldots, \alpha_{n+1})$. The points of the form $(0, \alpha_2, \ldots, \alpha_{n+1})$ are an $n - 1$ flat in $PG(n, q)$ and the remaining points form an $EG(n, q)$. We can therefore associate each point in $EG(n, q)$ with an n-tuple $(\alpha_2, \ldots, \alpha_{n+1})$ and there are q^n points in $EG(n, q)$.

We call the $n-1$ flat in $PG(n, q)$ that we delete to form $EG(n, q)$ the flat at infinity. The lines in $EG(n, q)$ are simply those lines in $PG(n, q)$ that are not in the flat at infinity with the points in the flat at infinity removed. The original $PG(n, q)$ is generated by the $n-1$ flat and a point A external to the flat. The lines collinear with A and each point of the $n-1$ flat are the lines of the $EG(n, q)$ when the points of the $n-1$ flat are removed. Similarly, to form an l flat in $EG(n, q)$ consider an l flat in $PG(n, q)$. If it lies entirely in the flat at infinity, ignore it. Otherwise we define an l flat in $EG(n, q)$ as the points of the l flat in $PG(n, q)$ with the points of the flat at infinity deleted. To see that each such l flat has the same number of points, we consider the properties of the underlying vector spaces and denote the deleted $n-1$ flat by V and the l flat not entirely contained in the $n-1$ flat by W. Then $V + W$ is the entire $PG(n, q)$ and

$$\dim(V \cap W) = \dim V + \dim W - \dim(V + W)$$

and so

$$\dim(V \cap W) = n - 1 + l - n = l - 1$$

which implies that the number of points in V in common with W is $(1 + q + \cdots + q^{l-1})$ and hence $W \setminus (V \cap W)$ contains q^l points and is an l flat in $EG(n, q)$.

By a similar type of reasoning we can see that any l flat in $EG(n, q)$ is an $EG(l, q)$. For suppose the l flat in $EG(n, q)$ is obtained from the residue of some $PG(l, q)$ in $PG(n, q)$. Let V be the $n-1$ flat deleted from $PG(n, q)$ to form $EG(n, q)$ and W the $PG(l, q)$. Then again we have

$$\dim(V \cap W) = l - 1$$

and hence $V \cap W$ is an $l-1$ flat in $PG(n, q)$. But $V \cap W$ is also an $l-1$ flat in $W = PG(l, q)$ and hence the l flat in $EG(n, q)$ is an $EG(l, q)$.

Example Consider the previous example of the Fano geometry $PG(2, 2)$ and delete the line (A, B, C) to form $EG(2, 2)$ (Fig. 5). The lines that are left

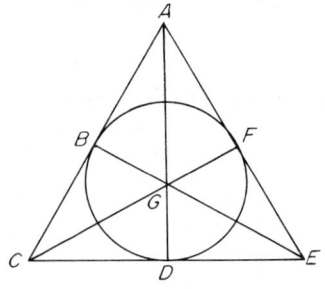

Fig. 5.

2.2 Finite Geometries: Their Collineation Groups and Codes

each have two points and are

$$(E, F) \quad (D, E) \quad (E, G) \quad (D, F) \quad (D, G) \quad (F, G)$$

In the $EG(2, 2)$ obtained we have the interesting property that it is no longer necessary for lines to intersect, and hence we have the idea of parallelism, which of course was lacking in the projective geometries. The lines of a Euclidean geometry can be divided into equivalence classes according to parallelism, since parallelism is an equivalence relation. For the above example these classes are

$$\{(E, F) \text{ and } (D, G)\}, \quad \{(E, G) \text{ and } (D, F)\}, \quad \{(D, E) \text{ and } (F, G)\}$$

As for projective geometries, we shall have need of certain information on l flats in $EG(n, q)$. To determine the number of l flats in $EG(n, q)$, we recall that every l flat in $EG(n, q)$ is the residue of an l flat in $PG(n, q)$ that is not contained in the $n - 1$ flat at infinity. The number of such flats is

$$\frac{(1 + q + \cdots + q^n) \cdots (q^l + \cdots + q^n)}{(1 + q + \cdots + q^l) \cdots (q^{l-1} + q^l)q^l}$$

$$- \frac{(1 + q + \cdots + q^{n-1}) \cdots (q^l + \cdots + q^{n-1})}{(1 + q + \cdots + q^l) \cdots (q^{l-1} + q^l)q^l}$$

which may be more conveniently written as

$$q^{n-l} \prod_{k=1}^{l} \frac{q^{n+1-k} - 1}{q^{l+1-k} - 1}$$

The number of m flats containing a given l flat in $EG(n, q)$ is just the number of m flats in $PG(n, q)$ containing a given l flat in $PG(n, q)$, which follows since the given l flat in $EG(n, q)$ cannot be in the flat at infinity and hence any m flat containing it cannot be in it either. Thus the number is

$$\frac{(q^{l+1} + \cdots + q^n)(q^{l+2} + \cdots + q^n) \cdots (q^m + \cdots + q^n)}{(q^{l+1} + \cdots + q^m)(q^{l+2} + \cdots + q^m) \cdots (q^{m-1} + q^m)q^m} = \prod_{k=l+1}^{m} \frac{q^{n-k+1} - 1}{q^{m-k+1} - 1}$$

(2.4)

From the previous equation we see that there are

$$q^{n-1} \frac{q^n - 1}{q - 1}$$

lines in $EG(n, q)$ and from the analytic description of the geometry each line contain q points. As seen before, each line of a $PG(n, q)$ is either entirely contained in a given $n - 1$ flat of $PG(n, q)$ or else has precisely one point in the flat.

Much as we did for projective geometries, we can obtain a representation of $EG(n, q)$ in terms of elements of $GF(q^n)$. We have already observed that for $n \geq 3$ the points of $EG(n, q)$ can be taken to be the points of $V_n(q)$. If α is a primitive element of $GF(q^n)$, then each n-tuple over $GF(q)$ can be uniquely identified with a power of α. In this representation a line through the point α^i is the set of all points of the form

$$\alpha^j = \alpha^i + \gamma \alpha^{i_1}, \qquad \gamma \in GF(q)$$

In a similar fashion an m flat through α^i is the set of points

$$\alpha^j = \alpha^i + \gamma_1 \alpha^{i_1} + \gamma_2 \alpha^{i_2} + \cdots + \gamma_m \alpha^{i_m}, \qquad \gamma_i \in GF(q)$$

where the points α^i, α^{i_1}, α^{i_2}, ..., α^{i_m} are independent and $i_j \in \{\infty, 0, 1, \ldots, q^n - 2\}$. Similarly a line through the (distinct) points α^i and α^{i_1} is the set of points

$$\{\beta \alpha^i + (1 - \beta)\alpha^{i_1}, \qquad \beta \in GF(q)\}$$

and an m flat through the $m + 1$ points α^i, α^{i_1}, ..., α^{i_m} is the set of points

$$\{\beta_0 \alpha^i + \beta_1 \alpha^i + \cdots + \beta_m \alpha^{i_m}, \qquad \beta_i \in GF(q)\}$$

where $\sum_{i=0}^{m} \beta_i = 1$.

We consider now some groups of transformations that it is possible to define on finite geometries. From Section 1.8 in Chapter 1 the general linear group $GL(n, q)$ (called the general linear homogeneous groups by some authors) is the set of $n \times n$ nonsingular matrices over $GF(q)$ and has order

$$|GL(n, q)| = q^{n(n-1)/2} \prod_{i=1}^{n} (q^i - 1)$$

Now consider the transformation

$$Ay_i = \sum_{j=0}^{n} \alpha_{ij} x_j, \qquad i = 0, 1, \ldots, n$$

defined on $V_{n+1}(q)$, where the transformation is nonsingular and we adopt the notation Ay_i to indicate that we identify transformations that are nonzero scalar multiples of one another. The set of such transformations forms a group, the projective group, which we denote by $P(n, q)$ and it is clear that

$$|P(n, q)| = \frac{|GL(n + 1, q)|}{q - 1}$$

The elements of $P(n, q)$ are, of course, intimately connected with the properties of $PG(n, q)$. Indeed, each element of $P(n, q)$ is a one-to-one transformation of $PG(n, q)$ onto itself. Furthermore, from the definition of a line in $PG(n, q)$ and the linearity (of a sort) of the transformation in $P(n, q)$ it follows that

2.2 Finite Geometries: Their Collineation Groups and Codes

each element of $P(n, q)$ takes a line of $PG(n, q)$ into a line. The elements of $P(n, q)$ permute the points of $PG(n, q)$, and hence the group has a natural representation as a permutation group on these points, and the degree of the permutation group is $(1 + q + \cdots + q^n)$.

For $n > 0$, $P(n, q)$ is doubly transitive, for consider the action of a group element on the two $(n + 1)$-tuples $(1, 0, \ldots, 0)$ and $(0, 1, 0, \ldots, 0)$. If the element is (α_{ij}), then these two $(n + 1)$-tuples are taken to be $(\alpha_{00}, \alpha_{10}, \ldots, \alpha_{n0})$ and $(\alpha_{01}, \alpha_{11}, \ldots, \alpha_{n1})$, respectively. Since these may be chosen independently, there exists an element of $P(n, q)$ that takes any two points of $PG(n, q)$ to any other two points of $PG(n, q)$, and thus, for $n > 0$, the group is doubly transitive. It can be verified that $P(1, q)$ is triply transitive.

By a collineation in $PG(n, q)$ we mean a point transformation that takes lines into lines. We have observed that elements of $P(n, q)$ are collineations and we ask if there are any others. Consider the transformation

$$\alpha y_i = x_i^p, \quad i = 0, 1, \ldots, n$$

of the point $(x_0, x_1, \ldots, x_n)$ in a $PG(n, q)$, where again we identify scalar multiples and the characteristic of $GF(q)$ is p, i.e., $q = p^k$. If $\mathbf{x}_1 = (\alpha_0, \alpha_1, \ldots, \alpha_n)$ and $\mathbf{x}_2 = (\beta_0, \beta_1, \ldots, \beta_n)$ are points of $PG(n, q)$ then the line through these points is defined by

$$\{\gamma \mathbf{x}_1 + \sigma \mathbf{x}_2, \quad \gamma, \sigma \in GF(q)\}$$

The above transformation, defined on the coordinates of each point, takes a point on this line to the point

$$(\gamma^p \alpha_0^p + \sigma^p \beta_0^p, \gamma^p \alpha_1^p + \sigma^p \beta_1^p, \ldots, \gamma^p \alpha_n^p + \sigma^p \beta_n^p)$$

since $\binom{p}{i} = 0$ in a field of characteristic p if $i \neq 0$ or p. Since the mapping $\gamma \to \gamma^p$ is an automorphism of $GF(q)$, the set of transformed points lies on the line $\alpha' \mathbf{x}_1' + \sigma' \mathbf{x}_2'$, where

$$\mathbf{x}_1' = (\alpha_0^p, \alpha_1^p, \ldots, \alpha_n^p) \quad \text{and} \quad \mathbf{x}_2' = (\beta_0^p, \beta_1^p, \ldots, \beta_n^p)$$

and the transformation in question is a collineation. It follows that a transformation of the form

$$\alpha y_i = \sum_{j=0}^{n} \alpha_{ij} x_j^{p^l}, \quad i = 0, 1, \ldots, n$$

where $l = 0, 1, \ldots, k - 1$ and $|\alpha_{ij}| \neq 0$, is a collineation. Clearly, the set of all collineations is a group. An important theorem states that every collineation of $PG(n, q)$ is of the above form and we denote the group of collineations by $C(n, q)$. The order of $C(n, q)$ is of course k times the order of $P(n, q)$ since the only new factors introduced into the argument are the k possible powers of p in the exponent. Collineation groups and their properties are rather more completely described by Carmichael (1956 Chapter 12), and the reader is referred there for more information.

Example The smallest example of a projective geometry over a non-prime field is $PG(2, 2^2)$, which we discuss here. Denote the elements of $GF(2^2)$ by $\{0, 1, \alpha, \alpha^2\}$, where, of necessity, $\alpha + 1 = \alpha^2$ and α is a primitive element. The geometry contains 21 points (Table V). There are five points on any line and the 21 lines are given by Table VI.

TABLE V

$P_1 = \{1, 0, 1\}$	$P_8 = \{1, 1, 0\}$	$P_{15} = \{1, \alpha^2, \alpha^2\}$
$P_2 = \{1, 0, \alpha\}$	$P_9 = \{1, \alpha, 1\}$	$P_{16} = \{1, \alpha^2, 0\}$
$P_3 = \{1, 0, \alpha^2\}$	$P_{10} = \{1, \alpha, \alpha\}$	$P_{17} = \{0, 0, 1\}$
$P_4 = \{1, 0, 0\}$	$P_{11} = \{1, \alpha, \alpha^2\}$	$P_{18} = \{0, 1, 1\}$
$P_5 = \{1, 1, 1\}$	$P_{12} = \{1, \alpha, 0\}$	$P_{19} = \{0, 1, \alpha\}$
$P_6 = \{1, 1, \alpha\}$	$P_{13} = \{1, \alpha^2, 1\}$	$P_{20} = \{0, 1, \alpha^2\}$
$P_7 = \{1, 1, \alpha^2\}$	$P_{14} = \{1, \alpha^2, \alpha\}$	$P_{21} = \{0, 1, 0\}$

TABLE VI

$L_1 = \{P_1, P_2, P_{17}, P_4, P_3\}$	$L_8 = \{P_2, P_7, P_{18}, P_{13}, P_{12}\}$	$L_{15} = \{P_4, P_6, P_{19}, P_{13}, P_{11}\}$
$L_2 = \{P_1, P_5, P_{21}, P_9, P_{13}\}$	$L_9 = \{P_2, P_8, P_{19}, P_{15}, P_9\}$	$L_{16} = \{P_4, P_7, P_{20}, P_{14}, P_9\}$
$L_3 = \{P_1, P_6, P_{20}, P_{15}, P_{12}\}$	$L_{10} = \{P_3, P_5, P_{19}, P_{14}, P_{12}\}$	$L_{17} = \{P_4, P_8, P_{21}, P_{16}, P_{12}\}$
$L_4 = \{P_1, P_7, P_{19}, P_{16}, P_{10}\}$	$L_{11} = \{P_3, P_6, P_{18}, P_{16}, P_9\}$	$L_{18} = \{P_5, P_6, P_{17}, P_7, P_8\}$
$L_5 = \{P_1, P_8, P_{18}, P_{14}, P_{11}\}$	$L_{12} = \{P_3, P_7, P_{21}, P_{15}, P_{11}\}$	$L_{19} = \{P_9, P_{10}, P_{17}, P_{11}, P_{12}\}$
$L_6 = \{P_2, P_5, P_{20}, P_{16}, P_{11}\}$	$L_{13} = \{P_3, P_8, P_{20}, P_{13}, P_{10}\}$	$L_{20} = \{P_{13}, P_{14}, P_{17}, P_{16}, P_{15}\}$
$L_7 = \{P_2, P_6, P_{21}, P_{14}, P_{10}\}$	$L_{14} = \{P_4, P_5, P_{18}, P_{15}, P_{10}\}$	$L_{21} = \{P_{18}, P_{19}, P_{17}, P_{20}, P_{21}\}$

Consider the points of the geometry containing only elements of $GF(2)$, i.e., $P_1, P_4, P_5, P_8, P_{17}, P_{18}, P_{21}$. The lines of this subgeometry are "sublines" of the previous geometry, i.e.,

$$L_1' = \{P_1, P_4, P_{17}\}, \quad L_2' = \{P_1, P_5, P_{21}\}, \quad L_3' = \{P_1, P_8, P_{18}\}$$
$$L_4' = \{P_4, P_5, P_{18}\}, \quad L_5' = \{P_4, P_8, P_{21}\}, \quad L_6' = \{P_5, P_8, P_{17}\}$$
$$L_7' = \{P_{17}, P_{18}, P_{21}\}$$

This, of course, is just $PG(2, 2)$, the Fano geometry. In general $PG(m, q)$ is contained in $PG(m, q^t)$ as a subgeometry. The points $P_{17}, P_{18}, P_{19}, P_{20}$, and P_{21} form a hyperplane and subtracting these points leaves the Euclidean geometry $EG(2, 2^2)$. The lines L_i', $i = 1, \ldots, 20$, of the geometry are those of

2.2 Finite Geometries: Their Collineation Groups and Codes

$PG(2, 2^2)$ with the line L_{21} and its points removed. The parallelism classes of $EG(2, 2^2)$ are

$$\{L_1', L_{18}', L_{19}', L_{20}'\}, \quad \{L_2', L_{12}', L_7', L_{17}'\}$$

$$\{L_3', L_6', L_{13}', L_{16}'\}, \quad \{L_4', L_{10}', L_9', L_{15}'\}, \quad \{L_5', L_8', L_{11}', L_{14}'\}$$

For the remainder of the section we will discuss majority logic decoding and the application of concepts of finite geometry to construct and analyze codes that are majority logic decodable. The forerunners of most of these ideas are the Reed–Muller codes discovered by Muller (1954) and provided with a decoding algorithm by Reed (1954). The work of Rudolph (1967) and Weldon (1969) has prompted the more recent interest in these codes.

To introduce majority logic decoding, we consider the following simple situation. Let $c(x) = \sum_{i=0}^{n-1} c_i x^i$ be a codeword in the (n, k) code $\mathscr{C}$ over $GF(q)$ with dual code $\mathscr{C}^\perp$ and let the error polynomial be $e(x) = \sum_{i=0}^{n-1} e_i x^i$ so that the received word polynomial is

$$r(x) = c(x) + e(x) = \sum r_i x^i = \sum (c_i + e_i) x^i.$$

Decoding the received word is equivalent to determining the error polynomial $e(x)$. The codeword $c(x)$ is orthogonal to every codeword in $\mathscr{C}^\perp$. Let $c'(x) = \sum_{i=0}^{n-1} c_i' x^i \in \mathscr{C}^\perp$; then

$$\langle r(x), c'(x) \rangle \triangleq \sum_{i=0}^{n-1} r_i c_i' = \sum_{i=0}^{n-1} e_i c_i'$$

Suppose that $\mathscr{C}$ has a distance three and that we can find two codewords $c'(x)$ and $c''(x)$ in $\mathscr{C}^\perp$ such that $c_0' = 1$ and $c_0'' = 1$ and if $c_i' \neq 0$, then $c_i'' = 0$ and if $c_i'' \neq 0$, then $c_i' = 0$. This leads to two equations such that e_0 appears in both and if e_i appears in one it cannot appear in the other, $i > 0$. Now suppose only one error occurs in transmission, i.e., the polynomial $e(x)$ has precisely one nonzero coefficient. If this error occurs in the zeroth position, then $e_0 = \alpha \in GF(q)$ and the two equations read

$$\langle r(x), c'(x) \rangle = \alpha = \langle r(x), c''(x) \rangle$$

If the error appears in the jth position, then by the assumptions on $c'(x)$ and $c''(x)$, c_j' and c_j'' cannot both be nonzero. If $c_j' \neq 0$, then the equations read

$$\langle r(x), c'(x) \rangle = \alpha c_j' \neq 0, \quad \langle r(x), c''(x) \rangle = 0$$

If both $c_j' = 0 = c_j''$, then both equations will be equal to zero. We draw the following conclusions if one error or less occurs:

(i) $\langle r(x), c'(x) \rangle = \langle r(x), c''(x) \rangle = \alpha$, then $e_0 = \alpha$.
(ii) $\langle r(x), c'(x) \rangle = \beta$, $\langle r(x), c''(x) \rangle = \gamma$, where one or both of β and γ are zero, then $e_0 = 0$.

Thus, contingent upon finding the polynomials $c'(x)$ and $c''(x)$, we have a criterion for correcting an error occurring in the zeroth position. If we can carry out a procedure for the other error coefficients e_j, $j = 1, \ldots, n - 1$, then we can correct any single error.

We generalize this simple procedure and consider $\mathscr{C}$ as an (n, k) code with distance d over $GF(q)$. Any sum of the form $\langle r(x), c'(x) \rangle$, where $c'(x) \in \mathscr{C}^\perp$, is called a parity check sum.

Definition A set of parity check sums $\langle r(x), c^{(i)}(x) \rangle$, $i = 1, 2, \ldots, J$, $c^{(i)}(x) \in \mathscr{C}^\perp$, is said to be orthogonal on the jth position (or on e_j) if:

 (i) $c_j^{(i)} = 1$, $i = 1, \ldots, J$.
 (ii) $c_k^{(i)} \neq 0$ in one sum, then $c_k^{(l)} = 0$, $\forall l \neq i$.

If $J = d - 1$ and $(d - 1)/2$ or fewer errors occur, then if $e_j = \alpha \neq 0$ since fewer than $(d - 1)/2$ other errors occur, fewer than $(d - 1)/2$ parity check sums will be different from α, i.e., in this case a majority of the sums will equal α. Similarly, if $e_j = 0$, then a majority (or at least half) of the sums will be zero and hence from a majority count of the values $\langle r(x), c^{(i)}(x) \rangle$ we can determine the value of e_j under the stated assumptions.

Definition If a set of $d - 1$ parity check sums orthogonal on e_i can be found for each $i = 0, 1, \ldots, n - 1$ for an (n, k, d) linear code, then the code is said to be completely or one-step orthogonalizable. If a set of $t - 1$ parity check sums orthogonal on e_i can be found for each $i = 0, 1, \ldots, n - 1$ for an (n, k, d) code, $t < d$, then we say the code is a one-step majority-logic-decodable code capable of correcting $\lfloor (t - 1)/2 \rfloor$ errors.

The term majority-logic-decodable code in general implies that majority logic decoding of the code may not correct all the errors that the code is capable of correcting. Massey (1963) uses the term *threshold decoding* to encompass methods using sets of orthogonal parity checks to achieve decoding.

There are few classes of codes that are completely one-step orthogonalizable. For example, the binary maximum length codes are completely one-step orthogonalizable.

We extend the notion of one-step majority logic decodability by first defining:

Definition A set of parity check sums is orthogonal on a set of positions $A = \{i_1, i_2, \ldots, i_k\}$ if:

 (i) For each $i_j \in A$ the corresponding error term e_j appears in each parity check sum, and with the same nonzero coefficient in each sum.

(ii) For each coordinate position $l \notin A$, e_l appears in at most one sum of the set with a nonzero coefficient.

Thus, rather than an estimate of a single error bit, we now obtain a majority vote on a weighted sum of the error positions in A. We now define an L-step majority-logic-decodable code recursively, using Berlekamp's (1968) definition, and restrict our attention to binary codes.

Definition A linear code with distance d is said to be L-step orthogonalizable if and only if the code contains sets of positions $A_1, A_2, \ldots$ such that

(i) For each i the code contains $d - 1$ parity check sums orthogonal on A_i.

(ii) The subcode of the original code that satisfies all these additional parity check sums is $(L - 1)$-step orthogonalizable.

When we talk of a nonbinary code being L-step majority logic decodable, the generalization will be intuitively straightforward. Basically, all coefficients of error terms in a set of parity check sums orthogonal on a set of error positions will be identical for these positions. This generalization of one-step majority logic decodability has led to the discovery of several interesting new classes of codes, including the finite geometry codes to be discussed in this section.

We begin the discussion of finite geometry codes with the projective geometry codes. It is convenient to consider the finite geometry $PG(m - 1, p^s)$ and describe its points as nonzero m-tuples over $GF(p^s)$ with scalar multiples identified. Let $n = (p^{ms} - 1)/(p^s - 1)$. The projective geometry code will be a code of length n, and we identify each coordinate position of an n-tuple with a point in the geometry. By an incidence vector of a t flat of $PG(m - 1, p^s)$ is meant a binary n-tuple with ones in those positions corresponding to points in the t flat and zeros elsewhere.

Definition A projective geometry code of order r and length $n = (p^{ms} - 1)/(p^s - 1)$ over $GF(p)$ is the largest cyclic code whose null space contains all the incidence vectors of all the r flats in $PG(m - 1, p^s)$.

The significance of the definition will become apparent later when majority logic decoding is discussed. At the moment it is important to establish the connection between these codes and the polynomial codes of Chapter 1.

Some authors define the rth-order p-ary projective geometry code to have length $(p^{(m+1)s} - 1)/(p^s - 1)$, containing all the flats of degree r of $PG(m, p^s)$ in its null space. While this relative difference of one in the dimension of the geometry affects the calculations, it makes little difference to the ideas. We choose our definition to preserve a certain symmetry with Euclidean geometry codes, whose definition will follow. With our definition the difference between the dimension of the geometry and the order of the flats contained in its null space will be the same.

That the coordinate positions can be labeled so that the incidence matrix of r flats in $PG(m, q)$ is cyclic can easily be demonstrated from the description of flats given in Equations (2.2) and (2.3). Thus if $\xi^{s_1}, \xi^{s_2}, \ldots, \xi^{s_{m+1}}$ are independent points in $PG(n, q)$, then so are $\xi\xi^{s_1}, \xi\xi^{s_2}, \ldots, \xi\xi^{s_{m+1}}$, and if the points

$$\alpha_i = \sum_{j=1}^{m+1} \varepsilon_{ij} \xi^{s_j}$$

form an m flat, then the points

$$\xi\alpha_i = \sum_{j=1}^{m+1} \varepsilon_{ij} \xi^{s_j+1}$$

form another m flat. If the coordinate positions are labeled with the first v distinct powers of ξ, the cyclic nature of the matrix is demonstrated.

Let L be a t flat in $PG(m-1, p^s)$ generated by the $t+1$ independent points $\{\mathbf{a}_0, \mathbf{a}_1, \ldots, \mathbf{a}_t\}$. Define $L^\perp$ to be the set of points $\mathbf{X}$ such that

$$\mathbf{a}_i \cdot \mathbf{X}^T = 0, \qquad 0 \leqslant i \leqslant t$$

which is, of course, an $m - t - 2$ flat in $PG(m-1, p^s)$. Consider the multivariable polynomial

$$f(\mathbf{X}) = \prod_{i=0}^{t} \{1 - (\mathbf{a}_i \cdot \mathbf{X}^T)^{p^s-1}\}$$

and observe that

$$f(\mathbf{X}) = \begin{cases} 1, & \mathbf{X} \in L^\perp \\ 0, & \mathbf{X} \notin L^\perp \end{cases}$$

Now consider an (n, m, s, μ, p) polynomial code with $b = p^s - 1$, implying that $n = (p^{ms} - 1)/(p^s - 1)$. Trivially the polynomial $f(\mathbf{X})$ is an element of $Q_m(\mu, p^s - 1)$ for $\mu \geqslant t + 1$ and hence the codeword $v(f)$ corresponding to this polynomial is in the polynomial code. It is clear that this codeword can be interpreted as the incidence vector of the $m - t - 2$ flat $L^\perp$. Thus the incidence vector of every $m - t - 2$ flat of $PG(m-1, p^s)$ is in the (n, m, s, μ, p) polynomial code if $0 \leqslant t < \mu$. Thus the dual of a $((p^{ms} - 1)/(p^s - 1), m, s, \mu, p)$ polynomial code is certainly contained in the projective geometry code of order $m - \mu - 1$.

The BCH bound on the (n, m, s, μ, p) code with $n = (p^{ms} - 1)/(p^s - 1)$ and $b = p^s - 1$ is given by Theorem 10.9 in Chapter 1:

$$d_{\min} \geqslant \frac{(R+1)p^{Qs} - 1}{b}$$

where R and Q are the remainder and quotient, respectively, obtained by dividing $m(p^s - 1) - \mu(p^s - 1)$ by $p^s - 1$. Clearly $R = 0$ and $Q = m - \mu$, to yield

$$d_{\min} \geq \frac{p^{(m-\mu)s} - 1}{p^s - 1}$$

Since the $m - \mu - 1$ flats are in the code and have precisely this weight, we conclude that this is the exact minimum distance of the code. Since this polynomial code is a subfield subcode of a GRM code, it follows from Delsarte (1970a, Theorem 12) that the minimum weight codewords generate the code. Thus the polynomial code is the minimal cyclic code containing the required flats, which implies that the dual code is the maximal code with the required flats in its null space, i.e., the projective geometry code. In general, the projective geometry code of order r has a null space generated by the incidence vectors of the r flats and this is the dual of the polynomial code $((p^{ms} - 1)/(p^s - 1), m, s, m - r - 1, p)$.

To see the relevance of this to majority logic decoding consider a projective geometry code of order r. There are

$$\frac{(1 + p^s + p^{2s} + \cdots + p^{(m-1)s})(p^s + \cdots + p^{(m-1)s}) \cdots (p^{rs} + \cdots + p^{(m-1)s})}{(1 + p^s + \cdots + p^{rs})(p^s + \cdots + p^{rs}) \cdots (p^{(r-1)s} + p^{rs})p^{rs}}$$

distinct r flats in $PG(m - 1, p^s)$. Now every $r - 1$ flat of $PG(m - 1, p^s)$ is contained in

$$\frac{p^{sr} + \cdots + p^{s(m-1)}}{p^{sr}} = 1 + \cdots + p^{s(m-r-1)} = \frac{p^{s(m-r)} - 1}{p^s - 1} = J$$

distinct r flats. If we fix an $r - 1$ flat L in $PG(m - 1, p^s)$, then every point of $PG(m - 1, p^s)$ is either in the $r - 1$ flat L or else in precisely one r flat containing L. Thus we use the J r-flats containing L to obtain a set of parity checks which are orthogonal on L. If e_i is the error digit on the ith coordinate position, then if $\lfloor J/2 \rfloor$ or fewer errors occur, the value of the sum

$$s_i = \sum_{j \in L} e_j$$

is given correctly by a majority of the J error sums on the r-flats containing L. By similar reasoning there are precisely J $(r - 1)$-flats containing a given $r - 2$ flat and, again assuming that $\lfloor J/2 \rfloor$ or fewer errors occur, we obtain correct estimates of error sums on the $r - 2$ flats. Clearly, in r steps we obtain correct error estimates, under the same conditions, of the 0 flats, i.e., the error digits themselves. The rth-order projective geometry code is thus r-step orthogonalizable and r-step majority logic decodable with distance at least $J + 1$. Interestingly, it can be shown that this is also the BCH bound on the minimum distance of the rth-order projective geometry code, which can be shown by considering the description of the roots of the associated polynomial code. These codes were originally considered by Rudolph (1964, 1967), who observed that the incidence matrix of r flats in a projective geometry can be made cyclic. Goethals and Delsarte (1968) approached the problem from a polynomial point of view and obtained an interesting characterization of the codeword zeros. Until recently the dimensions of these codes were known in only a few cases. The result of Hamada (1968), however, establishes the dimension for all cases.

The development of Euclidean geometry codes proceeds along lines similar to those of projective geometry codes. This time we label the $n = p^{ms} - 1$ coordinate positions with the nonzero points of the Euclidean geometry $EG(m, p^s)$.

Definition The Euclidean geometry code of order r and length $p^{ms} - 1$ over $GF(p)$ is the largest cyclic code whose null space contains the incidence vectors of all $r + 1$ flats of $EG(m, p^s)$.

Again, we first consider the relationship between these codes and polynomial codes. Let L be a t flat in $EG(m, p^s)$ and $\{\mathbf{a}_1, \mathbf{a}_2, \ldots, \mathbf{a}_t\}$ a basis. The polynomial

$$f(\mathbf{X}) = \prod_{i=1}^{t} \{1 - (\mathbf{a}_i \cdot \mathbf{X}^T)^{p^s - 1}\}$$

is of degree $t(p^s - 1)$. If $L^\perp$ is the null space of L, then $f(\mathbf{X})$ is the indicator polynomial of $L^\perp$, which is an $m - t$ flat. If $t(p^s - 1) \leq \mu$, then $f(\mathbf{X}) \in Q_m(\mu, 1)$ and the corresponding $(p^{ms} - 1)$-tuple is in the polynomial code $((p^{ms} - 1), m, s, \mu, p)$ for $b = 1$ and $\mu = D(p^s - 1)$ for some positive integer D. The flat $L^\perp$ goes through the origin since the origin is a solution to the equations $(\mathbf{a}_i \cdot \mathbf{X}^T) = 0$, $1 \leq i \leq t$. Notice that since coordinate positions are labeled only with the nonzero points of $EG(m, p^s)$, the incidence vector of an $m - t$ flat through the origin is of weight $p^{m-t} - 1$, since the origin is deleted. The weight of an $m - t$ flat not through the origin is p^{m-t}. Similarly, let $L'^\perp$ be the solutions to the set of equations

$$\mathbf{a}_i \cdot (\mathbf{X} - \mathbf{b})^T = 0, \quad 1 \leq i \leq t$$

The solutions form an $m - t$ flat through the point $\mathbf{b}$. The polynomial

$$f'(X) = \prod_{i=1}^{t} \{1 - [\mathbf{a}_i \cdot (\mathbf{X} - \mathbf{b})^T]^{p^s - 1}\}$$

is of degree $t(p^s - 1)$ and is the indicator function of the $m - t$ flat $L'^\perp$. The corresponding codeword $v(f')$ is the incidence vector of this flat.

2.2 Finite Geometries: Their Collineation Groups and Codes

For an (n, m, s, μ, p) polynomial code with $b = 1$ and $n = p^{ms} - 1$ suppose that

$$\mu = D(p^s - 1) + N, \qquad 0 \leqslant N < p^s - 1$$

From the previous considerations it follows that such a code contains every $m - t$ flat of $EG(m, p^s)$ for $0 \leqslant t \leqslant D$, i.e., all flats of order $m - D, m - D + 1, \ldots, m$.

Using precisely the decoding scheme as for projective geometry codes and noting that the number of $m - D$ flats that intersect a given $m - D - 1$ flat in $EG(m, p^s)$ is

$$J = \frac{p^{(D+1)s} - 1}{p^s - 1}$$

we conclude that the dual of the polynomial code is $(m - D)$-step majority logic decodable with minimum distance at least

$$J + 1 = \frac{p^{(D+1)s} - 1}{p^s - 1} + 1$$

Actually, with arguments similar to those used in deriving the BCH bound for an arbitrary polynomial code, Lin (1968, Theorem 4) showed that the dual code of a $(p^{ms} - 1, m, s, \mu, p)$ polynomial code has

$$d_{\min} \geqslant (\lambda + 1)p^{Ds} + p \cdot p^{(D-1)s}$$

where

$$\mu = D(p^s - 1) + N, \qquad 0 \leqslant N < p^s - 1$$

and

$$N = \lambda p^{s-1} + \sigma, \qquad 0 \leqslant \sigma < p^{s-1}$$

Choosing μ to be minimum for a given D, i.e., choosing $N = 0$, yields the bound

$$d_{\min} \geqslant p^{Ds} + p \cdot p^{(D-1)s}$$

The difference between this bound and the bound given by majority logic decoding is

$$(p^s - 2) + (p^s - 2)p^s + (p^s - 2)p^{2s} + \cdots + (p^s - 2)p^{(D-2)s} + (p - 2)p^{(D-1)s}$$

which indicates that for large s and D, majority logic decoding of the codes is falling far short of using the full error-correcting capabilities of the code.

It has been shown that a polynomial code of length $p^{ms} - 1$, $b = 1$, and $\mu = D(p^s - 1)$ contains all the $m - D$ flats of $EG(m, p^s)$. From Theorem 10.9 of Chapter 1 the BCH bound for such a code is given by

$$d_{min} \geq \frac{(R+1)p^{Qs} - 1}{b}$$

where $b = 1$, and R and Q are the remainder and quotient obtained by dividing $m(p^s - 1) - \mu$ by $p^s - 1$. For our situation $R = 0$ and $Q = m - D$, yielding

$$d_{min} \geq p^{(m-D)s} - 1$$

However, an $m - D$ flat through the origin has weight $p^{(m-D)s} - 1$ and we conclude that this is exactly the minimum weight of the polynomial code. Recalling that a polynomial code is a subfield subcode of a GRM code Delsarte (1970a, Theorem 11) shows that the minimum weight vectors actually generate the code. Thus the polynomial code is the smallest code that contains the required flats, and we conclude that the dual of the polynomial code of length $p^{ms} - 1$, $b = 1$, $\mu = (m - r - 1)(p^s - 1)$ over $GF(p)$ is the rth-order Euclidean geometry code.

Now let α be a primitive root of $GF(q^{ms})$ and let α^h be a zero of the polynomial code of length $q^{ms} - 1$, $b = 1$. Then, from a characterization of the roots it follows that

$$\min_{0 \leq l < s} \omega_{q^s}(hq^l) = j, \qquad 0 < j < m(q^s - 1) - \mu$$

Suppose the minimization is achieved with hq^l for some integer l and let

$$h = \sum \alpha_i q^{si} = \sum \beta_i p^i, \qquad 0 \leq \alpha_i \leq q^s - 1, \quad 0 \leq \beta_i \leq p - 1$$

Let $h' \in J(h)$, where $J(\cdot)$ is defined in Section 1.9 of Chapter 1. Then

$$h' = \sum \beta_i' p^i = \sum \alpha_i' q^{si}$$

and since $0 \leq \beta_i' \leq \beta_i$ we have that $0 \leq \alpha_i' \leq \alpha_i$ and hence that $\omega_{q^s}(h') \leq \omega_{q^s}(h)$. From the construction it follows that $\omega_{q^s}(h'q^l) \leq \omega_{q^s}(hq^l)$ and hence that $\alpha^{h'}$ is also a zero of the code. From Theorem 9.1 of Chapter 1 it follows that the extended polynomial code with $b = 1$ is invariant under the doubly transitive affine group of permutations.

Many authors refer to what we have defined as an rth-order Euclidean geometry code as a type 0, rth-order Euclidean geometry code. This is to distinguish them from the following:

Definition A type 1 Euclidean geometry code of order r and length $p^{ms} - 1$ over $GF(p)$ is the largest cyclic code whose null space contains the incidence vectors of all the $r + 1$ flats of $EG(m, p^s)$ that do not pass through the origin.

We have observed already the problem of flats which do or do not pass through the origin with respect to the fact that coordinate positions are labeled with only the nonzero elements of $EG(m, p^s)$. It is instructive to consider the problem a little further. Number p^{ms} coordinate positions with p^{ms} elements of $EG(m, p^s)$. Let H_1 be the incidence matrix of all the $r + 1$ flats in $EG(m, p^s)$ and H_2 the incidence matrix of all those $r + 1$ flats in $EG(m, p^s)$ that do not pass through the origin. Let $\mathscr{C}_1$ and $\mathscr{C}_2$ be the codes that have H_1 and H_2 as generator matrices, respectively. Let the first coordinate position be labeled with the origin of $EG(m, p^s)$ and denote by H_1', H_2', $\mathscr{C}_1'$, and $\mathscr{C}_2'$ the respective matrices and codes with the first position deleted. From the foregoing, the code $\mathscr{C}_1'$ is the polynomial code of length $p^{ms} - 1$, $\mu = (m - r - 1)(p^s - 1)$, and $b = 1$ over $GF(p)$. The dual of $\mathscr{C}_1'$ is the rth-order Euclidean geometry code if $g(X)$ is the generator polynomial of the dual of $\mathscr{C}_1'$, then α^h is a root iff

$$\max_{0 \leq l < s} \omega_{p^s}(hp^l) \leq (m - r - 1)(p^s - 1)$$

where α is a primitive element of $GF(p^{ms})$. $\mathscr{C}_2$ is obviously a proper subcode of $\mathscr{C}_1$ since it does not contain the incidence vector of any flat through the origin. Actually $\mathscr{C}_2$ has dimension precisely one less than that of $\mathscr{C}_1$. To see this, suppose F_0 is an $r + 1$ flat through the origin. It contains $p^{s(r+1)}$ points and it is possible to find $p^{(m-r-1)s} - 1$ other flats F_i, $1 \leq i \leq p^{(m-r-1)s} - 1$, not through the origin, which are parallel to F_0 so that, as sets,

$$EG(m, p^s) = \bigcup_{i=0}^{p^{(m-r-1)s}} F_i$$

If U is the all-ones vector and U_i the incidence vector of F_i, then

$$U = \sum_{i=0}^{p^{(m-r-1)s}-1} U_i$$

However U, which is a codeword of $\mathscr{C}_1$, cannot be a codeword of $\mathscr{C}_2$ since this would imply

$$U_0 = U - \sum_{i=1}^{p^{(m-r-1)s}} U_i$$

which says a flat through the origin is in $\mathscr{C}_2$. Since any flat through the origin can be obtained in this way, we conclude that the closure of U and $\mathscr{C}_2$ is $\mathscr{C}_1$ and hence the dimensions of $\mathscr{C}_1$ and $\mathscr{C}_2$ differ by exactly one.

Since an affine transformation of $EG(m, q^s)$ takes a t flat to a t flat, then $\mathscr{C}_1$ is invariant under the group of affine transformations. Similarly $\mathscr{C}_2$ is invariant under the subgroup that fixes the origin. This implies that $\mathscr{C}_2'$ is cyclic. Let $g_1(x)$ and $g_2(x)$ be the generator polynomials of $\mathscr{C}_1'$ and $\mathscr{C}_2'$

respectively. The generator polynomial of the dual of $\mathscr{C}_1'$ clearly does not have the unit element as a root since $\mathscr{C}_1'$ contains the all-ones vector. Thus the generator polynomial of $\mathscr{C}_1'$ does not have 1 as a root. On the other hand, every codeword polynomial corresponding to a row of H_2 has weight $p^{s(r+1)}$, which is zero mod p, implying that one is a zero of $g_2(x)$. We conclude that

$$g_2(x) = (x-1)g_1(x)$$

The dual of $\mathscr{C}_2'$ is a type 1 Euclidean geometry code of order r. From the characterization of the roots of the type 0 code we can conclude that α^h is a root of the generator polynomial of a type 1 Euclidean geometry code iff

$$0 < \max_{0 \leq l < s} \omega_{p^s}(hp^l) \leq (m-r-1)(p^s-1)$$

Notice that both the type 0 and type 1 Euclidean geometry codes are r-step orthogonalizable and, using r-step majority logic decoding, are capable of correcting

$$\left\lfloor \frac{p^{s(m-r+1)} - 1}{2(p^s - 1)} \right\rfloor$$

errors.

The original rth-order Reed–Muller codes (Reed, 1954; Muller, 1954) were binary, noncyclic (i.e., extended), type 0 Euclidean geometry codes. The null space of the codes of length 2^m contained every $r+1$ flat of $EG(m, 2)$. The dimension of the code was known to be

$$\sum_{i=0}^{r} \binom{m}{i}$$

and its distance 2^{m-r}. This approach to Reed–Muller codes is discussed in Exercise 6. The natural generalizations of these codes, the GRM codes, has been discussed in the previous chapter.

Slightly before the development of polynomial and geometry codes, Weldon (1966) considered planar difference set codes. These however, correspond to projective geometry codes with a parity check matrix containing the incidence vectors of lines in a projective plane. These p-ary codes, which have length $n = p^{2s} + p^s + 1$ and dimension (Graham and MacWilliams, 1966)

$$k = \binom{p+1}{2}^s + 1$$

will be considered in a later section. The more general class of difference set codes, whose parity check matrices contain all the incidence vectors of hyperplanes in $PG(m, p^s)$, will also be discussed there. The dimensions of these codes are

$$k = \binom{p+m-1}{p-1}^s + 1$$

where the codes are of length $(p^{ms} - 1)/(p^s - 1)$ (Smith, 1969; Goethals and Delsarte, 1968; MacWilliams and Mann, 1968). This latter reference also shows that the rank over $GF(p)$ of the hyperplanes in $EG(m, p^s)$ is

$$\binom{p+m-1}{p-1}^s - 1$$

A problem that has received attention recently is reducing the number of steps required in majority logic decoding of the geometry codes. Weldon(1968) described a scheme, applicable to RM and Euclidean geometry codes, that provided some improvement. He also demonstrated that nonorthogonal parity check sums could be used in decoding the geometry codes in fewer steps. Perhaps the most impressive results on this problem are those of Chen (1971, 1972), who showed that a type 1, rth-order Euclidean geometry code of length $p^{ms} - 1$ can be majority logic decoded in N steps, where

$$N = \begin{cases} 1, & r = 0 \\ 2, & \tfrac{1}{2}m \geqslant r > 0 \\ i+1 = 1 + \log_2\left\lceil \dfrac{m}{m-r} \right\rceil, & (1 - 2^{-i})m \geqslant r > (1 - 2^{-(i-1)})m \end{cases}$$

The result is obtained essentially by a flat counting argument. A similar result also holds for the projective geometry codes. Lin (1972a) was also able to give a shortening algorithm for the geometry codes, which resulted in majority logic decodable codes able to correct the same number of errors as the original code.

Delsarte (1969) generalizes the concept of a Euclidean geometry in an unusual way and obtains majority logic decodable codes. In another paper (Delsarte, 1971a) he defines a class of linear binary codes using the finite inversive planes [e.g., see Dembowski (1968)] and these codes are one-step orthogonalizable. Bose and Burton (1966) use projective geometries to show that the binary codes of maximum length n for a minimum distance $d = 4$ are Hamming codes and are unique.

The projective and Euclidean geometry codes are, of course, closely related and we consider this relationship further. Let α be a primitive element of $GF(p^{ms})$ and let $\beta = \alpha^n$, $n = (p^{ms} - 1)/(p^s - 1)$ be the corresponding primitive element of $GF(p^s)$. Let $\alpha^{s_1}, \ldots, \alpha^{s_{r+1}}$ be $r + 1$ independent points of $GF(p^{ms})$ over $GF(p^s)$. An r flat of $PG(m - 1, p^s)$ is the set of points α^j described by

$$\alpha^j = \sum_{k=1}^{r+1} \beta^{i_k} \alpha^{s_k} \tag{2.5}$$

as β^{i_k}, $k = 1, \ldots, r + 1$, assume all possible values subject to the condition that they not be simultaneously zero. In the projective geometry the set of $p^s - 1$ points $\{\alpha^j, \alpha^{j+n}, \alpha^{j+2n}, \ldots, \alpha^{j+(p^s-2)n}\}$ are identified since they are scalar multiples of one another over $GF(p^s)$. We say the set consists of $p^s - 1$ replications of the point α^j. On the other hand, the set of all points α^j satisfying Equation (2.5) for some set of β^{i_k}, $k = 1, \ldots, r + 1$, including the all-zero combination, forms an $r + 1$ flat through the origin in $EG(m, p^s)$. This flat has $p^{(r+1)s}$ points, while the r flat in $PG(m - 1, p^s)$ has $(p^{(r+1)s} - 1)/(p^s - 1)$. Clearly the $p^s - 1$ replications of points in an r flat in $PG(m - 1, p^s)$ together with the origin form an $r + 1$ flat in $EG(m, p^s)$ which passes through the origin. Since the number of r flats in $PG(m - 1, p^s)$ [by Eq. (2.1)] is

$$\frac{(1 + p^s + \cdots + p^{s(m-1)})(p^s + \cdots + p^{s(m-1)}) \cdots (p^{sr} + \cdots + p^{s(m-1)})}{(1 + p^s + \cdots + p^{sr})(p^s + \cdots + p^{sr}) \cdots (p^{s(r-1)} + p^{sr})p^{sr}}$$

is the same as the number of $r + 1$ flats through the origin in $EG(m, p^s)$ [by Eq. (2.4)] we conclude that the converse is also true, i.e., every $(r + 1)$ flat of $EG(m, p^s)$ that goes through the origin consists of $p^s - 1$ replications of an r flat of $PG(m - 1, p^s)$ plus the origin.

Notice, finally, that all the codes in this section were formed over $GF(p)$. The codes are the null spaces of certain incidence matrices containing only zeros and ones. The rank of this matrix is the same over $GF(p^l)$, l any positive integer, as it is over $GF(p)$, since all calculations are done in the prime subfield. The dimension of the code generated by the matrix, using $GF(p^l)$, is the same [over $GF(p^l)$] as that using $GF(p)$ [over $GF(p)$]. In particular the two codes have the same generator polynomial, which is a polynomial over $GF(p)$. Our main reason for using $GF(p)$ was that in this case it has been established that the respective polynomial codes described in this section are minimal, in the sense that they are the smallest cyclic codes containing the required flats. Their duals are then the geometry codes. If the codes are generated over $GF(p^l)$ rather than $GF(p)$, but using the same geometries, then the resulting code is closely related to the $GF(p)$ code. Specifically, if we view $GF(p^l)$ as a vector space over $GF(p)$, then an arbitrary linear combination of vectors over $GF(p)$ of the $GF(p)$ code yields a vector in the $GF(p^l)$ code, by associating with each of the code vectors a basis element in the vector space of $GF(p^l)$ over $GF(p)$. Alternatively, we could use the subfield subcode concept of the previous chapter to discuss the code over $GF(p)$.

2.3 Balanced Incomplete Block Designs and Codes

The arrangement of the elements of some set into blocks or subsets with certain properties is a problem that attracts much attention, both in mathematics and the applied sciences. Many of the applications lie in the theory of experimental design, but the problems encountered in the study of such systems are of considerable mathematical complexity. In this section we introduce the idea of a balanced incomplete block design (BIBD hereafter) and examine some of its more elementary properties.

A BIBD is a special, but very important, case of the more general incidence system, which may be described as a set of blocks B_i, $i = 1, \ldots, b$, of elements taken from the set $A = \{a_1, a_2, \ldots, a_v\}$ with an incidence relation indicating which elements are in which blocks. An element is, in general, in more than one block. If there are v elements in the set A, it is somewhat customary to refer to them by the integers one through v and we shall do so. The incidence system is a little too general to study and we define the following special case, a BIBD.

Definition A balanced incomplete block design is an incidence system on v elements with b blocks such that (i) each block contains the same number of objects $k < v$, (ii) each element appears in the same number of blocks r, and (iii) every unordered pair of distinct elements appears in precisely λ blocks.

Very often a BIBD is referred to as a (v, b, r, k, λ) configuration. In the following we will often refer to such a configuration simply as a design. The theory of such designs has a very intimate relationship with both number theory and group theory. As mentioned previously, BIBD's arise from problems in the design of experiments. The term balanced refers to the fact that each pair of elements occurs in precisely the same number of blocks, while the term incomplete implies that each block contains fewer than v elements. Before proceeding we give a simple example to verify the existence of such designs.

Example Consider the Fano geometry introduced in the previous section. We reproduce its diagram here with letters replaced by integers (Fig. 6). The lines of the geometry will be taken as the blocks of the design and are

(1, 2, 3), (3, 4, 5), (1, 5, 6), (1, 4, 7) (3, 6, 7), (2, 5, 7), (2, 4, 6)

Each element appears three times and each pair of elements appears in exactly one block. This system is a (7, 7, 3, 3, 1) configuration. The Fano geometry is the projective geometry $PG(2, 2)$, and we constructed the design from the incidence relations between points and lines. Later in the section we shall construct BIBD's from the incidence relations of the points in r flats of any finite geometry.

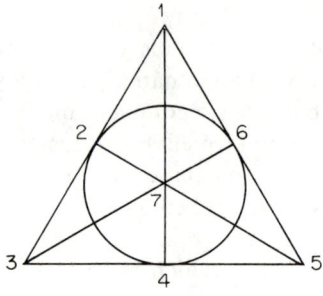

FIG. 6

To further verify the existence of such designs, we note that if one takes all possible combinations of k elements as the blocks, then a BIBD is obtained with parameters

$$b = \frac{v!}{k!(v-k)!}, \qquad r = \frac{(v-1)!}{(k-1)!(v-k)!}, \qquad \lambda = \frac{(v-2)!}{(k-2)!(v-k)!}$$

However, such a design is uninteresting, both mathematically and practically.

It is clear that the parameters of a design must satisfy certain relationships. Each of the b blocks contains k elements and the total number of elements in the array of blocks is bk. Since each element appears r times (the number of replications of the element) in the array and there are v elements, we have

(i) $bk = rv$.

Considering the r blocks containing a given element, we can form $r(k-1)$ pairs containing this element. Since there are $v-1$ possible sets of unordered pairs of elements containing the given element and each such pair occurs λ times in the array, we have

(ii) $r(k-1) = \lambda(v-1)$.

These are the basic necessary conditions for the existence of (v, b, r, k, λ) configuration. They are not, however, sufficient, as can be shown by example, and the determination of necessary and sufficient conditions for the existence of designs is an unsolved problem. For any given set of parameters of a design

2.3 Balanced Incomplete Block Designs and Codes

it must, of course, be true that $k < v$ and $\lambda < r$. In the case when $b = v$, then $k = r$ and we say that the design is symmetric.

The relationships (i) and (ii) imply that the design is specified by the three parameters (v, k, λ) (although any other set of three parameters can also be used to specify the design). It is also possible to show that $b \geq v$, an inequality due to Fisher, which is proven after the following theorem.

We define the incidence matrix $\Gamma = (\gamma_{ij})$ of a (v, b, r, k, λ) configuration as the $v \times b$ matrix of zeros and ones, where

$$\gamma_{ij} = \begin{cases} 1 & \text{if } i \in B_j \\ 0 & \text{otherwise} \end{cases}$$

where, for the ith element, we take the integer i. Some of the basic properties of designs can be derived using incidence matrices. It will also have significance in some problems of coding. We consider some of its properties in the following theorem.

Theorem 3.1 Let Γ be the $v \times b$ incidence matrix of a (v, b, r, k, λ) configuration. Then

(i) $\Gamma\Gamma^T = (r - \lambda)I_v + \lambda J_{vv}$;
(ii) $\det(\Gamma\Gamma^T) = [r + (v - 1)\lambda](r - \lambda)^{v-1}$;
(iii) $\Gamma J_{bb} = r J_{vb}$;
(iv) $J_{vv}\Gamma = k J_{vb}$;

where J_{mn} is the $m \times n$ matrix of all ones and I_n is the $n \times n$ identity matrix.

Proof To show (i), we notice that in the ith row of Γ a one appears in the jth position if and only if the element i is in the jth block. Thus, if $\Gamma\Gamma^T = \Sigma = (\sigma_{ij})$, then, since each row of Γ has r ones, we have $\sigma_{ii} = r$ for $i = 1, 2, \ldots, v$. The inner product of the ith row of Γ with the jth row, $i \neq j$, is the number of blocks in which both the elements i and j occur, which by definition is λ, and so $\sigma_{ij} = \lambda$, $i \neq j$. The resulting matrix is then most conveniently described by

$$\Gamma\Gamma^T = \Sigma = (r - \lambda)I_v + \lambda J_{vv}$$

To find the determinant of Σ, which has the form

$$\Sigma = \begin{bmatrix} r & \lambda & \lambda & \cdots & \lambda \\ \lambda & r & \lambda & \cdots & \lambda \\ \lambda & \lambda & r & \cdots & \lambda \\ \vdots & \vdots & \vdots & & \vdots \\ \lambda & \lambda & \lambda & \cdots & r \end{bmatrix}$$

we subtract the first column from each of the other columns and then add to the first row all rows below it, to give the matrix

$$\begin{bmatrix} r+(v-1)\lambda & 0 & 0 & \cdots & 0 \\ \hline \lambda & r-\lambda & & & \\ \lambda & & r-\lambda & & 0 \\ \vdots & & & \ddots & \\ \lambda & & 0 & & r-\lambda \end{bmatrix}$$

and the determinant of $\Gamma\Gamma^T$ is $[r+(v-1)\lambda](r-\lambda)^{v-1}$.

The third and fourth relationships in the theorem merely indicate that each row of Γ contains r ones and each column k ones. □

The fact that $r > \lambda$ implies that $\Gamma\Gamma^T$ is of rank v, by (ii). However, the rank of Γ is at most b and since rank $\Gamma \geqslant \text{rank}(\Gamma\Gamma^T) = v$, it follows that $b \geqslant v$.

We should perhaps mention that there are many interesting results on BIBD's which are pertinent to both number theory and matrix theory. For example, it can be shown [e.g., Ryser (1963, p. 111)] that in a symmetric BIBD with parameters $v = b$, $k = r$, and λ, if v is even, then $k - \lambda$ is a square, while if v is odd, then the Diophantine equation

$$x^2 = (k-\lambda)y^2 + (-1)^{(v-1)/2}\lambda z^2$$

has a nontrivial solution in integers for x, y, and z. Such a statement serves as a nonexistence theorem for symmetric BIBD's since for any v, k, and λ the foregoing conditions are certainly necessary for the existence of such a design. It is not known whether they are sufficient conditions.

If we are given a BIBD, there are several procedures for constructing other BIBD's from it and we consider three of them here. Perhaps we should first mention that two BIBD's are said to be isomorphic if one can be obtained from the other by a relabeling of the elements if the order in which the blocks are written down is immaterial.

If Γ is the incidence matrix of the given design, then the complement of Γ, which we denote by $\bar{\Gamma}$, obtained by replacing a zero in Γ by a one and a one by a zero, will be the incidence matrix of a design, which we call the complementary design. To verify that $\bar{\Gamma}$ is the incidence matrix of a design and to find its parameters, we proceed as follows. Each row of $\bar{\Gamma}$ has $b - r$ ones and each column has $v - k$ ones, indicating that for the new design $\bar{r} = b - r$ and $\bar{k} = v - k$. Furthermore, since two distinct rows of Γ have ones in precisely λ places at the same time, then they differ in $2r - \lambda$ places and hence any

2.3 Balanced Incomplete Block Designs and Codes

two rows of $\bar{\Gamma}$ have ones in $b - (2r - \lambda)$ places at the same time. Thus $\bar{\Gamma}$ is the incidence matrix of a $(v, b, b - r, v - k, b - 2r + \lambda)$ configuration if Γ is the incidence matrix of a (v, b, r, k, λ) configuration.

The residual design is another configuration that may be obtained from a given design, provided the given design is symmetric, i.e., $b = v$. If Γ is the incidence matrix of such a design, then, it can be shown that [see, for example, Ryser (1963, p. 103)].

$$\Gamma\Gamma^T = \Gamma^T\Gamma = (r - \lambda)I_v + \lambda J_{vv}$$

where Γ is a $v \times v$ matrix. The inner product of any two distinct rows of Γ^T is λ, implying that in a symmetric design any two distinct blocks have precisely λ elements in common. Now, from the given design with parameters (v, b, r, k, λ) ($b = v$ and $k = r$) delete a block and all elements occurring in that block from all other blocks. Thus we have $b - 1$ blocks remaining, and since we have removed precisely λ elements from each block, they contain $k - \lambda$ elements. This system clearly forms a design, called the residual design, and has parameters $(v - k, b - 1, r, k - \lambda, \lambda)$ where, again $b = v$ and $k = r$. It is important to note that the properties of the residual design may depend on which block of the original design was deleted and the deletion of a different block may not lead to an isomorphic design.

Suppose we start again with a symmetric (v, b, r, k, λ) design. To form the derived design, we choose a block and for the blocks of the derived design we take the sets of elements which each of the other blocks have in common with the chosen block. This gives a design with parameters $(k, b - 1, r - 1, \lambda, \lambda - 1)$ and again the properties of the derived design depend on the initial block chosen.

There are a wide variety of schemes available for the construction of BIBD's and the interested reader should consult Hall (1964, 1967a, b). We consider only one more technique since it uses the finite geometries which we have already investigated. Consider the case of a projective geometry $PG(n, q)$ first and for the elements of the design we take the points of the geometry. For the elements of a block we take the points of an m flat, where m is a constant for a given design. From the previous section we have that

$$v = 1 + q + q^2 + \cdots + q^n \quad \text{and} \quad k = 1 + q + q^2 + \cdots + q^m$$

where we are using the m flats in $PG(n, q)$ for the blocks of the design. The number of blocks is the number of distinct m flats in $PG(n, q)$, which we found to be

$$b = \frac{(1 + q + \cdots + q^n)(q + \cdots + q^n) \cdots (q^m + \cdots + q^n)}{(1 + q + \cdots + q^m)(q + \cdots + q^m) \cdots (q^{m-1} + q^m)q^m}$$

In the previous section we found that the number of $PG(m, q)$ in $PG(n, q)$ containing a given $PG(l, q)$ is

$$\frac{(q^{l+1} + \cdots + q^n)(q^{l+2} + \cdots + q^n) \cdots (q^m + \cdots + q^n)}{(q^{l+1} + \cdots + q^m)(q^{l+2} + \cdots + q^m) \cdots (q^{m-1} + q^m)q^m}$$

Thus, the number of m flats that are the blocks of the design containing a given point is, setting $l = 0$ in the above equation,

$$r = \frac{(q + \cdots + q^n)(q^2 + \cdots + q^n) \cdots (q^m + \cdots + q^n)}{(q + \cdots + q^m)(q^2 + \cdots + q^m) \cdots (q^{m-1} + q^m)q^m}$$

which is, of course, the number of replications of each point in the design. The parameter λ for the design is the number of m flats containing a given pair of points. However, any m flat containing the pair of points also contains the line joining the points and hence λ is the number of m flats containing a given line. This quantity is found by setting $l = 1$ in the above equation to give

$$\lambda = \frac{(q^2 + \cdots + q^n)(q^3 + \cdots + q^n) \cdots (q^m + \cdots + q^n)}{(q^2 + \cdots + q^m)(q^3 + \cdots + q^m) \cdots (q^{m-1} + q^m)q^m}$$

and $\lambda = 1$ for $m = 1$, and we have found all the parameters of the design.

In precisely the same manner we can use the m flats of $EG(n, q)$ to give a design with parameters

$$v = q^n, \qquad k = q^m, \qquad b = q^{n-m} \prod_{j=1}^{m} \frac{(q^{n+1-j} - 1)}{(q^{m+1-j} - 1)}$$

$$r = \frac{(q + \cdots + q^n)(q^2 + \cdots + q^n) \cdots (q^m + \cdots + q^n)}{(q + \cdots + q^m)(q^2 + \cdots + q^m) \cdots (q^{m-1} + q^m)q^m}$$

$$\lambda = \begin{cases} 1 & \text{if } m = 1 \\ \dfrac{(q^2 + \cdots + q^n)(q^3 + \cdots + q^n) \cdots (q^m + \cdots + q^n)}{(q^2 + \cdots + q^m)(q^3 + \cdots + q^m) \cdots (q^{m-1} + q^m)q^m} & \text{otherwise} \end{cases}$$

We have observed that when $b = v$, and hence $k = r$, the design is said to be symmetric and its incidence matrix is square. From the parameters of the designs derived from the projective geometries $PG(n, q)$ it follows that when $m = n - 1$, i.e., when the blocks of the design are determined by points in hyperplanes, then $b = v = 1 + q + q^2 + \cdots + q^n$ and hence these designs are symmetric with $\lambda = (q^{n-1} - 1)/(q - 1)$. In particular, the lines of any projective plane [not just $PG(2, q)$] determine a symmetric BIBD with $\lambda = 1$. The converse of this statement is also true, i.e., if we have a symmetric BIBD with $v = n^2 + n + 1$ and $k = n + 1$ and $\lambda = 1$, then we can form a projective plane of order n by identifying the elements of the designs with points of the

2.3 Balanced Incomplete Block Designs and Codes

geometry and the blocks of the design with the lines of the geometry. For a proof of this see Ryser (1963, p. 104).

In the next two sections of this chapter we shall consider some other combinatorial structures that may also be used to derive some combinatorial designs. However, rather than treat them as special cases of BIBD's, we prefer to present them in their original setting and deduce the associated design from them.

If it is possible to separate the b blocks of a BIBD into r sets of n blocks each such that each variety occurs exactly once in each set, then $b = nr$ and we say the design is resolvable (RBIBD). Such designs were studied by Bose (1942). It can be shown that for such designs

$$b \geqslant v + r - 1 \tag{3.1}$$

and since $r \geqslant 1$, this is a stronger version of Fisher's inequality. For an ordinary BIBD (i.e., not necessarily resolvable), if Fisher's inequality (i.e., $b \geqslant v$) holds with equality, then any two blocks in the design have the same number of elements in common. If Eq. (3.1) holds with equality for an RBIBD, then it can be shown that any two blocks belonging to different sets have exactly the same number of elements in common. A resolvable design with this property is called an affine resolvable BIBD (ARBIBD). The converse is also true, i.e., if the design is affine resolvable then Eq. (3.1) holds with equality. We state this as a theorem [due to Bose (1942)]:

Theorem 3.2 If a balanced incomplete block design D with parameters (v, b, r, k, λ) is resolvable, then $b \geqslant v + r - 1$, and equality holds if and only if D is affine resolvable.

Proof Since D is resolvable, then $v = nk$ and $b = nr$. Let R be a resolution class. Let B be a block of the class, and let $B_1, B_2, \ldots, B_t$, $t = (r-1)n$, be the blocks of the remaining classes. Let $x_i = |B \cap B_i|$, $i = 1, 2, \ldots, t$. Since each variety of B occurs $r - 1$ more times,

$$\sum_{i=1}^{t} x_i = k(r-1)$$

Also, since each pair of varieties of B occurs $\lambda - 1$ times elsewhere,

$$\sum_{i=1}^{t} x_i(x_i - 1) = (\lambda - 1)k(k-1)$$

If $\bar{x}$ denotes the mean of the x_i, then

$$\bar{x} = \frac{k(r-1)}{t} = \frac{k}{n} = \frac{k^2}{v}$$

Now

$$\sum_{i=1}^{t} (x_i - \bar{x})^2 = \sum x_i^2 - t\bar{x}^2$$

Since $v > k$, $b \geq v + r - 1$. If $b = v + r - 1$, each $x_i = k^2/v$, and the design is affine resolvable. Conversely, if the design is affine resolvable, each $x_i = \bar{x}$, and therefore $b = v + r - 1$. □

The following lemma will be used in the next chapter.

Lemma 3.1 Let D be a balanced incomplete block design with parameters (v, b, r, k, λ) which satisfy $b = v + r - 1$. If every pair of blocks intersect in either zero or k^2/v varieties, then D is affine resolvable.

Proof Since $b = v + r - 1$, the basic relations can be used to prove

$$v = \frac{(r-1)k}{r-k}, \quad b = \frac{r(r-1)}{r-k}, \quad \lambda = r - k$$

Let B be a block of D. Remove B and all varieties of B from D. The result is a configuration D' with $v' = v - k = k(k-1)/(r-k)$ varieties in which every variety of D' occurs in r blocks and every pair of distinct varieties of D' occurs in λ blocks of D'. The configuration D' has two block sizes, k and $k' = k - k^2/v$. Let x be a variety of D' and let $s(x)$ denote the number of blocks of size k' that contain x. Since every other variety of D' occurs with x exactly λ times, one obtains

$$s(x)(k'-1) + [r - s(x)](k-1) = \lambda(v'-1)$$

$$s(x) = \frac{\lambda k}{k_1 - k} = \frac{\lambda v}{k} = r - 1$$

Thus every variety of D' occurs in exactly one block of size k. But these blocks are the blocks of D that do not meet B. Therefore the block B and those that are disjoint from it form a resolution class. But B was arbitrary; hence every block can be assigned to a resolution class. □

Example An example of an ARBIBD with $b = 12$, $v = 9$, $r = 4$, $k = 3$, and $\lambda = 1$ is

$$\left.\begin{array}{l} b_1 = \{1, 2, 3\} \\ b_2 = \{4, 5, 6\} \\ b_3 = \{7, 8, 9\} \end{array}\right\} \text{set 1} \qquad \left.\begin{array}{l} b_7 = \{1, 5, 9\} \\ b_8 = \{2, 6, 7\} \\ b_9 = \{3, 4, 8\} \end{array}\right\} \text{set 3}$$

$$\left.\begin{array}{l} b_4 = \{1, 4, 7\} \\ b_5 = \{2, 5, 8\} \\ b_6 = \{3, 6, 9\} \end{array}\right\} \text{set 2} \qquad \left.\begin{array}{l} b_{10} = \{1, 6, 8\} \\ b_{11} = \{2, 4, 9\} \\ b_{12} = \{3, 5, 7\} \end{array}\right\} \text{set 4}$$

2.3 Balanced Incomplete Block Designs and Codes

The number of varieties in common with any two blocks of different sets of an ARBIBD is k^2/v, implying that $v|k^2$. For an arbitrary BIBD, if it has the property that any two blocks have the same number of elements in common, then $b = v$ and the design is symmetric. The converse also holds and any two blocks in a symmetric design must have precisely λ elements in common.

While three parameters specify the other two parameters of a BIBD, for an ARBIBD we only require two parameters. If an ARBIBD has parameters (v, b, r, k, λ), then these parameters must be of the form

$$v = n^2(n-1)t + n^2, \quad b = n(n^2t + n + 1), \quad r = n^2t + n + 1$$
$$k = n(n-1)t + n, \quad \lambda = nt + 1$$

A design that is associated with the BIBD's and also with the Latin squares to be considered in the next section is the orthogonal array, which we require later in this section for our discussion of equidistant codes.

Definition Let $\mathbf{A}$ be a $k \times N$ matrix with entries from the set $\{0, 1, \ldots, s-1\}$. We say that $\mathbf{A}$ is an orthogonal array of strength t, size N, with k constraints and s levels if each $t \times N$ submatrix of $\mathbf{A}$ contains all possible $t \times 1$ column vectors with the same frequency λ, where λ is called the index of the array. We denote such an array by $OA(N, k, s, t)$.

This is the definition given by Bose and Bush (1952). Hall (1967a) defines an orthogonal array as the special case of the foregoing definition for $t = 2$ and $\lambda = 1$, $N = n^2$, and $s = n$. Certain orthogonal arrays are closely related to transversal systems.

Definition Let S be a class of m mutually disjoint t-sets w_i, $i = 1, 2, \ldots, m$. A transversal system $\Gamma_0(m, t)$ consists of t^2 sets Y_j, $j = 1, 2, \ldots, t^2$, each containing m elements such that $|w_i \cap Y_j| = 1$, $i = 1, 2, \ldots, m$ and $j = 1, 2, \ldots, t^2$, and if $j \neq k$, then $|Y_j \cap Y_k| \leq 1$. The sets Y_j are called *transversals*.

Lemma 3.2 There exists a transversal system $\Gamma_0(m, t)$ if and only if there exists an orthogonal array $OA(t^2, m, t, 2)$ of index one.

Proof Let us assume that there exists an orthogonal array $OA(t^2, m, t, 2)$ on the set of symbols $1, 2, \ldots, t$. Let sets $w_i = \{a_{ij}, j = 1, 2, \ldots, t\}$ be given. Form an array A of the symbols of $V = \bigcup_{i=0}^{m} w_i$ by replacing the symbol j in the ith row of the $m \times t^2$ orthogonal array by the element a_{ij}. Let Y_k be the set of elements in the kth column of A. Then these sets form a transversal system $\Gamma_0(m, t)$. Indeed, it is clear that $|w_i \cap Y_k| = 1$, and if Y_k and Y_l, $k \neq l$, intersect in two or more elements, say $\{a_{ij}, a_{rs}\} \subseteq Y_k \cap Y_l$, then the pair of symbols (j, s) occurs twice in rows i and r of the array in columns k and l, contradicting the definition of orthogonal array. Conversely, given the transversal system, it is evident that the elements of the sets Y_k can be ordered to form the rows of an orthogonal array. □

Definition A set of t transversals of a transversal system $\Gamma_0(m, t)$ is a parallel class if no two have a common element.

If $\Gamma_0(m, t)$ contains e (or more) parallel sets of transversals, it is denoted by $\Gamma_e(m, t)$.

Lemma 3.3 There is a system $\Gamma_t(m - 1, t)$ if and only if there is a system $\Gamma_0(m, t)$.

Proof Let a transversal system $\Gamma_0(m, t)$ defined on sets $w_i = \{a_{ij}\}$ be given. Let P_j be the class of transversals containing the element a_{mj}, $j = 1, 2, \ldots, t$. The sets of P_j correspond to a parallel class in the transversal system $\Gamma_0(m - 1, t)$ obtained by removing the elements of a_m from $\Gamma_0(m, t)$. Hence this system has t parallel classes and can be denoted by $\Gamma_t(m - 1, t)$. Clearly the construction can be reversed to prove the converse. □

Theorem 3.3 There exists a projective plane or order k if and only if there exists an orthogonal array $OA(k^2, (k + 1), k, 2)$ of index one.

Proof Let us assume that there exists a projective plane of order k. Then there exists an affine plane A of order k. Let $l_1, l_2, \ldots, l_k$ be the lines of a parallel class of A. Then the remaining lines form a transversal system $\Gamma_0(k, k)$. Since these lines can be resolved into k parallel classes, this system is a system $\Gamma_k(k, k)$. Hence there exists a transversal system $\Gamma_0(k + 1, k)$ and therefore an orthogonal array $OA(k^2, (k + 1), k, 2)$. The argument can be reversed to establish the converse. □

Example The matrix

$$A = \begin{bmatrix} 0 & 1 & 0 & 0 & 0 & 1 & 1 & 1 \\ 0 & 0 & 1 & 0 & 1 & 0 & 1 & 1 \\ 0 & 0 & 0 & 1 & 1 & 1 & 0 & 1 \\ 0 & 1 & 1 & 1 & 0 & 0 & 0 & 1 \end{bmatrix}$$

is an $OA(8, 4, 2, 3)$ of strength three and index one.

We give a construction for such arrays, using the results of Bose and Bush (1952). Let C be a $k \times r$ matrix over $GF(q)$ with the property that every t rows are linearly independent. Label all the elements of $V_r(q)$, viewed as column vectors, by $X_1, X_2, \ldots, X_{q^r}$. We claim that the $k \times q^r$ matrix A,

2.3 Balanced Incomplete Block Designs and Codes

whose ith row is $\mathbf{CX}_i$, is an $OA(q^r, k, q, t)$ of index q^{r-t}. Let $\mathbf{A}'$ be a $t \times q^r$ submatrix of $\mathbf{A}$ and let $\mathbf{C}'$ be the corresponding submatrix of $\mathbf{C}$ consisting of rows of $\mathbf{C}$ corresponding to those chosen for $\mathbf{A}'$. Since $\mathbf{C}'$ has rank t, the equation $\mathbf{C}'\mathbf{X} = \mathbf{y}$ has q^{r-t} solutions for each $\mathbf{y} \, \varepsilon \, V_t(q)$, implying $\mathbf{A}$ is an $OA(q^r, k, q, t)$ of index q^{r-t}. Since nonzero scalar multiples of a row of $\mathbf{C}$ do not affect the independence of the rows, we can view the problem of constructing $\mathbf{C}$ as choosing k points of $PG(r - 1, q)$ so that no t of them lie in a μ flat for $\mu \leqslant t - 2$. Notice the similarity of this argument to that used in proving the Varshamov–Gilbert bound.

It follows immediately from the definition that an $OA(N, k, s, t)$ of index λ has $N = \lambda s^t$. It can be shown (Bose and Bush, 1952) that for $t = 2$ the number of constraints is bounded by

$$k \leqslant \left\lfloor \frac{\lambda s^2 - 1}{s - 1} \right\rfloor$$

while for $t = 3$ it must be true that

$$k \leqslant \left\lfloor \frac{\lambda s^2 - 1}{s - 1} \right\rfloor + 1$$

There are several close connections between codes and BIBD's and in the remainder of this section we discuss two of them. Let C be a binary, linear (n, k) code and suppose it is invariant under a doubly transitive permutation group. Let D_i be the set of vectors of weight i, $|D_i| = N_i$. Let $\mathbf{A}$ be the $N_i \times n$ matrix whose rows are the codewords of D_i. Since a permutation of coordinate positions does not affect the weight of a codeword, the set D_i is also invariant under the permutation group. There exists in the group a permutation which takes column one of $\mathbf{A}$ to column j and this permutation has the same effect as rearranging the rows of $\mathbf{A}$. Thus the number of ones in the first column is the same as the number in the jth column and we conclude that each column has $r = iN_i/n$ ones, since the total number of ones in $\mathbf{A}$ is iN_i. Similarly, let λ be the number of rows in which the first two columns of $\mathbf{A}$ both have ones. There exists a permutation that takes column one to column j and column two to column k. Again, this permutation can be effected by a rearrangement of the rows in which any two columns both have ones. Thus columns j and k both have ones in λ rows. Permute the rows of $\mathbf{A}$ so that the r ones in the first column of $\mathbf{A}$ are in the first r rows. From the preceding every other column of $\mathbf{A}$ must have exactly λ ones in the first r rows. Since each row has i ones, by counting the number of ones two ways, we obtain

$$(n - 1)\lambda + r = ri$$

or

$$\lambda = \frac{i(i - 1)}{n(n - 1)} N_i$$

Clearly, if we treat the column positions of **A** as the varieties, then **A** is the incidence matrix of a BIBD with parameters

$$v = n, \qquad b = N_i, \qquad k = i, \qquad r = \frac{iN_i}{n}, \qquad \lambda = \frac{r(i-1)}{n}$$

Thus the codewords of a given weight of any code that is invariant under a doubly transitive group yield a BIBD. We have encountered several classes of codes that are invariant under doubly transitive groups. These include the primitive BCH code (a simple consequence of Theorem 9.1 in Chapter 1), the GRM codes, and the type 0 Euclidean geometry codes. In later sections and the next chapter we will consider the relationship between t designs, which are particular types of BIBD's and codewords of constant weight in a code. Generally speaking, the conditions required to obtain t designs are far more severe than those required to obtain BIBD's.

For the remainder of the section we will consider the work of Semakov and Zinov'ev (1968) on the relationships between equidistant (ED) codes, RBIB and ARBIB designs, and orthogonal arrays of strength two. From Theorem 11.2 of Chapter 1 we have that in any code of length n with N codewords over an alphabet of q symbols and with minimum distance d [an (n, N, q, d) code]

$$d \leqslant nN(q-1)/(N-1)q \tag{3.2}$$

If we form the $N \times n$ matrix **A** whose rows are the codewords, then equality is achieved in (3.2) only if each of the q symbols appears the same number of times in each column, i.e., $N = qt$, where each symbol appears t times in each column. A code that attains equality in the bound of (3.2) is called a maximal ED code and denoted as an ED_m code. In an ED_m code we have $N = qt$ and

$$d = nqt(q-1)/(qt-1)q$$

Since $qt - 1 = (q-1)(t-1) + (q-1) + (t-1)$, we can reduce this by the quantity $(q-1, t-1)$. If we can write

$$n = c(qt-1)/(q-1, t-1)$$

then

$$d = ct(q-1)/(q-1, t-1) \qquad c \text{ an integer}, \quad c \geqslant 1$$

We exclude the case where $t = 1$ from consideration since this corresponds to the trivial situation of each codeword containing n repetitions of the same symbol. Notice that when $c = 1$ we must have that

$$(N-1) \geqslant n \geqslant \max\left\{\frac{(N-1)}{(q-1)}, \frac{(N-1)}{(t-1)}\right\}$$

2.3 Balanced Incomplete Block Designs and Codes

We now establish the equivalence between a (v, b, r, k, λ) RBIB, D, and an (n, N, q, d) ED_m code $\mathscr{C}$. Let $\mathbf{A}$ be the $N \times n$ code matrix and let $\mathbf{B}$ be the $v \times b$ incidence matrix of D with r ones in each row and k ones in each column. Suppose we are given the design D and let $b = qr$ and $v = qk$. Denote by $S_1, S_2, \ldots, S_r$ the sets of blocks with the property that each of the v elements appears precisely once in each set. Each set S_i contains q blocks and suppose the first q columns of the incidence matrix $\mathbf{B}$ of the design correspond to S_1. Let $\mathbf{B}'$ be the $v \times q$ submatrix of $\mathbf{B}$ containing its first q columns. Each row of $\mathbf{B}'$ contains one one while each column contains k ones. Label the rows of the $q \times q$ identity matrix by the elements of $X = \{0, 1, \ldots, q-1\}$ in some arbitrary but fixed manner. Then replace each row of $\mathbf{B}'$ by its corresponding element of X. Doing this for the other sets S_i yields a $v \times r$ matrix $\mathbf{C}$ with elements from X. Since in any design every pair of elements appears in precisely λ blocks, and since these λ blocks must appear in different sets, it follows that any two rows of $\mathbf{C}$ agree in exactly λ places. Viewing the rows of $\mathbf{C}$ as a code over X yields an ED code with

$$N = v = qk, \qquad d = r - \lambda, \qquad n = r$$

However, for any design

$$\lambda = r(k-1)/(v-1)$$

and it follows that

$$d = r - \lambda = r(v-k)/(v-1) = rk(q-1)/(qk-1) = nN(q-1)/q(N-1)$$

and the code is maximal.

The converse statement is also true and is shown simply by retracing these arguments, i.e., in the code matrix $\mathbf{A}$ replace each element of X by its associated row of I_q and the resulting matrix is the incidence matrix of an RBIBD. We have shown the following theorem.

Theorem 3.4 An ED_m code $\mathscr{C}$ with parameters $(n, N = qt, q, d)$ is equivalent to an RBIBD with parameters $(v = qk, b = qr, r, k, \lambda)$ and

$$t = k, \qquad N = qt = v, \qquad n = r, \qquad qn = b, \qquad d = n - \lambda = r - \lambda$$

Suppose now that we are given an ARBIBD with incidence matrix $\mathbf{B}$ and consider the ith column of the set S_1 and the jth column of the set S_2. The blocks corresponding to these two columns have precisely μ elements in common, implying that when we replace rows of $\mathbf{B}$ by r-tuples of elements of X the same ordered pair of elements in the first and second columns of the code will appear μ times. Since there are q^2 such ordered pairs, we must have that $v = q^2\mu$. Hence in a code corresponding to an ARBIBD in any two columns each ordered pair of elements from X appears μ times. Retracing the argument yields the converse, i.e., given a code with the property that in any two columns of its code matrix every ordered pair of elements appears μ times, then an ARBIBD can be obtained from it.

Let J be the transpose of the matrix over X of an $OA(N, k, S, 2)$ of index λ'. Thus J is an $N \times k$ matrix with the property that every ordered pair of elements of X appears λ' times in every pair of columns. But, using the correspondence between the rows of the identity matrix and elements of X, we see that an ARBIBD with parameters (v, b, r, k, λ) with $b = qr$ and $v = qk$ is equivalent to an $OA(v, r, q, 2)$ of index k^2/v.

Semakov and Zinov'ev (1968) further show that ED_m codes with $n = (N - 1)/(q - 1)$ are equivalent to ARBIBD's and hence to orthogonal arrays of appropriate parameters.

There are many other interesting relationships between balanced codes (all codewords have the same weight) and combinatorial configurations. The important t designs will be considered in the context of coding in the next chapter. We mention one other result on equidistant codes. Deza (1973) showed that a binary equidistant code of length n with m codewords at a mutual distance of $2k$ must have $m \leqslant k^2 + k + 2$ unless the code is trivial (i.e., each column of its code matrix contains m or $m - 1$ equal entries). Van Lint (1973) then showed that for $k > 1$ a code of length n, mutual distance $2k$, and with $m = k^2 + k + 2$ exists for sufficiently large n iff a $PG(2, k)$ exists. These results will be considered in the next chapter.

2.4 Latin Squares and Steiner Triple Systems

We shall refer to a symmetric BIBD with parameters $b = v$, $k = r$ as a (v, k, λ) configuration.

Definition Let x be a set of n elements $\{x_1, x_2, \ldots, x_n\}$. A Latin square of order n is then an $n \times n$ array of elements of x such that each row and each column of the array contains each element of x exactly once.

Example Let $x = \{1, 2, 3, 4\}$; then an example of a Latin square of order four is

$$\begin{array}{cccc} 1 & 3 & 2 & 4 \\ 4 & 2 & 3 & 1 \\ 2 & 4 & 1 & 3 \\ 3 & 1 & 4 & 2 \end{array}$$

2.4 Latin Squares and Steiner Triple Systems

Notice that we can permute rows and columns of the array without destroying the Latin square property. This implies that we can always permute the rows and columns of the array so that the elements in the first row and first column are ordered. Thus the foregoing array can be put into the form

$$\begin{array}{cccc} 1 & 2 & 3 & 4 \\ 2 & 1 & 4 & 3 \\ 3 & 4 & 1 & 2 \\ 4 & 3 & 2 & 1 \end{array}$$

Notice also that the group multiplication table of any group is, by definition, a Latin square.

Definition A pair of Latin squares of order n is said to be orthogonal if, when one square is superimposed on the other so that each cell of the $n \times n$ array contains an ordered pair, each of the n^2 possible ordered pairs occurs exactly once in the array.

Similarly, we say that a set of Latin squares is an orthogonal set if the squares are pairwise orthogonal.

Example The two Latin squares L_1 and L_2 of order four

$$L_1: \begin{array}{cccc} 1 & 2 & 3 & 4 \\ 2 & 1 & 4 & 3 \\ 3 & 4 & 1 & 2 \\ 4 & 3 & 2 & 1 \end{array} \qquad L_2: \begin{array}{cccc} \alpha & \beta & \gamma & \delta \\ \gamma & \delta & \alpha & \beta \\ \delta & \gamma & \beta & \alpha \\ \beta & \alpha & \delta & \gamma \end{array}$$

when superimposed give

$$\begin{array}{cccc} (1, \alpha) & (2, \beta) & (3, \gamma) & (4, \delta) \\ (2, \gamma) & (1, \delta) & (4, \alpha) & (3, \beta) \\ (3, \delta) & (4, \gamma) & (1, \beta) & (2, \alpha) \\ (4, \beta) & (3, \alpha) & (2, \delta) & (1, \gamma) \end{array}$$

and hence L_1 and L_2 are orthogonal.

The orthogonality of two Latin squares is not destroyed by a relabeling of the elements. Thus, if we have a set of r orthogonal Latin squares of order n, we can relabel the elements of each square so that the first row contains in order the elements $1, 2, \ldots, n$ and i appears in the ith column of each square. The set of elements in any given cell not in the first row of each square must be distinct since all identical pairs appear in the first row for any two given squares. Furthermore, the element i cannot appear in the ith column of any square, and we conclude that $r \leq n - 1$ and we can never have more than $n - 1$

pairwise orthogonal Latin squares (pols) of order n. A set of $n-1$ orthogonal Latin squares is said to be complete.

There are many techniques for constructing orthogonal Latin squares, but we limit ourselves to those using finite fields. The first result concerns complete sets of orthogonal squares.

Theorem 4.1 If $n \geq 3$ is a prime power, then there exists a complete set of orthogonal Latin squares of order n.

Proof We prove the theorem by actual construction of the complete set of squares using $GF(p^e)$, where $n = p^e$, p a prime. Let α be a primitive root of $GF(p^e)$ and by convention, let $\alpha^\infty = 0$. Define the $n \times n$ arrays

$$L_k = (l_{ij}^{(k)}), \qquad i, j = 1, 2, \ldots, p^e - 1, \infty, \quad k = 1, 2, \ldots, p^e - 1$$

where the last row and column are for convenience labeled ∞, and $l_{ij}^{(k)}$ is defined by the relation

$$l_{ij}^{(k)} = \alpha^k \cdot \alpha^i + \alpha^j, \qquad i, j = 1, 2, \ldots, p^e - 1, \infty, \quad k = 1, 2, \ldots, p^e - 1$$

We first verify that the L_k, $k = 1, 2, \ldots, p^e - 1$ are Latin squares. If two elements in the same row are identical, then we would have

$$\alpha^k \cdot \alpha^i + \alpha^j = \alpha^k \cdot \alpha^i + \alpha^{j'} \qquad \text{or} \qquad \alpha^j = \alpha^{j'}$$

implying that $j = j'$. Similarly, if two elements in the same column are identical, then

$$\alpha^k \cdot \alpha^i + \alpha^j = \alpha^k \cdot \alpha^{i'} + \alpha^j$$

implying that $i = i'$. Thus L_k is a Latin square. To show that L_{k_1} and L_{k_2} are orthogonal for any two distinct integers less than $n - 1$, suppose on the contrary that

$$(l_{ij}^{(k_1)}, l_{ij}^{(k_2)}) = (l_{i'j'}^{(k_1)}, l_{i'j'}^{(k_2)})$$

i.e., assume that the same ordered pair appears twice in the superposition of the squares. This equation implies that

$$\alpha^{k_1} \cdot \alpha^i + \alpha^j = \alpha^{k_1} \cdot \alpha^{i'} + \alpha^{j'}$$
$$\alpha^{k_2} \cdot \alpha^i + \alpha^j = \alpha^{k_2} \cdot \alpha^{i'} + \alpha^{j'}$$

which on subtraction implies that

$$(\alpha^{k_1} - \alpha^{k_2})(\alpha^i - \alpha^{i'}) = 0$$

and since k_1 and k_2 were assumed distinct, $i = i'$ and, consequently, $j = j'$. Thus we have a complete set of $n - 1$ orthogonal Latin squares of order n. □

2.4 Latin Squares and Steiner Triple Systems

It is not possible to construct a complete set of orthogonal Latin squares of order n for arbitrary n. However, the following is possible; let the prime factorization of n be

$$n = p_1^{e_1} p_2^{e_2} \cdots p_k^{e_k}$$

where the p_i are distinct primes and the e_i are positive integers. Then if

$$r = \min_i (p^{e_i} - 1), \qquad r \geq 2$$

it is possible to construct r pairwise orthogonal Latin squares of order n. We give the construction of these squares given by Hall (1964). Let M be the set of elements of the form $(\alpha_1, \alpha_2, \ldots, \alpha_k)$, $\alpha_i \in GF(p^{e_i})$ and define addition and multiplication of elements of the set componentwise, i.e.,

$$(\gamma_1, \gamma_2, \ldots, \gamma_k) + (\beta_1, \beta_2, \ldots, \beta_k) = (\gamma_1 + \beta_1, \gamma_2 + \beta_2, \ldots, \gamma_k + \beta_k)$$
$$(\gamma_1, \gamma_2, \ldots, \gamma_k)(\beta_1, \beta_2, \ldots, \beta_k) = (\gamma_1 \beta_1, \gamma_2 \beta_2, \ldots, \gamma_k \beta_k)$$

Notice that the set M contains precisely n elements. Now let α_i be a primitive element of $GF(p_i^{e_i})$, $i = 1, \ldots, k$, and let $z = (\alpha_1, \alpha_2, \ldots, \alpha_k)$. From the definition of r we see that the ith components of z^j, $j = 0, 1, \ldots, r - 1$, are distinct for $i = 1, \ldots, k$. We construct r arrays L_i, $i = 0, 1, \ldots, r - 1$, in the following manner. Designate the n elements of M by $a_0, a_1, a_2, \ldots, a_{n-1}$. The cell of each array will contain an element of M determined in the following manner: In the (i, j) square of the array L_l, $l = 0, 1, \ldots, r - 1$, we put the element b of M, where

$$a_i = z^l a_j + b, \qquad i, j = 0, 1, \ldots, n - 1, \quad l = 0, 1, \ldots, r - 1$$

We first check that L_l is a Latin square. If the element b occurs twice in the same row of L_l then

$$a_i = z^l a_j + b = z^l a_{j'} + b$$

which implies $j = j'$. Similarly if b occurs twice in the same column, then

$$a_i = z^l a_j + b, \qquad a_{i'} = z^l a_j + b$$

implying that $i = i'$. Thus L_l, $l = 0, 1, \ldots, r - 1$, is a Latin square. To show that the squares are orthogonal, consider L_{l_1} and L_{l_2}, l_1 and l_2 distinct, and suppose in their superposition that the ordered pair (b_1, b_2) occurs in cells (i, j) and (i', j'). This implies the equations

$$a_i = z^{l_1} a_j + b_1, \qquad a_{i'} = z^{l_1} a_{j'} + b_1$$
$$a_i = z^{l_2} a_j + b_2, \qquad a_{i'} = z^{l_2} a_{j'} + b_2$$

Using the first set of equations, we find that

$$(b_1 - b_2) + a_j(z^{l_1} - z^{l_2}) = 0$$

and from the second set

$$(b_1 - b_2) + a_{j'}(z^{l_1} - z^{l_2}) = 0$$

Subtracting these equations gives

$$(a_j - a_{j'})(z^{l_1} - z^{l_2}) = 0$$

and since $l_1 \neq l_2$, $j = j'$ and consequently $i = i'$ and the set is orthogonal. Notice the dependence of the argument on the form of z.

From a complete set of Latin squares of order $n \geq 3$ it is possible to construct a projective plane of order n and, conversely, from a projective plane of order n it is possible to construct a complete set of Latin squares of order n [cf. Ryser, (1963, p. 92)]. However, we have already seen that from a projective plane of order n we can construct an $(n^2 + n + 1, n + 1, 1)$ configuration, and in this manner we establish the connection between symmetric BIBD's and complete sets of orthogonal Latin squares. If we have a set of r orthogonal Latin squares of order n, $r < n$, then construction techniques exist for partially balanced designs, i.e., a design where the frequency of appearance of a given pair varies depending on the pair. The interested reader is referred to the work of Hall (1964). Many other construction techniques for sets of orthogonal Latin squares exist, and again the reader is referred to the work of Hall (1964, 1967a).

For many years there had been a question of the existence of orthogonal Latin squares. Euler in 1782 had conjectured that when n is equal to 2 mod 4 ("unevenly even" numbers) then no pair of orthogonal Latin squares of order n would exist. For $n = 6$ Tarry (1901) verified this by the complete enumeration of all possibilities. The falsity of the conjecture, however, was demonstrated by the result, due to Bose et al. (1960), that for n equal to 2 mod 4 and for $n > 6$ there exists a pair of orthogonal Latin squares of order n. The equivalence between complete sets of orthogonal Latin squares and projective planes might indicate that one method of constructing a projective plane of order ten, the lowest-order unknown case, is to try and find a set of nine pairwise-orthogonal Latin squares. This approach has not yet proven fruitful.

We briefly mention here Steiner triple systems, although we will not require them until the next chapter. These triple systems are special cases of t designs where a λ-(t, d, n) t-design D is a collection of d-subsets of a set X, $|X| = n$, such that any t-subset of X is in exactly λ subsets of D. There are some new and interesting relationships between t designs and codewords of certain fixed weights in particular codes which we will consider in a later section and the next chapter.

A Steiner system S of order n is a family of subsets of a set of elements of order n, such that each subset in the family contains k elements and each t-

2.4 Latin Squares and Steiner Triple Systems

element subset is contained in precisely one of the k-element subsets, i.e., it is a λ-(t, k, n) design where $\lambda = 1$. For convenience we designate a Steiner system by $S(t, k, n)$. Of particular interest are the Steiner triple systems for which $k = 3$ and $t = 2$. It follows immediately that a Steiner triple system is a BIBD with $k = 3$ and $\lambda = 1$, $v > 3$, and conversely. From the basic relationships between the parameters of a BIBD we find that

$$b = \tfrac{1}{6}v(v-1), \qquad r = \tfrac{1}{2}(v-1)$$

from which it follows that $v \equiv 1$ or $3 \mod 6$, i.e., v is of the form $6n + 1$ or $6n + 3$ for some positive integer n. Since Steiner triple systems can be constructed for all such values, these conditions are both necessary and sufficient for the existence of such systems.

We give two recursive techniques for constructing Steiner triple systems from given ones. Suppose we are given two Steiner triple systems S_1 of order v_1 defined on the set $\{a_1, a_2, \ldots, a_{v_1}\}$ and S_2 of order v_2 defined on the set $\{b_1, b_2, \ldots, b_{v_2}\}$. To construct a Steiner system S of order $v_1 v_2$ we take as the elements on which S is defined the $v_1 v_2$ ordered 2-tuples (a_i, b_j). A triple of these elements $\{(a_i, b_r), (a_j, b_s), (a_k, b_t)\}$ will be in S if it satisfies one of the following conditions:

(i) If $r = s = t$ and (a_i, a_j, a_k) is in S_1. Thus triples of the form $\{(a_i, b_r), (a_j, b_r), (a_k, b_r)\}$, $r = 1, 2, \ldots, v_2$, are in S if (a_i, a_j, a_k) is in S_1.

(ii) If $i = j = k$ and (b_r, b_s, b_t) is in S_2.

(iii) If (a_i, a_j, a_k) is in S_1 and (b_r, b_s, b_t) is in S_2.

That S is a Steiner triple system of order $v_1 v_2$ follows easily. Notice that isomorphic copies of both S_1 and S_2 are contained in S.

A second powerful construction of Steiner triple systems, given by Hall (1967a, p. 238), shows that given a system of order v_2 containing a subsystem of order v_3, and a system of order v_1, then we can construct a system of order $v = v_3 + v_1(v_2 - v_3)$. The first construction is just a special case of this last one, as is seen by choosing a subdesign, i.e., $v_3 = 0$. Finally, we note that the points on the lines of the Fano geometry, considered in the previous two sections, also form a Steiner triple system, and that any BIBD with $k = 3$ and $\lambda = 1$ forms a Steiner triple system.

Consider an orthogonal array $OA(n^2, k, n, 2)$ of index one and let **A** be the $k \times n^2$ array matrix with elements from the set $\{1, 2, \ldots, n\}$. From the ith row of **A** we form an $n \times n$ matrix $\mathbf{B}_i$ where the first n entries of the row form the first row of $\mathbf{B}_i$, the second n entries form the second row of $\mathbf{B}_i$, etc. From the definition of the orthogonal array the matrices $\mathbf{B}_i$, $i = 1, \ldots, k$, form a set of k pols. Hence an $OA(n^2, n-1, n, 2)$ of index one is equivalent to a projective plane of order n, $n \geq 3$.

We have already observed the equivalence between a set of $n - 1$ pols of order n and a projective plane of order n. Also, the relationship between coding, block designs, and projective planes has been considered. We consider further the relationship between sets of pols and coding and commence by proving a theorem given by Singleton (1964), attributed to Golomb and Posner.

Theorem 4.2 An $(n, q^2, n - 1)$ code over an alphabet on q symbols is equivalent to a set of $n - 2$ pols of order q.

Proof Suppose we are given a set of $n - 1$ pols of order q. Number the rows and columns in the same fashion with the q alphabet symbols on which the squares are defined. Let $L_1, L_2, \ldots, L_{n-2}$ be the set of pols and L the set of pols superimposed, so that each of the q^2 cells of L contains an ordered $n-2$-tuple the ith position of which contains the alphabet symbol in the same cell of L_i. Suppose cell (a_i, a_j) of L contains the $(n-2)$-tuple $(a_{ij}^1, a_{ij}^2, \ldots, a_{ij}^{n-2})$, where a_i, a_j, and a_{ij}^k are elements of the q-ary alphabet. Form the code C with q^2 codewords:

$$C = \{X_{ij} = (a_i, a_j, a_{ij}^1, a_{ij}^2, \ldots, a_{ij}^{n-2}), \quad 1 \leq i, j \leq q\}$$

Consider the distance between two distinct codewords X_{ij} and X_{rs}. These two codewords can agree in at most one place in the last $n - 2$ places since if they agreed in two places, the corresponding Latin squares when superimposed would contain a repeated ordered 2-tuple, contrary to the assumed orthogonality. However, if X_{ij} and X_{rs} agree in either the first or second position, then they can agree in none of the last $n - 2$ places since if they did, it would imply that a symbol is repeated in either a row or a column of one of the squares, depending on whether the first or second symbols of the two codewords are the same. Thus the resulting code has distance $n - 1$ and is maximum distance separable.

For the converse, suppose we are given a code of block length n with q^2 codewords, distance $n - 1$, over a q-ary alphabet. Reversing the above process, we identify a cell in L, with rows and columns labeled as before, by the first two coordinate positions and place the ordered $(n - 2)$-tuple consisting of the last $n - 2$ positions in that cell. The ith coordinates yield a Latin square L_i, $1 \leq i \leq n - 2$, and their pairwise orthogonality follows from the distance properties of the code. □

The method of the above theorem can be extended to Latin hypercubes, yielding codes with higher dimension. Golomb and Posner (1964) laid to rest a question of Golay (1958) concerning the existence of a perfect single-error-correcting code of length seven over an alphabet of six symbols containing 6^5 codewords. The sphere-packing condition is, of course, satisfied for these parameters. Since $d = n - k + 1$, the code is maximum distance separable and any five coordinate positions contain all 6^5 possible 5-tuples over the alphabet. Take, for convenience, the first five positions and consider the 36 codewords that begin with a fixed triple. If we delete the (identical) first three positions from each of the 36 codewords, a code of length four and distance three over an alphabet of six symbols results. From the previous theorem this implies the existence of two pols of order six. However, as previously mentioned, Tarry (1901), by exhaustive search, showed that two pols of order six do not exist. Hence no perfect single error-correcting code of length seven over an alphabet of six symbols exists.

2.5 Quadratic Residues and Codes

Let p be an odd prime and consider the integers mod p. We say that a is a quadratic residue of p if $a \not\equiv 0 \bmod p$ and there is a solution to the equation

$$x^2 \equiv a \bmod p$$

If no solution exists, a is called a quadratic nonresidue of p. Consider the $(p - 1)/2$ integers, taken mod p,

$$1^2, 2^2, 3^2, \ldots, [(p - 1)/2]^2$$

These must be distinct integers mod p, for if $a^2 = b^2$, mod p, then $a = b$ or $a = p - b$ and since $a, b \leq (p - 1)/2$ by assumption, $a = b$. Since $a^2 = (p - a)^2$ mod p, it follows that there are precisely $(p - 1)/2$ quadratic residues mod p, which are the above set of numbers. Consequently, there must be $(p - 1)/2$ quadratic nonresidues of p.

The Legendre symbol is a useful notation in the investigation of quadratic residues. If a is a quadratic residue of p, $p \nmid a$, then we denote this fact by writing $(a/p) = +1$. If a is a quadratic nonresidue of p, then $(a/p) = -1$. There are many interesting properties that can be proved about this symbol and we state only a few of them. For any odd prime p, it is true that

$$(a/p) = a^{(p-1)/2} \bmod p$$

To see this, we observe that from Fermat's theorem, for any nonzero integer a and prime p such that $p \nmid a$, $a^{p-1} \equiv 1 \bmod p$. It follows that $a^{(p-1)/2} \equiv \pm 1$ mod p. If a is a quadratic residue, then $a \equiv x^2$ for some x and hence $a^{(p-1)/2} \equiv 1 \bmod p$. But the equation $y^{(p-1)/2} \equiv 1$ cannot have more than $(p - 1)/2$ distinct solutions mod p and hence these solutions are precisely the quadratic residues. Since every nonzero element satisfies the equation

$$y^{(p-1)} - 1 = (y^{(p-1)/2} - 1)(y^{(p-1)/2} + 1)$$

it follows that every quadratic nonresidue satisfies $a^{(p-1)/2} \equiv -1$ mod p. From this relationship it readily follows that

$$(a/p)(b/p) = (ab/p) \quad \text{and} \quad (-1/p) = (-1)^{(p-1)/2}$$

implying that $-1 \equiv (p-1) \pmod{p}$ is a quadratic residue if $p \equiv 1$ mod 4 and a nonresidue if $p \equiv 3$ mod 4. Thus if a and b are two quadratic residues or two quadratic nonresidues, then their product will be a quadratic residue. Similarly, the product of a residue and a nonresidue is a nonresidue.

The law of quadratic reciprocity states that if p and q are distinct odd primes, then

$$(p/q)(q/p) = (-1)^{(p-1)(q-1)/4}$$

Since the proof is quite lengthy and can be found in many books on number theory, we omit it.

There is an interesting property of factorials using quadratic residues, which we consider. Let a be a fixed element mod p and suppose it is a quadratic residue and $x_1^2 \equiv a$ mod p. Also, $p - x_1$ is a solution since $(p - x_1)^2 \equiv a$ mod p and $x_1(p - x_1) \equiv -a$ mod p. The elements x_1 and $p - x_1$ are the only solutions to the congruence $x^2 \equiv a$ mod p. For any other element x' mod p there is precisely one element x'' such that $x'x'' \equiv a$ mod p. Thus the elements $1, 2, \ldots, p - 1$ can be divided into x_1 and $p - x_1$ [for which $x_1(p - x_1) \equiv -a$ mod p] and $(p - 3)/2$ classes each of two elements x' and x'' such that $x'x'' \equiv a$ mod p. Thus if a is a quadratic residue, we conclude that

$$(p - 1)! = \Pi x \equiv - a \cdot a^{(p-3)/2} \equiv -a^{(p-1)/2}$$

If a is a quadratic nonresidue, however, we can divide the elements $1, 2, \ldots, p - 1$ into $(p - 1)/2$ classes x' and x'' such that $x'x'' \equiv a$ mod p for each class, from which it follows that

$$(p - 1)! \equiv -(a/p)a^{(p-1)/2} \text{ mod } p$$

Since unity is always a quadratic residue, we have

$$(p - 1)! \equiv -1 \text{ mod } p$$

which is just Wilson's theorem. From these relations it also follows directly that

$$(a/p) \equiv a^{(p-1)/2} \text{ mod } p$$

a fact shown previously.

For another interesting interpretation of the Legendre symbol, we say that the (ordinary) residue of an integer k mod p is minimal if it lies between

2.5 Quadratic Residues and Codes

$-p/2$ and $+p/2$. For some fixed integer k, $p \nmid k$, consider the set of minimal residues of the set

$$k, 2 \cdot k, 3 \cdot k, \ldots, [(p-1)/2] \cdot k$$

which we denote by $r_1, r_2, \ldots, r_a, -s_1, -s_2, \ldots, -s_b$, where $0 < r_i$, $s_j < p/2$ and $a + b = (p-1)/2$. Since $p \nmid k$, if $r_i = r_j$, $i \neq j$, then we would have $u \cdot k \equiv v \cdot k \pmod{p}$, which would imply $p \mid (u-v)k$, which is impossible since $0 < u, v < p/2$. Similarly, no two of the s_i are equal. If $r_i = s_j$, then, for some u, v we have that

$$uk \equiv r_i \bmod p \quad \text{and} \quad vk \equiv -s_j \bmod p$$

or

$$(u+v)k \equiv 0 \bmod p$$

Since $p \nmid k$, this equation implies that

$$u + v \equiv 0 \bmod p$$

which, again, is impossible since $0 < u, v < p/2$. Thus the minimal residues are distinct and the set $\{r_1, r_2, \ldots, r_a, s_1, s_2, \ldots, s_b\}$ accounts for all the integers $\{1, 2, \ldots, (p-1)/2\}$. It follows that

$$k \cdot 2k \cdot 3k \cdots [(p-1)/2]k \equiv (-1)^b \cdot 1 \cdot 2 \cdots [(p-1)/2] \bmod p$$

or equivalently

$$k^{(p-1)/2} \equiv (-1)^b \bmod p$$

However, since

$$(k/p) = k^{(p-1)/2} \bmod p$$

we conclude that

$$(k/p) \equiv (-1)^b \bmod p$$

a result that is referred to as Gauss's lemma. For $k = 2$ we can find $(2/p)$ by considering the minimal residues of the set

$$\{2, 4, 6, \ldots, p-1\}$$

It is easily established that $(2/p) = 1$ if p is a prime of the form $8n \pm 1$ and $(2/p) = -1$ if p is a prime of the form $8n \pm 3$. Since $\frac{1}{8}(p^2 - 1)$ is even if p is of the form $8n \pm 1$ and odd otherwise,

$$(2/p) = (-1)^{(p^2-1)/2}$$

Notice that from the law of reciprocity

$$(p/q)(q/p) = (-1)^{(p-1)(q-1)/4}$$

and thus if either $(p-1)/2$ or $(q-1)/2$ is an even integer, then $(p/q) = (q/p)$. If they are both odd, then $(p/q) = -(q/p)$ and this occurs only when the odd primes p and q are of the form $4n+3$.

The theory of quadratic residues has been put to some interesting uses in the solution of certain congruences. Berlekamp (1968) uses the law of reciprocity in factoring the polynomial $x^n - 1$ over $GF(p)$, where p and n are distinct odd primes (see Exercise 16).

Recall that there are $(p-1)/2$ quadratic residues of the odd prime p, which are $1^2, 2^2, \ldots, [(p-1)/2]^2$ and which we denote by R_0. We now use these to construct some $(p, (p-1)/2)$ codes over $GF(q)$ that have good distance properties.

Let η be a primitive pth root of unity in some extension field of $GF(q)$, $(q, p) = 1$, and denote the set of quadratic nonresidues by R_1. Assume that the order of q mod p divides $(p-1)/2$, i.e., $q \in R_0$, and notice that char$(GF(q))$ is not related to p except that it is relatively prime to p. Define the two polynomials

$$f_1(x) = \prod_{i \in R_0} (x - \eta^i), \qquad f_2(x) = \prod_{i \in R_1} (x - \eta^i)$$

and observe that these polynomials will have coefficients in $GF(q)$ and that

$$x^p - 1 = (x-1)f_1(x)f_2(x)$$

The polynomials $f_i(x)$ will be irreducible over $GF(q)$ iff the multiplicative order of q mod p is $(p-1)/2$. With these polynomials we define the following codes.

(i) $\mathscr{C}_1$ is the quadratic residue code with parameters $(p, (p+1)/2)$ generated by $f_1(x)$.

(ii) $\mathscr{C}_2$ is the expurgated quadratic residue code with parameters $(p, (p-1)/2)$ generated by $(x-1)f_1(x)$.

(iii) $\mathscr{C}_3$ is the extended quadratic residue code with parameters $(p+1, (p+1)/2)$ obtained by annexing an overall parity check to $\mathscr{C}_1$.

The code $\mathscr{C}_2$ is, of course, a subcode of $\mathscr{C}_1$ consisting of all those codewords $c(x) = \sum_{i=0}^{p-1} c_i x^i$ such that $\sum_{i=0}^{p-1} c_i = 0$, whence the name expurgated. The original investigation of Mattson and Solomon (1961) was restricted to the binary case and $f_i(x)$ irreducible. In this case if the roots of $f_i(x)$ expressed as powers of a primitive pth root of unity, then these powers are either all quadratic residues or all quadratic nonresidues. The BCH bound on the minimum distance of these codes is, generally speaking, not very good. We proceed to obtain a better bound. We do not restrict the discussion to $f_i(x)$ irreducible.

2.5 Quadratic Residues and Codes

Lemma 5.1 Let $c(x) \in \mathscr{C}_1 \backslash \mathscr{C}_2$ be a codeword of weight d; then $d^2 \geq p$ and if -1 is a quadratic nonresidue mod p, i.e., $p \equiv -1 \bmod 4$, then this bound can be improved to $d^2 - d + 1 \geq p$.

Proof Since multiplication of a quadratic residue by a quadratic nonresidue is a quadratic nonresidue, the codes generated by $f_1(x)$ and $f_2(x)$ are equivalent, as can be seen by applying Theorem 7.5 of Chapter 1. Let $c(x)$ by a codeword of weight d in $\mathscr{C}_1 \backslash \mathscr{C}_2$ and let $c'(x)$ be a codeword of weight d in the code $\mathscr{C}_1'$ generated by $f_2(x)$ but not in the expurgated code generated by $(x-1)f_2(x)$. Then $f_1(x)|c(x)$, $(x-1) \nmid c(x)$, $f_2(x)|c'(x)$, and $(x-1) \nmid c'(x)$ and we consider the product $c(x)c'(x)$. We write

$$c(x)c'(x) = q(x)(x^p - 1) + r(x), \qquad \deg r < p$$

Since $(x-1) \nmid c(x)c'(x)$ and $f_1(x)f_2(x)|c(x)c'(x)$ it follows that $f_1(x)f_2(x)|r(x)$ and hence, since $\deg(f_1 f_2) = p - 1$,

$$\alpha f_1(x)f_2(x) = r(x) = \alpha(1 + x + \cdots + x^{p-1}), \qquad \alpha \in GF(q), \quad \alpha \neq 0$$

However, the Hamming weight of the product of these two polynomials $\omega_H(c(x)'c(x))$ cannot be greater than $\omega_H(c(x))\omega_H(c'(x))$ and we conclude that

$$p \leq d^2$$

If $p \equiv 3 \bmod 4$, then we have seen that -1 is a quadratic nonresidue, in which case the above bound can be sharpened as follows. For $c'(x)$ we take the codeword $c(x^{-1})$, which is in the desired code and of weight d. In the product $c(x)c(x^{-1})$, d of the terms collapse into unity, of which $d - 1$ were not taken into account by the above argument. Hence, the weight of $c(x)c(x^{-1})$ is at most $d^2 - (d-1)$ and we conclude that

$$p \leq d^2 - d + 1 \qquad \square$$

Berlekamp (1968) shows that every extended binary QR code is invariant under the doubly transitive projective unimodular group and uses this fact to show that the minimum weight of an augmented binary QR code of length n is an odd number d bounded by

$$d^2 > n \qquad \text{if} \quad n \equiv +1 \bmod 8$$
$$d(d-1) \geq n - 1 \qquad \text{if} \quad n \equiv -1 \bmod 8$$

Two particularly interesting examples of QR codes are the so-called Golay codes. One is a binary (23,12) code of weight seven and the other is an (11, 6) code over $GF(3)$ of weight five. They are both perfect codes and will be considered further in Section 2.8.

There is a close connection between quadratic residue codes and the quasicyclic codes, introduced in Section 2.7. Also, certain extended quadratic residue codes are self-dual, a concept also introduced in Section 2.7. The relationships between quadratic residue codes, quasicyclic codes, and self-dual codes are interesting and not yet fully investigated. Assmus and Mattson (1972a, b) investigate some of these relationships in determining minimum weights of quadratic residue codes. Automorphism groups of quadratic residue codes are studied extensively by Assmus and Mattson (1969). In this regard notice that the permutation

$$\eta: \quad X \to X^r \quad r \quad \text{a quadratic residue}$$

certainly preserves the code C_1. In fact, the extended quadratic residue codes are invariant under certain projective special linear groups and this can be used to give information on the minimum weight of the code [see Assmus and Mattson (1969)].

2.6 Hadamard Matrices, Difference Sets, and Their Codes

Let H be an $m \times m$ matrix with elements of plus one and minus one such that

$$HH^T = mI$$

where I is the $m \times m$ identity matrix. Equivalently, H is a matrix of plus ones and minus ones whose rows are orthogonal. It follows directly from the definition that the determinant $|H|$ of such a matrix is

$$|H| = \pm m^{m/2}$$

For any real, square, $n \times n$ nonsingular matrix $A = (a_{ij})$ Hadamard's inequality states that

$$\det A \leq \left[\prod_{i=1}^{n} \left(\sum_{j=1}^{n} a_{ij}^2 \right) \right]^{1/2}$$

Since a matrix H of the form described satisfies the bound with equality, we refer to it as a Hadamard matrix. Complex Hadamard matrices and Hadamard matrices over the set of pth roots of unity have also been considered in the literature, but they will not be considered here. Multiplying a row or a column by minus one does not destroy the Hadamard property, and, similarly, permuting rows and columns does not destroy the property. Multiplying the appropriate rows and columns by minus one, we can always obtain a matrix H with the first row and first column all plus ones, and we call this a normalized Hadamard matrix. Except for the trivial Hadamard matrix

$$H_1 = [1]$$

this implies immediately that the order of a Hadamard matrix must be even, since for orthogonality, the second row must contain an even number of plus and minus ones. The normalized Hadamard matrix of order two is

$$H_2 = \begin{bmatrix} 1 & 1 \\ 1 & -1 \end{bmatrix}$$

Now suppose H_m is a normalized Hadamard matrix of order m, $m \geqslant 3$, and consider the first three rows of this matrix. Rows two through m must have an equal number of plus and minus ones and permute the columns of H_m so that the first $m/2$ elements of the second row are plus one and the remaining $m/2$ elements are minus one. Suppose there are x elements of $+1$ in the first $m/2$ elements of the third row and permute the columns so that the first x elements of the third row are plus one and the elements in the third row in columns $\frac{1}{2}m + 1$ to $\frac{1}{2}m + \frac{1}{2}(m - x/2)$ are plus ones. Now, for the second and third row to be orthogonal, we must have

$$2[x - (\tfrac{1}{2}m - x)] = 0$$

which implies that $m = 4x$. We conclude that the order of any Hadamard matrix of order greater than two is divisible by four. It remains an open question as to whether a Hadamard matrix of order $4t$ exists for all positive integers t, despite the many construction techniques known. We mention that if H_{m_1} and H_{m_2} are Hadamard matrices of orders m_1 and m_2, respectively, then the Kronecker product matrix $H_{m_1} \otimes H_{m_2}$ is easily shown to be a Hadamard matrix of order $m_1 m_2$.

We now consider the relationship between Hadamard matrices and designs. Suppose H is a normalized Hadamard matrix of order $4t$, $t \geqslant 2$. We form an incidence matrix Γ by replacing the minus ones by zeros and delete the first row and column. As we have seen, in a normalized Hadamard matrix H each row and each column contains the same number of plus and minus ones. This implies that the $(4t - 1) \times (4t - 1)$ matrix Γ is such that each row and each column contains $2t$ zeros and $2t - 1$ ones. It readily follows that

$$\Gamma\Gamma^T = tI_{4t-1} + (t - 1)J_{(4t-1)(4t-1)}$$

since in the normalized matrix H, plus ones in any two distinct rows match up in precisely t places and hence the inner product of any two distinct rows (or columns) of Γ is $t - 1$. Thus Γ is the incidence matrix of a $(4t - 1, 4t - 1, 2t - 1, 2t - 1, t - 1)$ symmetric **BIBD**. Conversely, a Hadamard matrix of order $4t$ can be obtained from a symmetric design with the above parameters by replacing the zeros of its incidence matrix by minus ones and adding a row and column of all ones. There are other methods of obtaining BIBD's from Hadamard matrices. For example, the complement of the above design is also a symmetric BIBD.

Example Consider again the BIBD formed from the Fano geometry as given in the previous section. The blocks are

$$B_1 = (1, 2, 3), \quad B_2 = (3, 4, 5), \quad B_3 = (1, 5, 6), \quad B_4 = (1, 4, 7)$$
$$B_5 = (3, 6, 7), \quad B_6 = (2, 5, 7), \quad B_7 = (2, 4, 6)$$

and using the integers to label the columns and rows, this design has the incidence matrix

$$\Gamma = \begin{bmatrix} 1 & 1 & 1 & 0 & 0 & 0 & 0 \\ 0 & 0 & 1 & 1 & 1 & 0 & 0 \\ 1 & 0 & 0 & 0 & 1 & 1 & 0 \\ 1 & 0 & 0 & 1 & 0 & 0 & 1 \\ 0 & 0 & 1 & 0 & 0 & 1 & 1 \\ 0 & 1 & 0 & 0 & 1 & 0 & 1 \\ 0 & 1 & 0 & 1 & 0 & 1 & 0 \end{bmatrix}$$

Changing the zeros to minus ones and adding a column and a row of all ones gives the normalized Hadamard matrix of order eight

$$H_8 = \begin{bmatrix} 1 & 1 & 1 & 1 & 1 & 1 & 1 & 1 \\ 1 & 1 & 1 & 1 & -1 & -1 & -1 & -1 \\ 1 & -1 & -1 & 1 & 1 & 1 & -1 & -1 \\ 1 & 1 & -1 & -1 & -1 & 1 & 1 & -1 \\ 1 & 1 & -1 & -1 & 1 & -1 & -1 & 1 \\ 1 & -1 & -1 & 1 & -1 & -1 & 1 & 1 \\ 1 & -1 & 1 & -1 & -1 & 1 & -1 & 1 \\ 1 & -1 & 1 & -1 & 1 & -1 & 1 & -1 \end{bmatrix}$$

Thus the Fano geometry is a projective plane, a Steiner triple system, a BIBD, and also yields an Hadamard matrix.

If one Hadamard matrix can be obtained from another by multiplying a row or a column by minus one or by permuting any set of the rows or the columns, we call them equivalent. It is not true that any two Hadamard matrices of the same order are equivalent.

The matrix

$$H = \begin{bmatrix} -1 & 1 & 1 & 1 \\ 1 & -1 & 1 & 1 \\ 1 & 1 & -1 & 1 \\ 1 & 1 & 1 & -1 \end{bmatrix}$$

2.6 Hadamard Matrices, Difference Sets, and Their Codes

is both Hadamard and a circulant. It is conjectured that a Hadamard matrix of order $n > 4$ cannot be circulant, although it is possible that a normalized Hadamard matrix with its first row and column removed is circulant. A Hadamard matrix is said to be regular if the sum of all the elements in each row or column is constant. It is possible to show that a regular, symmetric Hadamard matrix of order $4s$ exists only if s is an integer square. The existence of a regular Hadamard matrix H, of order $4k'^2$, where $HJ = \pm 2k'J$, is equivalent to the existence of an SBIBD with parameters $v = 4k^2$, $k = 2k'^2 \pm k'$, $\lambda = k^2 + k$ (Wallis et al., 1972; Hall, 1967a).

Since we have observed that the existence of a Hadamard matrix of order $4t$ is equivalent to the existence of a symmetric, balanced, incomplete block design with parameters $v = b = 4t - 1$, $r = k = 2t - 1$, $\lambda = t - 1$, all statements referring to codes and Hadamard matrices refer equally well to codes and SBIBD's with the appropriate parameters. In Section 1.11 we proved the bounds, due to Plotkin (1960),

(i) $A(4t - 2, 2t) \leqslant 2t$
(ii) $A(4t - 1, 2t) \leqslant 4t$ (6.1)
(iii) $A(4t, 2t) \leqslant 8t$

where $A(n, d)$ is the maximum number of binary codewords of length n and minimum distance d, and t is some positive integer. If equality in (iii) holds, then it must also hold in (i) and (ii) and equality will hold in (iii) when $4t - 1$ is a prime.

Suppose we are given a Hadamard matrix H of order $4t$. Assume it to be in normalized form so that its first row and column are plus ones. Each row has $2t$ plus ones and $2t$ minus ones. Form the matrix H' by replacing the minus ones with zeros. Form the $8t \times 4t$ matrix

$$H_1 = \begin{bmatrix} H' \\ H_c' \end{bmatrix}$$

where H_c' is the complement of H' obtained by interchanging zeros and ones. Any two rows of H' are a distance $2t$ apart, as are any two rows of H_c'. A row of H' is also a distance $2t$ from a row of H_c' unless the two are complements, in which case the two are a distance $4t$ apart. In this manner the existence of a Hadamard matrix of order $4t$ implies the existence of a $(4t, 8t, 2t)$ code, i.e., a code of length $4t$, distance $2t$, and with $8t$ codewords. The rows of H' with the first column removed form a $(4t - 1, 4t, 2t)$ code. In this code half the words begin with a one and half with a zero, since in H' each column except the first has $2t$ zeros and $2t$ ones. Choosing the $2t$ words that begin with a one and deleting these ones yields a $(4t - 2, 2t, 2t)$ code. Thus the existence of a Hadamard matrix of order $4t$ implies the existence of codes that satisfy the bounds of (6.1) with equality.

Conversely, suppose $\mathscr{C}$ is a binary $(4t, 8t, 2t)$ code and let M be the $8t \times 4t$ matrix which has the codewords as rows. Again half of the rows have a first coordinate of zero and half have one. Permute the rows until the first $4t$ rows begin with a one. If in the first row a column has a zero, interchange zeros and ones in that column, so that the first row will be all ones. Let H be the $4t \times 4t$ matrix obtained from the first $4t$ rows of the modified matrix M by replacing zeros by minus ones. From the properties of the code it follows immediately that H is a Hadamard matrix. Using similar reasoning, we can conclude that the existence of either a $(4t, 8t, 2t)$ code or a $(4t - 1, 4t, 2t)$ code is entirely equivalent to a Hadamard matrix of order $4t$ and hence to a certain SBIBD.

The role of Hadamard matrices in coding is far deeper then these elementary results would indicate [see, for example, the work of Semakov and Zinov'ev (1969a, b)]. Since, at the moment, many of the applications of Hadamard matrices involve particular rather than general constructions, we will not report on other work here.

As mentioned previously, a Steiner system is a λ-(t, d, n) design with $\lambda = 1$. Many of the interesting connections between codes and t designs will be discussed in the next chapter, where Steiner systems will arise as particular cases. However, there is one interesting observation to be made with regard to Steiner systems.

Definition An (n, w, λ) binary balanced code is a set of binary n-tuples, each of weight w, such that any two words agree in at most λ positions.

Let the maximum number of words in a balanced code be N. It follows directly from the bound on the quantity $R(n, d, e)$ given in Theorem 11.4 of Chapter 1 that

$$N \leq \left\lfloor \frac{n}{w} \left\lfloor \frac{n-1}{w-1} \left\lfloor \cdots \left\lfloor \frac{n-\lambda}{w-\lambda} \right\rfloor \cdots \right\rfloor \right\rfloor \right\rfloor \leq \binom{n}{\lambda+1} \Big/ \binom{w}{\lambda+1}$$

where the right-hand bound follows trivially from the left-hand bound.

A balanced (n, w, λ) code with $\binom{n}{\lambda+1}/\binom{w}{\lambda+1}$ codewords is called a complete balanced code. There are $\binom{n}{\lambda+1}$ binary n-tuples of weight $\lambda + 1$ possible. An n-tuple of weight w "contains" $\binom{w}{\lambda+1}$ such n-tuples. Since in a balanced (n, w, λ) code no two words can agree in more than λ positions, the equivalence between an (n, w, λ) complete balanced code and a 1-$((\lambda + 1), w, n))$ Steiner system is established, since such a Steiner system also contains the requisite number of $\binom{n}{\lambda+1}/\binom{w}{\lambda+1}$ blocks.

We consider now another interesting combinatorial structure, the difference set, which will also have significant applications in coding.

2.6 Hadamard Matrices, Difference Sets, and Their Codes

Definition If the set of k integers mod v, $D = \{a_1, a_2, \ldots, a_k\}$, is such that for any nonzero integer a mod v there exist precisely λ pairs (a_i, a_j) where

$$a_i - a_j = a \bmod v$$

then we call D a (v, k, λ) or perfect difference set. If $\lambda = 1$, D is called a planar difference set. In such a difference set $\lambda = k(k-1)/(v-1)$.

Once again, an important problem is the enumeration of all difference sets and this problem is unsolved, although several plausible conjectures exist. As with other structures, we consider the connection between difference sets and configurations and give two important methods for constructing difference sets.

A circulant matrix is an $n \times n$ matrix whose $(i+1)$th row is a cyclic shift of the ith row, i.e., one of the form

$$\begin{bmatrix} \alpha_1 & \alpha_2 & \alpha_3 & \cdots & \alpha_n \\ \alpha_n & \alpha_1 & \alpha_2 & \cdots & \alpha_{n-1} \\ \vdots & \vdots & \vdots & & \vdots \\ \alpha_2 & \alpha_3 & \alpha_4 & \cdots & \alpha_1 \end{bmatrix}$$

We now show that there is a one-to-one correspondence between the difference sets defined above and those (v, k, λ) configurations whose incidence matrices are circulants.

Let D be a (v, k, λ) configuration defined on the integers mod v. Clearly, adding any integer mod v to each element of D also yields a (v, k, λ) difference set and hence from D we can form $v - 1$ other difference sets

$$D_i = \{a_1 + i, a_2 + i, \ldots, a_k + i\}, \qquad i = 1, \ldots, v-1$$

Each integer mod v appears in precisely k of the sets $D, D_1, \ldots, D_{v-1}$. To show that every pair of integers (a, b) mod v appears in λ of these sets, let $a - b \equiv c$ mod v, where $c \not\equiv 0$ mod v. Then for precisely λ pairs (a_i, a_j), $a_i, a_j \in D$, we have

$$a - b \equiv a_i - a_j \bmod v$$

Thus, if we let $d \equiv a - a_i \equiv b - a_j$ mod v, then $a \equiv a_i + d$ and $b = a_j + d$ and hence both a and b belong to D_d. Since we can do this for each pair (a_i, a_j), the pair (a, b) belong to λ of the sets $D, D_1, D_2, \ldots, D_{v-1}$. Consequently these sets form a (v, k, λ) configuration, and its incidence matrix is a circulant.

Conversely, suppose we are given a (v, k, λ) configuration whose incidence matrix is a circulant. Suppose the elements are the integers $0, 1, \ldots, v-1$ and denote the blocks by $B_0, B_1, B_2, \ldots, B_{v-1}$. If block B_0 contains the elements $\{a_1, a_2, \ldots, a_k\}$ then, from the circulant property of the incidence matrix, block B_i contains the elements $a_1 + i, a_2 + i, \ldots, a_k + i$, mod v. There are λ blocks that contain the pair of elements $(0, d)$ $d \not\equiv 0$ mod v, which means there are precisely λ triples (j, a_k, a_l) such that $a_k + j \equiv d$ mod v and $a_l + j \equiv 0$ mod v, i.e., there are λ pairs (a_k, a_l) such that

$$a_k - a_l \equiv d \text{ mod } v$$

which, of course, implies that B_0 is a (v, k, λ) difference set. We have shown that a (v, k, λ) difference set is equivalent to a (v, k, λ) configuration that has a circulant incidence matrix.

There is also an interesting relationship between quadratic residues and difference sets, and hence between quadratic residues and (v, k, λ) configurations with circulant incidence matrices. Recall that in the discussion of quadratic residues we showed that $(a/p) = a^{(p-1)/2}$ and hence if p is an odd prime of the form $p = 4n + 3$, then $-1 \equiv p - 1$ is a quadratic nonresidue of p. Also, if r is a quadratic residue of p, p of the form $4n + 3$, then $-r \equiv p - r$ is a quadratic nonresidue and if r runs through the quadratic residues of p, then $-r$ runs through the quadratic nonresidues. We wish to show that when p is a prime of this form then its quadratic residues $R = \{r_1, r_2, \ldots, r_{(p-1)/2}\}$ form a difference set with parameters $v = p = 4n + 3$, $k = 2n + 1$, and $\lambda = n$. This is often called a Hadamard difference set, or equivalently, a Hadamard design, since it is an SBIBD with the right parameters that yields a $4(n + 1) \times 4(n + 1)$ Hadamard matrix.

Let r_i and r_j be two distinct quadratic residues of p, and suppose $r_i - r_j \equiv r$ mod p, where r is a quadratic residue. Then $r^{-1} r_i - r^{-1} r_j = r_i' - r_j' \equiv 1$ mod p, where $r_i' = r^{-1} r_i$ and $r_j' = r^{-1} r_j$, and r_i' and r_j' are also quadratic residues, i.e., elements of R. Also, if r_i and r_j are two elements of R such that $r_i - r_j \equiv 1$ mod p, then $rr_i - rr_j = r$ mod p for any quadratic residue of p and both rr_i and rr_j are in R. Thus the number of ordered pairs of elements of R, (r_i, r_j), whose difference equals a given quadratic residue is the same as for any other quadratic residue. If $r_i - r_j \equiv r$ mod p, r a quadratic residue, then $r_j - r_i \equiv -r$ mod p, where $-r$ is a quadratic nonresidue, and precisely the same argument holds. Thus, for any integer m mod p, the number of pairs of R whose difference equals m is the same for any m, and this, of course, is the definition of a difference set. To determine the parameters of the set, v is clearly equal to p and k to $(p - 1)/2$. Since there are $\frac{1}{2}(p-1)\frac{1}{2}(p-3)$ pairs of elements from R, we must have

$$\lambda = \frac{\frac{1}{2}(p-1)\frac{1}{2}(p-3)}{p-1} = \frac{p-3}{4} = n$$

It was shown that the incidence matrix of the Hadamard design associated with this difference set can be made circulant. Adding the row and column of ones to form the Hadamard matrix, however, destroys the circulant property.

2.6 Hadamard Matrices, Difference Sets, and Their Codes

Example The quadratic residues mod 19 are

$$\{1, 4, 5, 6, 7, 9, 11, 16, 17\}$$

and it is readily verified that these form a (v, k, λ) difference set with $v = 19$, $k = 9$, and $\lambda = 4$.

It was seen in an earlier section that the hyperplanes in $PG(n, q)$ can be used to form a symmetric block design. We show now that this design has an incidence matrix that is a circulant and hence the design determines a (v, k, λ) difference set, where

$$v = \frac{q^{n+1} - 1}{q - 1}, \quad k = \frac{q^n - 1}{q - 1}, \quad \lambda = \frac{q^{n-1} - 1}{q - 1}$$

if the hyperplanes in $PG(n, q)$ are used. Recall the representation of $PG(n, q)$ given earlier, where, if ξ is a primitive root of $GF(q^{n+1})$, then the first $v = (q^{n+1} - 1)/(q - 1)$ powers of ξ can be taken as the points of $PG(n, q)$. The elements ξ^i and ξ^j represent the same point iff $i \equiv j \mod v$ and the elements of the form ξ^{iv}, $i = 1, \ldots, q - 1$, with the zero element, form $GF(q)$. Consider the mapping on $GF(q^{n+1})$ defined by

$$\tau: \quad 0 \longmapsto 0$$
$$\xi^i \longmapsto \xi^{i+1}$$

This is a pointwise map of $GF(q^{n+1})$ which takes m flats into m flats since it preserves the independence of points. Now suppose τ^j maps a hyperplane onto itself. Then it must permute the points of a hyperplane in cycles of length, say, t, where $t | k$, k the number of points in a hyperplane. Since $\tau^v = 1$, the identity map, $t | v$ and since v and k are relatively prime, $t = 1$. Thus the map τ maps the hyperplanes in a chain of length v, i.e., if $V = PG(n - 1, q)$ is a particular hyperplane, then the other hyperplanes are $\tau(V)$, $\tau^2(V), \ldots, \tau^{v-1}(V)$. This, of course, is precisely the condition needed to verify that the incidence matrix of the design is a circulant. There are many other constructions known for difference sets. Since they tend to use concepts that are not important for our purpose, we will not discuss them here.

Example Consider the Fano geometry (Fig. 7). The lines of the geometry are

$$\{0, 1, 3\}, \quad \{1, 2, 4\}, \quad \{2, 3, 5\}, \quad \{3, 4, 6\}, \quad \{4, 5, 0\}, \quad \{6, 5, 1\}, \quad \{6, 0, 2\}$$

and in this representation the associated design is cyclic. The set or line $\{0, 1, 3\}$ is a planar difference set.

An important and useful notion in the investigation of a difference set is that of a multiplier. If $D = \{d_1, d_2, \ldots, d_k\}$ is a difference set, then every translation, mod v, is also a difference set, where the translation is $D' = \{d_1 + r, d_2 + r, \ldots, d_k + r\}$ for some integer r. If there exists an integer S with $(S, v) = 1$ such that the sets of elements, mod v, $D' = \{d_1 + r, d_2 + r, \ldots, d_k + r\}$ and $D'' = \{d_1 S, d_2 S, \ldots, d_s S\}$ are identical, then we call S a multiplier of the difference set D. It can be shown that if D is a (v, k, λ) difference set and p a prime such that $p | (k - \lambda)$, $p \nmid v$, and $p > \lambda$, then p is a multiplier of D.

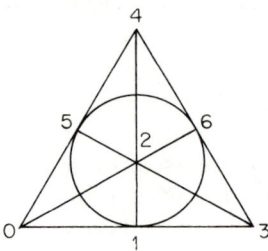

Fig. 7

A planar difference $(v, k, \lambda = 1)$ set can be shown to be equivalent to a projective plane of order $n = k - 1$. Singer (1938) constructed such difference sets using projective planes of order $n = p^s$, p a prime and s a positive integer. These sets are cyclic in the sense discussed previously and it is conjectured that every finite cyclic projective plane is Desarguesian and hence must have order $n = p^s$ for some prime p and integer s.

Weldon (1966) defined a difference-set cyclic code to be the null space of the incidence matrix of the associated design. However, the only cyclic difference sets used are those generated from the hyperplanes of a projective geometry, hence the only difference-set codes that have been considered are the projective geometry codes of the appropriate order. The dimensions of these codes were given by Graham and MacWilliams (1966), Smith (1969), and MacWilliams and Mann (1968). Since these codes are subsumed by the projective geometry codes, we consider them no further.

2.7 Self-Dual and Quasicyclic Codes

Let $\mathscr{C}$ be a linear code with block length n over $GF(q)$. As defined previously, the dual of $\mathscr{C}$ is

$$\mathscr{C}^\perp = \{u \in V_n(q); \quad (u, v) = 0, \quad \forall v \in \mathscr{C}\}$$

2.7 Self-Dual and Quasicyclic Codes

We say that $\mathscr{C}$ is self-dual if $\mathscr{C} = \mathscr{C}^\perp$. Notice that $\mathscr{C}^\perp$ is a subspace even if $\mathscr{C}$ is not a linear code; thus a self-dual code is automatically linear. If $\mathscr{C}$ is self-dual, then its block length n must be even since its dimension is $n/2$. Furthermore, the weight of each codeword is even if $q = 2$ since an odd-weight codeword cannot be orthogonal to itself. Thus over $GF(2)$ the all-ones vector is always in a self-dual code.

MacWilliams *et al.* (1972b) define $\mathscr{C}$ to be a weakly self-dual code over $GF(2)$ if it is linear, $\mathscr{C} \subset \mathscr{C}^\perp$, and the all-one vector is in $\mathscr{C}$. Clearly every self-dual code is weakly self-dual.

There are many known self-dual codes. For example, the extended binary (24, 12) Golay code is self-dual, as is the extended (12, 6) ternary Golay code. Both these codes are related to quadratic residue codes and the latter code is a symmetry code. In this section we consider classes of codes that are self-dual. We begin with the symmetry codes of Pless (1969, 1970, 1972a).

The symmetry codes will be a class of self-dual codes over $GF(3)$ with parameters $(2q + 2, q + 1)$ where $q \equiv -1 \mod 3$ and q is the power of an odd prime. We first construct a $(q + 1) \times (q + 1)$ matrix S_q over $GF(3)$ and will write its entries as either 0 or ± 1. Let a be an arbitrary (but fixed) mapping of the integers $0, 1, 2, \ldots, q - 1$ onto $GF(q)$ with the condition that $a(0) = 0$ and $a(1) = 1$. Let a^{-1} be the inverse mapping. To $GF(q)$ we adjoin an exceptional element ∞ and let $a(\infty) = \infty$. Label the rows and columns of $S_q = (s_{ij})$ with the elements $\infty, 0, 1, 2, \ldots, q - 1$ consecutively.

Define the function χ on $GF(q)$ such that $\chi(\eta) = 1$ if η is a square, -1 if η is a nonsquare, and 0 if $\eta = 0$. The matrix entries are then defined by

$$s_{\infty,\infty} = 0, \quad s_{i,\infty} = \chi(-1), \quad s_{\infty,i} = 1$$
$$s_{i,j} = \chi(a(j) - a(i)), \quad i, j = 0, 1, \ldots, q - 1$$

Definition The code with generator matrix $[I, S_q]$, I the identity matrix of order $q + 1$, is a symmetry code which we denote by $C(q)$.

Notice that if q is a prime, then we can take the map a to be the identity map. For example, if $q = 5$, then the matrix S_q is, with a the identity map,

	∞	0	1	2	3	4
∞	0	1	1	1	1	1
0	1	0	1	−1	−1	1
1	1	1	0	1	−1	−1
2	1	−1	1	0	1	−1
3	1	−1	−1	1	0	1
4	1	1	−1	−1	1	0

The following theorem gives the basic properties of the matrix S_q.

Theorem 7.1 If q is an odd power of a prime and S_q is as described above, then

(a) $S_q S_q^T = qI_{q+1}$ over $\mathbb{R}$ and hence $S_q S_q^T = -I_{q+1}$ over $GF(3)$. It follows that S_q is nonsingular and $S_q^{-1} = -S_q^T$ over $GF(3)$.
(b) If $q \equiv 1 \bmod 4$, then $S_q = S_q^T$.
(c) If $q \equiv 3 \bmod 4$, then $S_q = -S_q^T$.

Proof Notice that -1 is a square if $q \equiv 1 \bmod 4$ and is a nonsquare if $q \equiv 3 \bmod 4$. Also, if α and β are any two field elements, then $\chi(\alpha\beta) = \chi(\alpha)\chi(\beta)$. For q the power of an odd prime, there are precisely $(q-1)/2$ squares and nonsquares alike, implying that

$$\sum_{\alpha \in GF(q)} \chi(\alpha) = 0$$

Consider now the sum

$$\sum_{\alpha \in GF(q)} \chi(\alpha)\chi(\alpha + \beta)$$

for a fixed element $\beta \in GF(q)$. For any $\alpha \neq 0$ we can find an element $\gamma \in GF(q)$ such that $\alpha + \beta = \alpha\gamma$ and as α ranges through the nonzero elements of $GF(q)$ γ ranges through the nonunity elements, $\beta \neq 0$. Thus we can write

$$\sum_{\alpha \in GF(q)} \chi(\alpha)\chi(\alpha + \beta) = \sum_{\alpha \in GF(q)} \chi^2(\alpha)\chi(\gamma) = \sum_{\substack{\gamma \in GF(q) \\ \gamma \neq 1}} \chi(\gamma)$$

$$= \sum_{\gamma \in GF(q)} \chi(\gamma) - 1 = -1$$

since $\chi^2(\gamma) = 1$ for any nonzero γ. Applying these facts to the matrix S_q, we observe that the (i,j) element of $S_q S_q^T$ is the inner product of the ith and jth rows of S_q. Since each row of the $(q+1) \times (q+1)$ matrix has precisely one zero in it, the inner product of a row with itself is q (over $\mathbb{R}$), and since by assumption $q \equiv -1 \bmod 3$, this is -1. The inner product of the first row with any other row is simply $\sum_{\alpha \in GF(q)} \chi(\alpha) = 0$. The inner product of the ith and jth rows is

$$\chi(-1)^2 + \sum_{k=0}^{q-1} \chi(a(k) - a(i))\chi(a(k) - a(j))$$

$$= 1 + \sum_{k=0}^{q-1} \chi(a(k) - a(i))\chi(a(k) - a(i) + (a(i) - a(j)))$$

$$= 1 - 1 = 0$$

2.7 Self-Dual and Quasicyclic Codes

by the previous result. Hence $S_q S_q^T = -I_{q+1}$, over $GF(3)$, and $S_q^{-1} = -S_q^T$. From the definition $\chi(-1)S_q = S_q^T$, implying that if $q \equiv 1 \mod 4$ (-1 is a square), then $S_q = S_q^T$, while if $q \equiv 3 \mod 4$ (-1 not a square), then $S_q = -S_q^T$. □

Theorem 7.2 Every symmetry code $C(q)$ is self-dual and the weight of every codeword in it is divisible by three.

Proof Consider the generator matrix $G = [I, S_q]$ of $C(q)$. Notice that since the elements of this matrix are either 0 or ± 1, the scalar product of a row with itself is simply the weight of that row. Thus, since every row of S_q has exactly one zero element, the weight of any row of G is $q + 1$, which is 0 mod 3 since, by assumption, $q \equiv -1 \mod 3$. Similarly, since $S_q S_q^T = -I$, the rows of S_q are orthogonal to each other and it follows that the rows of G are orthogonal to each other. Thus $C(q)$ is self-dual and, since every codeword is orthogonal to itself, its weight must be divisible by three. □

Corollary If $q \equiv 1 \mod 4$, then $G' = [-S_q, I_q]$ is also a basis of $C(q)$. If $q \equiv 3 \mod 4$, then $G'' = [S_q, I_q]$ is also a basis of $C(q)$.

Proof In general, the matrix $[-S_q^T, I]$ is a basis of $C(q)^\perp$ and hence of $C(q)$. The result follows from Theorem 7.1. □

The general weight enumerator of these codes is not as yet known. It is not difficult to show that the weight of any two rows of the generator matrix is $(q + 7)/2$ while the weight of any three rows is bounded below by this quantity. In fact for the first five possible values of q the code $C(q)$ has minimum distance exactly $(q + 7)/2$. For $C(47)$, however, the minimum weight is not of this form, being strictly less than the predicted value. The weight enumerators of these five codes are known and, by applying the theorem of Assmus and Mattson (Theorem 6.1 in Chapter 3), various new 5-designs have been determined (Table VII). Note that $d(C(q))$ denotes the minimum distance of $C(q)$ and A_i the number of codewords of weight i. If the codewords of weight r yield a t-design, then the number of blocks in the design is $A_r/2$ since the codes are over $GF(3)$. From this we can calculate

$$\lambda = \frac{A_r}{2} \frac{\binom{r}{t}}{\binom{n}{t}}$$

A great deal of information on the groups that leave these codes invariant is given by Pless (1972a).

TABLE VII

q	$d(C(q))$	λ	Design parameters
5	6	1	(5, 6, 12)
11	9	6	(5, 9, 24)
		576	(5, 12, 24)
		8,580	(5, 15, 24)
17	12	45	(5, 12, 36)
		5,577	(5, 15, 36)
		209,685	(5, 18, 36)
		2,438,973	(5, 21, 36)
23	15	364	(5, 15, 48)
		50,456	(5, 18, 48)
		2,957,388	(5, 21, 48)
		71,307,600	(5, 24, 48)
		749,999,640	(5, 27, 48)
29	18	3,060	(5, 18, 60)
		449,820	(5, 21, 60)
		34,337,160	(5, 24, 60)
		1,271,766,600	(5, 27, 60)
		24,140,500,956	(5, 30, 60)
		239,329,029,060	(5, 33, 60)

Clearly any code that has a generator matrix of the form

$$[I, S]$$

will be self-dual over $GF(q)$ iff $SS^T = I$, where the matrices I and S have the same dimension. If S is a circulant matrix (each row a cyclic shift of the previous row), then this is an example of a code equivalent to a quasicyclic code to be considered next [see Chen et al. (1969) and Townsend and Weldon (1967) but note that the term self-orthogonal in the title of this last paper does not mean self-dual]. Some interesting results on circulants and their application in binary coding are given by Karlin (1969). MacWilliams (1971) considers the problem of constructing orthogonal circulant matrices over a finite field.

A quasi-cyclic code is one in which every codeword shifted by n_0 digits is also a codeword. By permuting the column positions an (mn_0, mk_0) quasi-cyclic code of rate k_0/n_0 can be assumed to have a generator matrix of the form

$$G = \begin{bmatrix} C_{11} & C_{12} & \cdots & C_{1n_0} \\ \vdots & \vdots & & \vdots \\ C_{k_01} & C_{k_02} & \cdots & C_{k_0n_0} \end{bmatrix}$$

2.7 Self-Dual and Quasicyclic Codes

where C_{ij} is an $m \times m$ circulant matrix. In particular, for a rate $\frac{1}{2}$ quasi-cyclic code the generator matrix has the form

$$G = [C_{11}, C_{12}]$$

and if either C_{11} or C_{12} is nonsingular, this is equivalent to

$$G' = [I, C]$$

where C is again a circulant matrix. The case where C_{11} and C_{12} are both singular matrices is largely uninteresting since it leads to nonuniquely decodable codes (Tavares et al., 1974). Thus a rate $\frac{1}{2}$ quasi-cyclic code will be assumed in this systematic form.

Chen et al. (1969) are able to show that these codes have many interesting properties. The quadratic residue codes, considered in Section 5, with one digit removed are equivalent to rate $\frac{1}{2}$ quasi-cyclic codes. Karlin (1969) also investigates the quasi-cyclic structure of such codes. Chen et al. (1969) also show that there exist very long, but not arbitrarily long, quasi-cyclic codes that meet the Gilbert bound. Perhaps the class of quasi-cyclic codes deserves more attention.

Consider, for a moment, the set of all $n \times n$ circulant matrices over $GF(q)$. With the circulant matrix

$$A = \begin{bmatrix} a_0 & a_1 & \cdots & a_{n-1} \\ a_{n-1} & a_0 & \cdots & a_{n-2} \\ \vdots & \vdots & & \vdots \\ a_1 & a_2 & \cdots & a_0 \end{bmatrix}$$

we identify the polynomial

$$a(x) = \sum_{i=1}^{n} a_i x^i$$

Denote by P_n the polynomial algebra $GF(q)[x]/(x^n - 1)$. If C_n is the algebra of $n \times n$ circulant matrices over $GF(q)$, it is readily verified that the mapping

$$\eta: P_n \longrightarrow C_n$$
$$a(x) \longmapsto A$$

is an isomorphism. The transpose of the circulant matrix corresponding to $a(x) = \sum a_i x^i$ will correspond to the polynomial

$$a(x)^T = \sum_{i=1}^{n} a_{n-i} x^k$$

which we call the transpose polynomial. Of course $a(x)^T = a(x^{n-1})$ mod $(x^n - 1)$. If $a(x) = a(x)^T$, we say it is a symmetric polynomial since it implies $A = A^T$. Suppose B is the circulant matrix that is the inverse of A and the corresponding polynomials are $b(x)$ and $a(x)$, respectively. It is not difficult to show that if $AB = I$, then $a(x)b(x) \equiv 1 \mod x^n - 1$. We call $b(x)$ the inverse polynomial of $a(x)$ and write it as $a^{-1}(x)$. If $AA^T = I$, then A is an orthogonal matrix and the corresponding polynomial $a(x)^T$ is such that $a(x)a(x)^T \equiv 1 \mod x^n - 1$ and is said to be an orthogonal polynomial. An orthogonal polynomial is necessarily invertible. MacWilliams (1971) determines the orders and some of the structure of the groups G_n (invertible circulants), S_n (symmetric invertible), and O_n (orthogonal circulants). Berlekamp (1965) determines the number of $n \times n$ circulant matrices of a given rank.

We have previously observed that a code over $GF(2)$ with a generator matrix of the form
$$G = [I, A]$$
is self-dual if A is orthogonal and this is one reason for looking at techniques to construct orthogonal and orthogonal circulant matrices.

There are a number of operations on the polynomial of a circulant matrix which yield equivalent codes. Specifically, consider the binary quasi-cyclic codes with generator matrices
$$[I, C_1] \quad \text{and} \quad [I, C_2]$$
where C_1 and C_2 are circulant matrices with polynomials $C_1(x)$ and $C_2(x)$, respectively. Then Tavares et al. (1974) are able to show that the binary codes will be equivalent if any of the five following conditions hold: (i) $C_2(x) = x^i C_1(x) \mod x^m - 1$, $0 < i < m$; (ii) $C_2(x) = C_1^{-1}(x)$; (iii) $C_2(x) = C_1^*(x)$; (iv) $C_2(x) = C_1(x)^T$; (v) $C_2(x) = C_1(x)^{2^r}$, r an integer and m odd. The polynomial $C_1^*(x)$ is the reciprocal polynomial of $C_1(x)$ discussed in the previous chapter.

Some very significant advances in the weight enumeration of self-dual codes have recently been made. For the remainder of the section we review a few of these results.

Let $\mathscr{C}$ be a linear code of length n over $GF(q)$ and $\mathscr{C}^\perp$ its dual. If A_i is the number of codewords of weight i in $\mathscr{C}$, then we call the bivariate polynomial
$$W(x, y) = \sum_{i=0}^{n} A_i x^{n-i} y^i$$
the weight enumerator of $\mathscr{C}$. If the weight enumerator of the dual is given by
$$W(x, y) = \sum_{i=0}^{n} A_i^\perp x^{n-i} y^i$$
then the MacWilliams identities yield
$$\sum_{i=0}^{n} A_i x^{n-i} y^i = \frac{1}{|\mathscr{C}^\perp|} \sum_{i=0}^{n} A_i^\perp [x + (q-1)y]^{n-i}(x - y)^i$$

2.7 Self-Dual and Quasicyclic Codes

If the code is self-dual, then the right-hand side of this equation reduces to

$$\sum_{i=0}^{n} A_i \left[\frac{x + (q-1)y}{\sqrt{q}} \right]^{n-i} \left(\frac{x-y}{\sqrt{q}} \right)^i$$

since, in this case, $|\mathscr{C}^\perp| = q^{n/2}$. We conclude that for a self-dual code

$$W(x, y) = W\left(\frac{x + (q-1)y}{\sqrt{q}}, \frac{x-y}{\sqrt{q}} \right)$$

and call any homogeneous polynomial with complex coefficients that has this property formally self-dual. Clearly the weight enumerator of a self-dual code is formally self-dual but the converse to this statement is not true. For example, the polynomial $x^2 y^2 (x^2 - y^2)^2$ is formally self-dual but is not the weight enumerator of any code since it has a negative coefficient.

Theorem 7.3 (Gleason, 1971) Let

$$W(x, y) = \sum_{i=0}^{n} A_i x^{n-i} y^i$$

be a formally self-dual weight enumerator of length n over $GF(q)$ in which every weight is divisible by c.

(i) If $q = 2$ and $c = 2$, then $W(x, y) = W(y, x)$, n is even, and $A_{n-i} = A_i$. In this case define

$$f(x, y) = x^2 + y^2, \qquad g(x, y) = x^2 y^2 (x^2 - y^2)^2$$

and let $n = 2r + 8s$.

(ii) If $q = 2$ and $c = 4$, then n is divisible by eight. In this case define

$$f(x, y) = x^8 + 14 x^4 y^4 + y^8, \qquad g(x, y) = x^4 y^4 (x^4 - y^4)^4$$

and let $n = 8r + 24s$.

(iii) If $q = 3$ and $c = 3$, then n is divisible by four and define

$$f(x, y) = x^4 + 8xy^3, \qquad g(x, y) = y^3 (x^3 - y^3)^3$$

and let $n = 4r + 12s$.

In each of these three cases the weight enumerator is given by

$$W(x, y) = \sum_{r, s} k_{rs} f(x, y)^r g(x, y)^s$$

where r and s are nonnegative integers satisfying the conditions given and k_{rs} are complex numbers.

Proofs of these statements are given by Berlekamp et al. (1972).

Example Consider the extended Golay codes, which, as we have already observed, are self-dual. The weight enumerator of the (24, 12) Golay code over $GF(2)$ is given by

$$W(x, y) = x^{24} + 759x^{16}y^8 + 2576x^{12}y^{12} + 759x^8y^{16} + y^{24}$$

Since $q = 2$ and $c = 4$, part (ii) of the theorem applies and we can express $W(x, y)$ as

$$W(x, y) = \alpha f(x, y)^3 + \beta g(x, y)$$

from which it is readily determined that $\alpha = 1$ and $\beta = -672$.

The (12,6) Golay code over $GF(3)$ has a weight enumerator

$$W(x, y) = x^{12} + 264x^6y^6 + 440x^3y^9 + 24y^{12}$$

and since part (iii) of the theorem applies, we can write this as

$$W(x, y) = \alpha f(x, y)^3 + \beta g(x, y)$$

A calculation reveals that $\alpha = 1$ and $\beta = -24$.

Using classical invariant theory, MacWilliams *et al.* (1972a) generalized the above theorem of Gleason to different weight enumerators. Mallows and Sloane (1973) show that if the minimum distance d of a self-dual code is maximized, then, for case (i) of Gleason's theorem $d \leq 2\lfloor n/8 \rfloor + 2$, for case (ii) $d \leq 4\lfloor n/24 \rfloor + 4$, and for case (iii) $d \leq 3\lfloor n/12 \rfloor + 3$. An explicit formula for the weight enumerator of such extremal codes is given and the existence of such codes is considered.

MacWilliams *et al.* (1972b) show that there exist binary self-dual codes with a distance that is asymptotically the same as that given by the Varshamov–Gilbert bound. Pless and Pierce (1973) extend the result to show that there exist self-dual codes over $GF(q)$, q an arbitrary power of a prime, that satisfy a modified Varshamov–Gilbert bound, which is asymptotically the same as the usual bound. Pless (1972b) catalogs a great deal of information on self-dual codes over $GF(2)$, including a complete listing of all $(n, n/2)$ self-dual codes, n even, $2 \leq n \leq 20$, and $(n, (n-1)/2)$, n odd, $3 \leq n \leq 19$. The number of codes in the equivalence class of each code, the complete group leaving the code invariant, and the weight enumeration of each code are given. Assmus and Mattson (1972a, b) also attempt to develop tools to investigate the weight structure and in particular the minimum weight of self-dual codes.

There seem to be at least four reasons for the recent interest shown in the class of self-dual codes. As mentioned previously, there exist codes in the class that asymptotically meet the Gilbert bound. The amount of information on the form of the weight enumerators has also been an encouraging development and will foster further work along these lines. It turns out that there are some very important connections between some of the new simple groups found and automorphism groups of self-dual codes. Thus (Pless, 1972b) the Conway group (Conway, 1968b, 1969; Higman and Sims, 1968) is closely related to the (24, 12) extended Golay code. This group is, in turn, related to a remarkable lattice in 24-dimensional Euclidean space discovered by Leech (1967). Finally, MacWilliams et al. (1973) approach the problem of the existence of a projective plane of order ten by considering the information known about the weight distribution of the (111, 56) binary code that the incidence matrix of such a plane would yield. Clearly the study of self-dual codes is reaching into some deep and interesting areas of mathematics.

2.8 Perfect Codes

We have seen examples of a perfect linear code over a finite field. Specifically, we have:

(i) Hamming codes: single-error-correcting codes over $GF(q)$ with length $n = (q^m - 1)/(q - 1)$ and $k = [(q^m - 1)/(q - 1)] - m$. When $(n, q - 1) = 1$ a cyclic version of such a code will exist.

(ii) Binary repetition codes of odd length: codes with two codewords, the all-ones and all-zeros codewords.

(iii) The Golay codes: the (11, 6) code over $GF(3)$ of distance five and the (23, 12) code over $GF(2)$ of distance seven. As mentioned in Section 2.6, both of these are quadratic residue codes.

Golay (1949) described all these codes, except that he treated the Hamming codes only over prime alphabets, in a half-page article. It has taken a quarter of a century to prove that, with one exception, there are no others. In this section we will give the construction of the only other class of perfect codes known and state some of the results found along the way to the nonexistence result.

Suppose that C is a code consisting of M codewords of n-tuples over an arbitrary alphabet of q-symbols. If C is a perfect e-error-correcting code, then the usual sphere packing condition must hold, i.e.,

$$M \left[\sum_{i=0}^{e} \binom{n}{i}(q-1)^i \right] = q^n$$

If $e = 1$, the condition reduces to

$$[1 + n(q - 1)] | q^n$$

If, in addition, q is a power of a prime, say p^r, then the length of the code n must be of the form $n = (q^m - 1)/(q - 1)$. For suppose

$$[1 + n(q - 1)] = p^\alpha q^a, \qquad 0 \leqslant \alpha < r$$

then we must have that

$$n = \frac{p^\alpha q^a - 1}{q - 1} = p^\alpha \frac{q^a - 1}{q - 1} + \frac{p^\alpha - 1}{q - 1}$$

and since the first term is an integer, the last must be also, implying $\alpha = 0$. Thus a perfect single-error-correcting code over an alphabet of q letters must have a length expressible in the form $(q^m - 1)/(q - 1)$ for some integer m. It follows immediately that the number of codewords in the code must be q^{n-m}.

There are examples of codes over alphabets of q symbols, q a power of a prime, where the alphabet is not a finite field (Lindström, 1969). There are also semilinear codes (Schönheim, 1969) and nonlinear codes (Vasil'ev, 1962; Schönheim, 1968) over finite fields, which are perfect. We will give the construction of these last codes by Schönheim shortly. It is not known whether perfect codes over nonprime power alphabets exist.

The approach of Golay in finding his nontrivial codes was simple. For the binary case he merely summed elements in the rows of the Pascal triangle and observed the parameters when he obtained a power of two. He found two cases in his search,

$$\sum_{i=0}^{3} \binom{23}{i} = 2^{11} \quad \text{and} \quad \sum_{i=0}^{2} \binom{90}{i} = 2^{12}$$

The first case corresponds to the (23, 12) Golay code.

As shown in the next chapter, it is a simple consequence of the t-design properties of perfect binary codes that no (90, 78) perfect code over $GF(2)$ with $d = 5$ can exist. The (11, 6) Golay code over $GF(3)$ was found in a similar manner.

We will consider the Golay codes further shortly. We now give Schönheim's generalization of Vasilev's construction of perfect, single-error-correcting, binary, nonlinear codes. We order the nonzero elements of $GF(q)$ in some arbitrary but fixed way as $\alpha_1, \alpha_2, \ldots, \alpha_{q-1}$ and let $v_i = (x_{i1}, x_{i2}, \ldots, x_{in})$, $i = 1, 2, \ldots, q - 1$, be an arbitrary set of vectors of $V_n(q)$. Define a parity check function on these $q - 1$ vectors as

$$p(v_i) = \sum_{i=1}^{q-1} \alpha_i \left(\sum_{j=1}^{n} x_{ij} \right)$$

Let $\mathscr{C}_n$ be a linear, single-error-correcting code of length $n = (q^M - 1)/(q - 1)$ and label the codewords in some fashion as c_i, $i = 0, 1, 2, \ldots, q^{n-M} - 1$. Define the code $\mathscr{C}_N$ as the set of vectors

$$\mathscr{C}_N = \{(v_1, v_2, \ldots, v_{q-1}, c + \sum_{i=1}^{q-1} v_i, p(v_i) + f(c)); \ v_i \in V_n(q), \ c \in \mathscr{C}_n\}$$

2.8 Perfect Codes

where $f: V_n(q) \to GF(q)$ is an arbitrary (but fixed) function such that $f(0) = 0$. Clearly the length of the code is $N = nq + 1 = (q^{M+1} - 1)/(q - 1)$ and $|\mathscr{C}_N| = q^{n(q-1)} q^{n-M} = q^{N-(M+1)}$. Thus, by induction, we can get codes of any length of the form $(q^m - 1)/(q - 1)$ with q^{n-m} vectors. From previous considerations, if we can show the code is single-error-correcting, then it must be perfect.

We break each codeword of $\mathscr{C}_N$ into three parts and denote the distances of corresponding parts of the two codewords

$$\mathbf{u} = (v_1, v_2, \ldots, v_{q-1}, c + \sum v_i, p(v_i) + f(c))$$

and

$$\mathbf{v} = (v_1', v_2', \ldots, v_{q-1}', c' + \sum v_i', p(v_i') + f(c'))$$

by

$$W_1 = d((v_1, v_2, \ldots, v_{q-1}), (v_1', v_2', \ldots, v_{q-1}'))$$

$$W_2 = d((c + \sum_{i=1}^{q-1} v_i), (c' + \sum_{i=1}^{q-1} v_i'))$$

and

$$W_3 = w(p(v_i) + f(c) - p(v_i') - f(c'))$$

To show that $W_1 + W_2 + W_3 \geq 3$, we consider the various possibilities. If $W_1 \geq 3$, we are finished. If $W_1 = 0$, then $v_i = v_i'$, $i = 1, \ldots, q-1$, and if $u \neq v$, $W_2 \geq 3$. If $W_1 = 1$, then $\sum v_i$ and $\sum v_i'$ differ in precisely one place, and if $c \neq c'$, then $W_2 \geq 2$. Thus $W_1 + W_2 \geq 3$. If $W_1 = 1$ and $c = c'$, then $W_2 = 1$ and $w(p(v_i) - p(v_i')) = 1$ and thus $W_1 + W_2 + W_3 = 3$. If $W_1 = 2$, there are two ways this can happen; either $d(v_i, v_i') = 2$ for some i or $d(v_i, v_i') = 1$ for two values of i. In the first instance $W_2 \geq 1$ if $c \neq c'$, while $W_2 = 2$ for $c = c'$, and in either case $W_1 + W_2 + W_3 \geq 3$. In the second instance if $c \neq c'$, then $W_2 \geq 1$ and $W_1 + W_2 \geq 3$, while if $c = c'$, then we have to show either $W_2 \geq 1$ or $W_3 \geq 1$. Suppose that

$$w(v_i - v_i') = 1 \quad \text{and} \quad w(v_j - v_j') = 1 \quad \text{for two indices} \quad i, j, \quad i \neq j$$

This implies that $x_{i\beta} - x_{i\beta}' \neq 0$ and $x_{j\beta} - x_{j\beta}' \neq 0$. Now if $W_2 = W_3 = 0$ then

$$x_{i\beta} - x_{i\beta}' + x_{j\beta} - x_{j\beta}' = 0$$

$$\alpha_i(x_{i\beta} - x_{i\beta}') + \alpha_j(x_{j\beta} - x_{j\beta}') = 0$$

But these equations imply that

$$(\alpha_i - \alpha_j)(x_{i\beta} - x_{i\beta}') = 0$$

which implies that either $\alpha_i = \alpha_j$ or $x_{i\beta} = x_{i\beta}'$ contrary to assumption. Notice that the all-zeros vector is in the code and the code is linear or nonlinear according as $f(\cdot)$ is a linear functional on $V_n(q)$ or not.

Thus the result of Tietäväinen indicates that the only perfect codes over a finite field are the single-error-correcting codes, binary repetition codes, and two Golay codes. As has been mentioned, some of the results of finite fields can be extended to codes over other structures and in these cases the idea of a linear code is less meaningful. For example, as an Abelian group, $V_n(q)$ has the structure of $GF(q)^+ \times GF(q)^+ \times \cdots \times GF(q)^+$, where $GF(q)^+$ denotes the additive group of $GF(q)$. Herzog and Schönheim (1971) modify this structure to $G_1 \times G_2 \times \cdots \times G_n$. where G_i is an arbitrary Abelian group of order p^{α_i}, p prime, and the groups need not be of the same type. A linear code is then a subgroup of the product group and Herzog and Schönheim construct perfect linear and nonlinear codes for the Hamming metric.

Before the result of Tietäväinen great efforts were expended in proving the nonexistence of perfect codes for various sets of parameters and in finding necessary conditions for their existence. Among the latter conditions found are the following theorems of Lloyd (1957) and van Lint (1971), which we state without proof.

Theorem 8.1 (*Lloyd*) If a perfect e-error-correcting code of length n over $GF(q)$ exists, then the polynomial

$$P_e(x) = \sum_{i=0}^{e} (-1)^i \binom{n-x}{e-i}\binom{x-1}{i}(q-1)^{e-i}$$

has e distinct integral zeros among the integers $1, 2, \ldots, n-1$.

Lenstra (1972) generalized this result by showing that it holds for more general alphabets than $GF(q)$.

Theorem 8.2. (*van Lint*) The zeros of the polynomial $P_e(x)$, $\alpha_1, \alpha_2, \ldots, \alpha_e$, satisfy the equations

(a) $\quad \displaystyle\sum_{i=1}^{e} \alpha_i = \frac{e(n-e)(q-1)}{q} + \frac{e(e+1)}{2}$

(b) $\quad \displaystyle\prod_{i=1}^{e} \alpha_i = e!\, q^{k-e}$

where $\sum_{i=0}^{e} \binom{n}{i}(q-1)^i = q^k$.

It is appropriate to mention here some relationships between the Mathieu groups, certain Steiner systems, and the Golay codes, although the combinatorial aspects of these relationships will be considered in some detail in the next chapter. The Mathieu groups play such an important role that some information on them has been gathered as Appendix C. The Mathieu groups M_{12} and M_{24}, which are the only 5-transitive permutation groups known, are the automorphism groups of the $S(5, 6, 12)$ and $S(5, 8, 24)$ Steiner systems. The Mathieu groups M_{11}, M_{23} (4-transitive), and M_{22} (3-transitive) are the automorphism groups of the Steiner systems $S(4, 5, 11)$, $S(4, 7, 23)$, and $S(3, 6, 22)$, respectively. Witt (1938a, b) has shown that, up to isomorphism, these five Steiner systems are unique. Now the minimal weight codewords in a linear perfect code span the code [see, e.g., Assmus and Mattson (1969)]. Also, the minimal weight codewords in the extended (24, 12) and (12, 6) Golay codes form the Steiner systems $S(5, 8, 24)$ and $S(5, 6, 12)$, while those of the codes (23, 12) and (11, 6) form the Steiner systems $S(4, 7, 23)$ and $S(4, 5, 11)$, respectively (Pless, 1968). The following two interesting theorems, which we state without proof, are due to Pless (1968) and summarize the extent to which the Golay codes are unique.

Theorem 8.3 Let $\mathscr{C}$ be an (11, 6) [resp. (23, 12)] code over $GF(3)$ [resp. $GF(2)$]. Then the following five statements are equivalent:

(i) $\mathscr{C}$ has the same weight distribution as the Golay code.

(ii) The nonzero coordinate indices of the minimum weight vectors of $\mathscr{C}$ form an $S(4, 5, 11)$ [resp. $S(4, 7, 23)$] Steiner system.

(iii) $\mathscr{C}$ is perfect.

(iv) $\mathscr{C}$ is self-dual and its minimum weight is five (resp. seven).

(v) $\mathscr{C}$ is equivalent to the Golay code.

The corresponding theorem for the extended codes is:

Theorem 8.4 Let $\mathscr{C}$ be a (12, 6) [resp. (24, 12)] code over $GF(3)$ [resp. $GF(2)$]. Then the following five statements are equivalent:

(i) $\mathscr{C}$ has the same weight distribution as the extended Golay code.

(ii) The nonzero coordinate indices of the minimum weight vectors of $\mathscr{C}$ form an $S(5, 6, 12)$ [resp. $S(5, 8, 24)$] Steiner system.

(iii) The minimum weight of $\mathscr{C}$ is six (resp. eight).

(iv) $\mathscr{C}$ is self-dual and has minimum weight six (resp. eight).

(v) $\mathscr{C}$ is equivalent to an extended Golay code.

2.9 Comments

In this chapter an attempt has been made to present a few of the interesting and natural connections between the fields of combinatorics and coding. Many others have not been mentioned. The subject is further considered in the next chapter, where the connections of coding with t designs are treated. The books by Ryser (1963) and Hall (1967a) were used liberally and are the standard combinatorial references. An excellent source of material on finite geometries and their collineation groups is Carmichael (1956). All material relating to coding is referenced in the text.

Exercises

1 Let $g(x)$ be the generator polynomial of the dual of the rth-order projective geometry code of length $(p^{ms} - 1)/(p^s - 1)$. Show that the code of this length generated by $(x - 1)g(x)$ is $(m - r - 1)$-step majority logic decodable.

2 Let $g(x)$ be the generator polynomial of the dual of the rth-order Euclidean geometry code of length $p^{ms} - 1$ over $GF(p)$. Show that the code of this length generated by $(x - 1)g(x)$ is $(m - r - 1)$-step majority logic decodable.

3 Let d' be the minimum distance of the dual code of a linear (n, k) code. Show that the number of errors that can be corrected by one-step majority logic decoding is upper-bounded by the quantity $(n - 1)/[2(d' - 1)]$. Extend this result to the L-step majority-logic-decoding case assuming all codes are binary.

4 Show that the rth-order projective geometry code has a BCH bound on its minimum distance the same as the bound achieved by majority-logic-decoding arguments.

5 Show that if $r < (m + 1)/2$, an rth-order projective geometry code obtained from $PG(m, p^s)$ is two-step majority logic decodable. Similarly, show that if $r \leq m/2$, an rth-order Euclidean geometry code obtained from $EG(m, p^s)$ is two-step majority logic decodable.

6 (Reed–Muller Codes) Let G' be the generator matrix for the cyclic binary maximum length code with parameters $(2^m - 1, m)$ and distance 2^{m-1}. To G' we first add a column of all zeros and then a row of all ones to obtain

$$G = \begin{bmatrix} 1 & 1 & 1 & \cdots & 1 \\ \hline 0 & & & & \\ \vdots & & G' & & \\ 0 & & & & \end{bmatrix}$$

The codewords of the code with generator matrix G are those of the original code with a zero added in the first position, plus the complements of words in the original code with a one added in the first position. Label the rows of G by $v_0, v_1, \ldots, v_m$ and let $n = 2^m$. If $u = (u_1, u_2, \ldots, v_n)$ and $v = (v_1, v_2, \ldots, v_n)$ are two binary n-tuples, we define a componentwise product as

$$uv = (u_1 v_1, u_2 v_2, \ldots, u_n v_n)$$

It follows trivially that $u^2 = u$ for any binary vector u.

Exercises

Definition The rth-order Reed–Muller (RM) code is the subspace of $V_n(2)$ spanned by the set of vectors

$$v = v_0^{i_0} v_1^{i_1} \cdots v_m^{i_m}, \quad i_j = 0 \text{ or } 1 \quad \text{and} \quad \sum_{j=0}^{m} i_j \leq r$$

Show that:

(i) The rth-order RM code of length 2^m has dimension $k = \sum_{j=0}^{r} \binom{m}{j}$ and distance 2^{m-r}.

(ii) The dual of the rth-order RM code of length 2^m is the dual of the $(m-r-1)$th-order RM code.

(iii) The vector v_i, $i > 0$, may be interpreted as the incidence vector of a hyperplane in $EG(m, 2)$.

(iv) The vector $v_i + v_j$, $i, j > 0$, using the interpretation of (iii), is the incidence vector of the symmetric difference of the flats corresponding to v_i and v_j, i.e., if v_i and v_j are incidence vectors of the $m-1$ flats F_i and F_j, respectively, then $v_i + v_j$ is the incidence vector of the symmetric difference $(F_i \backslash (F_i \cap F_j)) \cup F_j \backslash (F_i \cap F_j))$, where unions and intersections are in the set-theoretic sense. Show that this symmetric difference is also an $m-1$-flat.

(v) The vector $v_i v_j$ is the incidence vector of $F_i \cap F_j$, which is always an $m-2$ flat.

(vi) The complement of an $m-1$ flat is an $m-1$ flat.

(vii) The rth-order RM code of length 2^m contains all the t flats of $EG(m, 2)$, $t \geq m - r$. Conclude that the code is $(r + 1)$-step orthogonalizable.

7 (Bose and Bush, 1952) The solutions to the equation $\sum_{i=1}^{r} x_i = 0$ form an $r - 1$ flat in $PG(r-1, 2)$ containing $2^{r-1} - 1$ points. Show that the set of 2^{r-1} points not in this flat consists of those elements of $V_r(2)$ containing an odd number of ones. Show that if we use these points as the row vectors for a $2^{r-1} \times r$ matrix $\mathbf{C}$, then the $2^{r-1} \times 2^r$ matrix $\mathbf{A}$ whose ith row is $\mathbf{C} x_i$, $x_i \in V_r(2)$, $i = 1, \ldots, 2^r$, is an $OA(2^r, 2^{r-1}, 2, 3)$ of strength three.

8 Show that if there exist k orthogonal Latin squares of orders n_1 and n_2, respectively, then there exists a set of k orthogonal Latin squares of order $n_1 n_2$.

9 (Hall, 1967a) Given two orthogonal arrays $OA(n_1^2, k, n_1, 2)$ and $OA(n_2^2, k, n_2, 2)$, both of index one, construct an $OA((n_1 n_2)^2, k, n_1 n_2, 2)$ of index one.

10 (Lin, 1969) It was shown in the previous chapter that a GRM code of length q^m over $GF(q)$ is invariant under the doubly transitive permutation group consisting of transformations

$$\mathbf{X} \to \mathbf{AX} + \mathbf{B}$$

where $\mathbf{X} \in V_m(q)$, $\mathbf{A}$ is an $m \times m$ nonsingular matrix, and $\mathbf{B} \in V_m(q)$. Each coordinate position is labeled with an element of $V_m(q)$ and the group acts on these elements. Show that when $q = 2$ this group is actually triply transitive. Conclude that the shortened code obtained by deleting the coordinate position labeled with the zero vector is invariant under a doubly transitive group and hence that its constant weight vectors yield a BIBD. The minimum weight vectors in the cyclic binary rth-order RM code correspond to the BIBD's formed from the $m - r$ flats in $EG(m, 2)$ through the origin, and those of the next highest weight correspond to the BIBD's formed from the $m - r$ flats not through the origin.

11 (Pless, 1972b) An (n, k) code $\mathscr{C}$ is said to be the direct sum of two codes $\mathscr{C}_1$ and $\mathscr{C}_2$, denoted by $\mathscr{C} = \mathscr{C}_1 \oplus \mathscr{C}_2$, if: (i) $\mathscr{C}_1$ and $\mathscr{C}_2$ are vector subspaces of $\mathscr{C}$; (ii) every vector $c \in \mathscr{C}$ can be expressed as $c = c_1 + c_2$, $c_i \in \mathscr{C}_i$; (iii) a projection of $\mathscr{C}_1$ onto n_1 coordinate places is an (n_1, k_1) subspace and a projection of $\mathscr{C}_2$ onto n_2 coordinate places is an (n_2, k_2) subspace, where $n = n_1 + n_2$ and $k = k_1 + k_2$. Show that if $\mathscr{C}_1$ and $\mathscr{C}_2$ are left invariant by the permutation groups G_1 and G_2 respectively, then the direct product group $G_1 \times G_2$ is in the group leaving $\mathscr{C}$ invariant. Show also that if there are A_i vectors of weight i in $\mathscr{C}_1$ and B_i in $\mathscr{C}_2$, then the number of vectors of weight k in $\mathscr{C}_1 \oplus \mathscr{C}_2$ is $\sum_{i+j=k} A_i B_j$.

12 (Hsiao et al., 1970) Recall that if we can form at least $d - 1$ check sums orthogonal on each digit, then the code has minimum distance at least d. Construct a binary parity check matrix:

$$H = \begin{bmatrix} M_1 \\ M_2 \\ \vdots \\ M_{2t} \end{bmatrix} \quad I_{2tm} \end{bmatrix}$$

where the $m \times m^2$ matrices M_1 and M_2 are given by

$$M_1 = \begin{bmatrix} \overbrace{1\ 1\ \cdots\ 1}^{m} & & & \\ & \overbrace{1\ 1\ \cdots\ 1}^{m} & & 0 \\ & & \ddots & \\ & & & \overbrace{1\ 1\ \cdots\ 1}^{m} \\ 0 & & & \end{bmatrix}$$

$$M_2 = [I_m \ \vdots \ I_m \ \vdots \ \cdots \ \vdots \ I_m]$$

and the M_i, $3 \leqslant i \leqslant 2t$, are constructed from a set $L_i = (l_{rs}^i)$, $3 \leqslant i \leqslant 2t$, of pols in the following manner: Define the incidence matrix Q_i^μ of the element μ in L_i in the usual fashion, i.e., $(q_i^\mu)_{rs} = 1$ if $l_{rs}^i = \mu$ and 0 otherwise. For a

Exercises

given L_i the rows of each of the m incidence matrices are concatenated to form an m^2-tuple, e.g.,

$$V_i^\mu = [(q_i^\mu)_{11}, (q_i^\mu)_{12}, \ldots, (q_i^\mu)_{1m}, (q_i^\mu)_{21}, \ldots, (q_i^\mu)_{2m}, \ldots, (q_i^\mu)_{mm}]$$

Finally, the matrix M_i is formed by using V_i^j as its jth row. Show that the resulting parity check matrix yields a code of length $m^2 + 2tm$, dimension m^2, and is capable of correcting t errors.

13 An $n \times n$ matrix whose ith row is the $(i-1)$th row shifted cyclically to the left one position is called a back circulant. Show that such a matrix is symmetric. Show that the product of a back circulant matrix with a circulant matrix yields a symmetric matrix. Show that any two circulant matrices of the same order commute.

14 Show that -3 is a quadratic residue of primes of the form $6n + 1$ and a quadratic nonresidue of primes of the form $6n + 5$.

15 Show that the number of solutions to the equation

$$ax^2 + bx + c \equiv 0 \mod p$$

is given by $1 + [(b^2 - 4ac)/p]$.

16 (Berlekamp, 1968) Show that the cyclotomic polynomial of order 241 is irreducible over $GF(17)$. [*Hint*: It is sufficient to show that the multiplicative order of 17 mod 241 is 240 and this follows from the fact that $(17/241) = -1$. Use the law of reciprocity to establish this.]

3 Coding and Combinatorics

3.1 Introduction

Being concerned with problems of a finite nature, coding theory itself can be considered as a part of combinatorial theory. For example, certain coding problems themselves were treated under the title of packing problems before the concept of coding theory arose. [See, for example, Paige (1956).] In fact, the well-known Hamming–Golay codes were discussed by Zaremba (1950, 1952) before the results of Hamming and Golay appeared in print. We shall concern ourselves in this chapter, however, with the relationship between coding theory and certain aspects of experimental design and matroid theory.

3.2 General t Designs

Recall from the preceding chapter that a *t design* or $\lambda\text{-}(t, k, v)$ *configuration* is an ordered pair (S, B), where S is a set of cardinality v, and B is a family of k-subsets (called blocks) of S with the property that each t-subset of S is contained in precisely λ blocks of B. For nondegeneracy, we shall always assume that $0 < t < k < v$. While not much is known about general t designs, much work has been done in certain special cases. For each triple t, k, v satisfying $0 < t < k < v$ there are many t designs that may be obtained

3.2 General t Designs

trivially as follows. Let S be any v-set. Form C, the set of all possible k-subsets of S. The result (S, C) is a t design that we call the full combinatorial design. In this design $\lambda = \binom{v-t}{k-t}$. Let B be the family of k-subsets of S that includes each member of C exactly n times. (The term family is used here to indicate that each k-subset of S may occur several times as a block of B.) The result is a λ-(t, k, v) configuration in which $\lambda = n\binom{v-t}{k-t}$. In fact, given any λ-(t, k, v) configuration, we can *replicate* it n times to get an $n\lambda$-(t, k, v) configuration by employing each of its blocks n times.

Nontrivial t designs have been known for some time for each t, $0 < t < 6$. However, it is only recently that it has been shown that t designs other than full combinatorial designs and their replications also exist for all $t \geqslant 6$.

Proposition 2.1 Every λ_t-(t, k, v) configuration is a λ_{t-1}-$(t-1, k, v)$ configuration in which $\lambda_{t-1} = \lambda_t(v - t + 1)/(k - t + 1)$.

Proof Let $D = (S, B)$ be a λ_t-(t, k, v) design. Let X be any $(t-1)$-subset of S, and $\lambda(X)$ be the number of blocks of D that contain X. Clearly X is contained in $v - t + 1$ t-subsets of S. Let $C(X)$ denote the collection of such subsets. Each of these t-subsets occurs in λ_t blocks of D, these being the $\lambda(X)$ blocks that contain X. Now each of these blocks contains $k - t + 1$ members of $C(X)$. Thus $\lambda(X)(k - t + 1) = \lambda_t(v - t + 1)$. Since $\lambda(X)$ does not depend on the particular $(t-1)$-subset chosen, the result follows. □

Corollary 1 Let $\lambda_0 = b$ be the number of blocks (which contain the null subset), $\lambda_1 = r$ denote the number of blocks containing a given element, and λ_i be the number of blocks that contain a given i-subset, $0 \leqslant i \leqslant t$. Then $\lambda_i = \lambda_t\binom{v-i}{t-i}/\binom{k-i}{t-i}$.

Proof This follows from repeated application of the proposition. □

Corollary 2 Let u be an integer such that $0 \leqslant u \leqslant t$; then the number of subsets in B intersecting a given t-subset in u elements is independent of the t-subset chosen.

Proof This is equivalent to Corollary 1. □

Corollary 3 The complement of a t design is a t design.

Proof A straightforward application of the principle of inclusion–exclusion shows that the complementary configuration of a λ-(t, k, v) design is a λ'-$(t, v - k, v)$ design, where $\lambda' = \lambda\binom{v-k}{t}/\binom{k}{t}$. □

Proposition 2.2 The existence of a λ_t-(t, k, v) configuration implies the existence of a λ_t-$(t - i, k - i, v - i)$ configuration, $0 < i < t$.

Proof Let $C(x)$ be the set of blocks of $D = (S, B)$ that contain a given element of x of S. Every $(t - 1)$-subset of S-$\{x\}$ occurs with x in exactly λ_t blocks of D, these blocks being those of $C(x)$. Thus $(S - \{x\}, B')$, where B' is obtained from the blocks of $C(x)$ by removing x, is the required configuration for $i = 1$. Repeated application establishes the required result. □

Corollary The existence of a λ_t-(t, k, v) configuration implies the existence of a λ^t-$(t - 1, k, v - 1)$ design, where $\lambda^t = \lambda_t - \lambda_{t-1}$.

Proof This design consists of the blocks other than those containing a given variety x_1, that is, the set of blocks $B - C(x)$ in Proposition 2.2. □

Steiner Systems A 1-(t, k, v) design is called a *Steiner system*, which we denote by $S(t, k, v)$. The following is a brief catalogue of results on the existence of Steiner systems. For $t = 2$ (balanced incomplete block designs) various families have been found. Among the best-known families of such designs are:

(a) Steiner triple systems $S(2, 3, v)$, which exist for all $v \equiv 1$ or $3 \mod 6$.
(b) Hanani systems I, $S(2, 4, v)$, which exist for all $v \equiv 1$ or $4 \mod 12$.
(c) Hanani systems II, $S(2, 5, v)$, which exist for all $v \equiv 1$ or $5 \mod 20$.
(d) Projective and Euclidean systems, which exist for $n > 1$ and q a prime power: (i) $S(2, q, q^n)$ and (ii) $S(2, q + 1, (q^n - 1)/(q - 1))$.
(e) Hyperbolic systems $S(2, 2^n, 2^{2n+1} - 2^n)$ for n an integer > 1.
(f) Hermitian systems $S(2, q + 1, q^3 + 1)$, which exist for q a prime power.

In this regard we point out that it is known that for any given integer $k > 2$ there is a constant $c(k)$ such that $S(2, k, v)$ exists for $v > c(k)$ and $v \equiv k \mod k(k - 1)$.

For $t = 3$ the following systems are known:

(a) Quadruple systems $S(3, 4, v)$ exist for all positive $v \equiv 2$ or $4 \mod 6$.
(b) Inversive systems $S(3, q + 1, q^2 + 1)$ exist for q a prime power.

For $t = 4$ and $t = 5$ the only known Steiner systems are the Mathieu systems, $S(5, 8, 24)$, $S(5, 6, 12)$, $S(4, 7, 23)$, $S(4, 5, 11)$ [see Witt (1938b)].

Before investigating some of the connections between t designs and coding theory we digress as follows. With every linear code there is an associated combinatorial structure called a *matroid*. Many of the investigations concerning t designs and codes are in fact investigations of the properties of the matroids of linear codes, and we discuss briefly the concept of a matroid. Excellent sources of information on matroids are the publications by Tutte (1965, 1971). Matroids can be axiomatized in many ways; here we basically follow Tutte's (1965) work, which is ideally suited for our purposes.

3.3 Matroids

Definition A matroid M is a pair (S, F) where S is a finite set of elements called cells and F is a family of nonnull subsets of S such that (i) no member of F is a proper subset of another, and (ii) if a and b are two members of S, and X and Y are members of F such that $a \in X \cap Y$ and $b \in X - Y$, then there exists a member Z of F such that $a \notin Z$, $b \in Z$, and $Z \subseteq X \cup Y$.

The members of F are called the circuits of the matroid, and we refer to these axioms as the circuit axioms of the matroid. The *girth* $\alpha = \alpha(M)$ of the matroid is defined by the formula

$$\alpha(M) = \min_{X \in F} \{|X|\}$$

Definition The *weight of a circuit* X is its cardinality $|X|$.

Examples of Matroids We assume that the reader is familiar with the concept of a linear graph [see, for example, Tutte (1966)]. We denote the set of vertices of the graph by V or $V(G)$ and the edges by E or $E(G)$. A polygon of G is a connected subgraph of G in which each vertex is incident with exactly two edges. A bond in G is a nonnull subset B of $E(G)$ that satisfies the following condition: The graph H obtained from G by deleting the edges of B has two connected components C_1 and C_2 such that each edge of B has one end in each component.

Example 1 If F is the family of polygons of a graph G, then $(E(G), F)$ is a matroid, as is readily verified. This is called the polygon matroid of G.

Example 2 If F is the family of bonds of a graph G, then $(E(G), F)$ is a matroid, as is readily verified.

Definition A matroid is *graphic* if it is the bond matroid of a graph; it is *cographic* if it is the polygon matroid of a graph.

Proposition 3.1 Let F' be a family of nonnull subsets of a set S. If F' satisfies matroid axiom (ii), then the minimal members of F' form a matroid on S, where a minimal member is one which contains no other under set inclusion.

Proof We first show that if $a \in X$ where $X \in F'$, then there is a minimal member Y of F' such that $a \in Y \subseteq X$. Indeed, take Y as a member of F' with least cardinality, which satisfies $a \in Y \subseteq X$. If Y is not minimal, then there exists $Z \in F'$ such that $Z \subseteq Y - \{a\}$. Choose $b \in Z$; then there is a member Z' of F' such that $a \in Z' \subset Y - \{b\} \subset X$, by axiom (ii), which contradicts the choice of Y. Now let a and b be members of S, and X and Y be minimal members of F' such that $a \in X \cap Y$ and $b \in X - Y$. Then there is a member W of F' such that $a \notin W$, $b \in W$, and $W \subseteq X \cup Y$, and therefore there is a minimal member Z of F' such that $b \in Z \subseteq W$. Then $a \notin Z$, $b \in Z$, and $Z \subseteq X \cup Y$. □

Definition A dendroid of a matroid M is a subset D of S that meets every circuit of M and is minimal with respect to this property.

By definition, there is a circuit of M such that $X \cap D = \{a\}$. X is uniquely determined by this condition. Indeed, if Y is another circuit such that $Y \cap D = \{a\}$, there is by (ii) a circuit $Z \subseteq X \cup Y$ that does not meet D. The circuit X is denoted by $J(D, a)$.

Lemma 3.1 Let D be a dendroid of a matroid $M = (S, F)$. Let a and b be elements of S such that $a \in D$ and $b \notin D$. Then $D' = (D - \{a\}) \cup \{b\}$ is a dendroid if and only if $b \in J(D, a)$.

Proof If D' is also a dendroid of M, then, since it meets $J(D, a)$, it must do so in b. Conversely, if $b \in J(D, a)$, then D' meets every circuit of M, since $D - \{a\}$ meets every member of $F - \{J(D, a)\}$. Consider any element $c \in D' - \{b\}$. Then $I = D' \cap J(D, c)$ is either $\{c\}$ or $\{b, c\}$. If $I = \{b, c\}$, then by (ii) there is a circuit W such that $c \in W$, $b \notin W \subseteq J(D, a) \cup J(D, c)$, hence $D \cap W = \{c\}$. In either case there is a circuit that meets D' only in $\{c\}$, so that no proper subset of D' meets all circuits of M. □

Corollary All dendroids of M have the same cardinality.

Proof Let D_1 and D_2 be two distinct dendroids of M. Let D_3 be a dendroid of M such that $|D_3| = |D_1|$ and $|D_2 \cap D_3|$ is as large as possible. If $D_2 \not\subseteq D_3$, choose $c \in D_3 - D_2$. Then $J(D_3, a)$ meets D_2 in a cell b, which is not in D_3. Since $D_4 = (D_3 - \{a\}) \cup \{b\}$ is a dendroid of M, and $|D_4 \cap D_2| = |D_3 \cap D_2| + 1$, we have a contradiction of the definition of D_3 and therefore conclude that $D_2 \subseteq D_3$. By definition, no proper subset of a dendroid is a dendroid; thus $D_2 = D_3$ and the corollary is established. □

Definition The *rank* of a matroid M is the cardinality of any dendroid of M. This quantity is denoted by $r(M)$.

3.4 Chains and Chain Groups

Definition A subset C of the set S of a matroid $M = (S, F)$ is spanning if C contains a dendroid of M.

A matroidal property that has direct implication in coding is given in the following lemma.

Lemma 3.2 If M is a matroid of girth α defined on a set S of cardinality n, then every subset of cardinality $n - \alpha + 1$ is spanning and $r(M) \leqslant n - \alpha + 1$.

Proof Since every circuit has at least α elements, every subset S' of S which has cardinality of at least $n - \alpha + 1$ must meet every circuit, and any subset $T \subseteq S'$ that is minimal with respect to this latter property will form a dendroid. The lemma follows. □

To show the relationships between matroids and linear algebra, and thus coding theory, we turn our attention now to chain groups.

3.4 Chains and Chain Groups

Let S be a finite set and R a commutative ring with a unit element and no divisors of zero. A *chain* on S is a function f which maps S into R. If $a \in S$, we call the element $f(a)$ the *coordinate* of a in f. The *domain* dom f of the chain f is the set of all $a \in S$ such that $f(a) \neq 0$. A chain is a *zero* chain iff dom f is null. Clearly the zero chain is unique. We define a new chain on S as the sum of two chains f and g by the rule $(f + g)a = f(a) + g(a)$, $a \in S$.

The product λf of an element λ of R and a chain f on S is defined by $\lambda f(a) = \lambda(f(a))$, $a \in S$.

A *chain group* on S over R is defined as any subset of the set of all chains on S that is closed with respect to both addition and multiplication. Let G be a chain group on S. A chain f on G is *elementary* if it is nonzero and there is no nonzero chain $g \in G$ such that dom $g \subsetneq$ dom f.

With every chain group an S we can associate a matroid M on S as follows.

Lemma 4.1 Let N be a chain group on S over R. Then the family F of domains of the elementary chains of N is a matroid on S.

Proof Let F' be the family of domains of the nonzero chains of N. Let f and g be two such chains and a and b be elements of S such that $a \in$ dom f $\cap$ dom g and $b \in$ dom $g -$ dom f. Multiply f by $g(a)$ and add the resulting chains to obtain the chain $h \in N$. Then $a \notin \text{dom } h$, $b \in \text{dom } h$, and dom $h \subseteq \text{dom } f \cup \text{dom } g$. Thus F' satisfies axiom (ii). Since the elementary chains have minimal domains, the result follows by Proposition 3.1. □

We call the above matroid the matroid of N and denote it by $M(N)$.
For later reference we note the following definition.

Definition A chain group over GF(2) is called binary.

The concept of a dendroid for matroids can be extended as follows to a chain group N on S over a commutative ring as above.

Definition A *dendroid* of a *chain group* N on a set S is a minimal subset of S that meets the domain of every nonzero chain of N.

Since each such domain contains that of an elementary chain, the dendroids of N are identical with those of $M(N)$.

If D is a dendroid of N, for each $a \in D$ there is a chain f_a of N such that dom f_a is the circuit $J(D, a)$ of $M(N)$.

Definition A set of chains f_a, one for each $a \in D$ as in the foregoing, is a *dendroid basis* of N.

Definition The weight of a dendroid basis $\{f_a\}$ is the product $\Pi_{a \in D} f_a(a)$, which is clearly nonzero.

Proposition 4.1 Let $B = \{f_a | a \in D\}$ be a dendroid basis of weight w of a chain group N over R. Let g be any nonzero chain of N. Then wg can be expressed as a linear combination of members of B, with coefficients in R.

Proof By adding suitable members of B to wg, one can reduce the coordinate of each $a \in D$ to zero. The resulting chain is zero, since its domain does not meet D. □

Corollary If dom g meets D in a single cell a, then dom $g = $ dom f_a, and hence g is an elementary chain.

The rank of $M(N)$ in the sense of the theory of linear independence is therefore the rank of N.

In the case where R is a field the chains of the dendroid basis can be chosen so that $f_a(a) = 1$ for each $a \in D$. In this case a dendroid basis becomes a true basis for N.

3.4 Chains and Chain Groups

Proposition 4.2 Let $M = (S, F)$ be the matroid of a chain group N over a field K. Let L be a nonempty subset of F. Let T be a subset of S that meets every member of L. If T is minimal with respect to this property, then T can be extended to a dendroid of M.

Proof Let D be a dendroid of M such that $|T \cap D|$ is as large as possible. Let L' be the set of members of L that do not meet $T \cap D$. If L' is not empty, choose $x \in T$ such that x belongs to some member l of L'. Then $x \notin D$. Then $x \in J(D, a)$ for some $a \in T \cap D$; otherwise, since every chain on $l \in N$ can be written as a linear combination of chains on $J(D, a)$, $a \in [D-(T \cap D)]$, every chain on l would vanish on x, violating the choice of x. Then $D' = (D \cup \{x\}) - \{a\}$ is a dendroid meeting T in a set of cardinality $|D \cap T| + 1$, violating the choice of D. Therefore L' must be empty. Thus every member of L meets $T \cap D$. But T is minimal with respect to this property, thus $T \cap D = T$, that is, $T \subseteq D$. □

Corollary If L contains the domains of a basis of N, then T is a dendroid.

Proof If T is a proper subset of D, then the chains on $J(D, a)$, $a \in T$, generate a $|T|$-dimensional subspace that includes all chains whose domains lie in L, contradicting the fact that L contains the domains of a basis of N. □

Chain groups over S are studied because of the fact that structures on S have properties that are reflected in certain of the chain groups of S. For example, if there is a linear graph whose edge set is S, then its polygon matroid has certain properties that must be reflected in certain chain groups that share the same matroid, which allows one to determine whether or not a given matroid is *cographic*, that is, the polygon matroid of a linear graph.

Our main concern here will be the relationship between t designs and matroids, as reflected in the linear codes they generate. For our purposes, we require yet the following considerations.

Let N be a chain group on a set S over a commutative ring R having a unit element and no divisors of zero.

Definition If S' is a subset of S, then the *restriction* of the chain f of N to S' is the chain g on S' such that $g(a) = f(a)$ for each $a \in S$.

Definition The *reduction of a chain group* N to S' is the chain group consisting of the restrictions of the chains of N to S'. (It is evident that the set of such chains forms a chain group under the usual definitions of addition and scalar multiplication; we refer to this chain group as $N \cdot S'$.)

Similarly we have the following definition.

Definition The contraction of a chain group N to S' (denoted by $N \times S'$) is the chain group consisting of the restrictions to S' of those chains f of N such that $\text{dom } f \subseteq S'$.

It is evident that $N \times S' \subseteq N \cdot S'$.

Definition The *reduction of a matroid* M to a subset $S' \subset S$ is the matroid consisting of the minimal members of the class of nonnull intersections of S' with the circuits of M. (It is evident that these subsets form a matroid, which is referred to as $M \cdot S'$.)

Definition The contraction of M to S' (denoted by $M \times S'$) is the matroid of the class of all circuits X of M such that $X \subset S'$.

The reader can verify that $M(N \times S) = M(N) \times S$ and that $M(N \cdot S) = M(N) \cdot S$.

3.5 Dual Chain Groups

Definition Two chains f and g on S over R are orthogonal if

$$\sum_{a \in S} f(a) \cdot g(a) = 0$$

Let N^* be the class of all chains h on S over R such that h is orthogonal to every member of N. Then N^* is a chain group on S over R.

Definition The *dual chain group* N^* of the chain group N is the chain group of all chains orthogonal to the chains of N.

Clearly $N \subset N^{**}$, and if R is a field, then $N = N^{**}$.

Lemma 5.1 *If N is a chain group over R, then $(N \cdot S')^* = N^* \times S'$ for any $S' \subseteq S$. Further, if R is a field, then $(N \times S')^* = N^* \cdot S'$.*

Proof Let f be a chain on S. Then f belongs to $(N \cdot S')^*$ if and only if it is orthogonal to every chain of $N \cdot S'$, that is, if and only if the chain g on S that satisfies $\text{dom } g = \text{dom } f$ and has f as its restriction to S' is orthogonal to every member of N. This condition is valid if and only if f belongs to $N^* \times S'$. Thus $(N \cdot S')^* = N^* \times S'$. Now if R is a field, we write

$$(N \times S')^* = ((N^*)^* \times S')^* = ((N^* \cdot S')^*)^* = N^* \cdot S'. \quad \square$$

3.6 Matroids, Graphs, and Coding

Before returning to coding theory, we remark that there is a corresponding theory of matroid duality, which is reflected in the theory of planar graphs. The theory of matroids is a rich and beautiful area of combinatorial mathematics, encompassing the area of finite geometries as well, but these considerations are beyond the scope of the present book. The reader is referred to Tutte (1971).

3.6 Matroids, Graphs, and Coding

Clearly every code C over a field K may be viewed as a set of chains defined on the set of coordinate positions in its concrete realization. For the remainder of this section we adopt such a vantage point. We refer to the chains of C as code chains or code vectors. We note that the Hamming weight of a chain f denoted by $w(f)$ is $|\text{dom } f|$ and the Hamming distance between two chains f and g [denoted by $d(f, g)$] is $w(f - g)$. A code over a field K is linear if and only if its sets of chains form a chain group.

Let us turn our attention for a moment to linear graphs.

Definition A *cycle* in a graph G is the edge set of a subgraph C of G such that each vertex of C is of even degree.

Definition By the mod 2 sum (or symmetric difference) of two sets A and B we mean the set $(A \cup B) - (A \cap B)$.

The cycles of G form a vector space $C(G)$ over GF(2) under mod 2 addition and scalar product $1 \cdot C = C$ for all cycles C and $0 \cdot C = \varnothing$, the subgraph with no edges.

Definition The *cycle space* of a graph G is the vector space of cycles of G under mod 2 addition. (Thus any linear graph yields a binary code.) Clearly the matroid of $C(G)$ is the polygon matroid of G.

The girth of a graph is the number of edges contained in the "smallest" polygon of the graph. Since the cycle space codes of graphs are linear, the distance d of such a code is the girth of the corresponding graph. Graphs of given girth and degree with as few vertices as possible are called *cages* by Tutte and have been investigated by Tutte (1966) and others. We do not pursue this interesting topic here.

Since by definition of girth we have the fact that in any linear (n, k) code the minimum mutual distance d of the code is the girth α of the matroid $M(C(G))$, Lemma 3.2, when interpreted for linear codes, is the well-known inequality $d \leqslant n - k + 1$ (Theorem 7.6 of Chapter 1).

Definition A linear code is maximum distance separable if $d = n - k + 1$.

Proposition 6.1 Every maximum-distance-separable (linear) binary code can be viewed as the cycle space of a linear graph.

Proof This follows from the fact that all maximum-distance-separable binary codes are of a trivial nature, as will be seen in Proposition 6.2. It will be seen that the graph in question either is a polygon or consists of a pair of vertices joined by several edges. □

In the following we refer to the circuits of least weight α of a matroid as the belts of the matroid.

Proposition 6.2 (*Assmus and Mattson* 1967a) An (n, k) code is maximum distance separable if and only if the belts of its matroid form a full combinatorial design.

Proof Let us assume that the belts of the matroid $M = (S, F)$ of an (n, k) code N form a full combinatorial design. Since in a linear code the minimum weight of the nonzero vectors is d, the design consists of the set of all d-subsets of the n-set S. By Lemma 3.2 every $(n - d + 1)$-subset of S is spanning. Moreover, each is a dendroid, since no subset of smaller size can meet all the belts of M; thus $k = r(M) = n - d + 1$.

Conversely, if $k = n - d + 1$, then $n - d = k - 1$. Now let us assume that some d-subset T of S is not a belt of the matroid M. Then, if T' is the complementary subset of T in S, T' meets all subsets of S with cardinality greater than or equal to d with the exception of T, and in particular all circuits of the matroid M. Thus T' is a dendroid and rank $r(M) < k$, a contradiction. □

Corollary Any binary maximum-distance-separable code is trivial.

Proof Let C be a binary optimal (n, k) code. If $k = 0$, the code is trivial. We now assume $k > 0$. Let $M = (S, F)$ be the matroid of C. Then the belts of C are the d-subsets of S. If $d \neq n$, then the mod 2 sum of any two d-subsets X and Y that intersect in a $(d-1)$-set has weight two, and therefore $d \leqslant 2$. Hence $d = 1, 2$, or n, and $k = n, n - 1$, or 1, respectively. □

Let us say that a code C of distance d (over some alphabet that we may take to be a ring) is a matroid code if its set of minimal domains form a matroid M whose girth $\alpha(M) = d$. Then if the rank (or dimension) of the code is defined to be the rank of its matroid, we may say that a matroid code is optimal

3.6 Matroids, Graphs, and Coding

if and only if $r(M) = n - \alpha(M) + 1$. Then the above proposition is valid for matroid codes as well. A similar statement applies to many of the results that follow.

In the further study of the relationship between a code and its set of minimal domains the following concept is useful.

Definition A code C over a field K is *scalar complete to level l* if for every code chain f of weight l or less: (i) λf is also a code chain for each nonzero scalar λ of K, and (ii) for all $g \in C$ such that dom g = dom f, the relation $g = \lambda f$ is valid for some scalar λ of K.

Definition A code C over a field K is scalar complete if it is scalar complete to level j, where j is the weight of the codeword of least nonzero weight in the code. Clearly, if a code is scalar complete to level $l > j$, it is scalar complete. Evidently every binary (not necessarily linear) code on an n-set is scalar complete to level n and every linear code is scalar complete. This result can be strengthened as follows.

Lemma 6.1 Every linear code C of distance d over $GF(q)$ is scalar complete to level $u = u(q, d)$, where u is the largest integer satisfying

$$u - \left[\frac{u}{q-1}\right] < d$$

where $[x]$ denotes the least integer that is greater than or equal to x. In particular, every linear code is scalar complete.

Proof It is evident that if $f \in C$, then $\lambda f \in C$ for all scalars $\lambda \in GF(q)$. Moreover, if f and g are any pair of weight w (>0) vectors of C such that dom f = dom g, then there is a nonzero scalar λ of K such that f and $\lambda_0 g$ agree on at least $[w/(q-1)]$ members of dom f. Indeed, if we consider the average agreement $\bar{a}$ of f with the vectors $\{\lambda g : \lambda \in GF(q), \lambda \neq 0\}$, this average is $[w/(q-1)]$, and therefore the vector with greatest agreement agrees in at least $[w/(q-1)]$ places. Now, if $w \leq u(q, d)$, then $w(f - \lambda_0 g) < d$, which implies that $f - \lambda_0 g = 0$, that is, $f = \lambda_0 g$, which establishes the lemma. □

Theorem 6.1 (*Assmus and Mattson 1969*) Let A and B be linear dual (n, k) and $(n, n-k)$ codes over $GF(q)$ with minimum distances d and e, respectively, and which are scalar complete to levels $v_0 \geq d$ and $w_0 \geq e$, respectively. Let t be an integer less than d. Suppose that the number of nonzero weights of B that are less than or equal to $n - t$ is itself less than or equal to $d - t$. Then for each weight v with $d \leq v \leq v_0$ the domains of weight v of the code A form a t design, and for each weight w with $e \leq w \leq \{n - t, w_0\}$ the domains of weight w of code B form a t design.

Proof Let S be the common coordinate set of A and B. Choose a fixed value of t satisfying the conditions of the hypothesis. Let T be any t-subset of S. We consider the codes $A' = A \cdot (S - T) = A'(T)$, the reduction of A to $S - T$, and $B' = B \times (S - T) = B'(T)$, the contraction of B to $S - T$. By Lemma 5.1, A' and B' are dual codes. Moreover, by Lemma 3.2, every $(n - d + 1)$-subset of S is spanning; thus $S - T$ contains a dendroid of $M(A)$ that is also a dendroid of $M(A) \cdot (S - T) = M(A')$; therefore A' is an $(n - t, k)$ code and B' is an $(n - t, n - k - t)$ code. Let $0 < w_1 < w_2 < \cdots < w_{d-t} \leq n - t$ be the possible nonzero weights less than or equal to $n - t$ appearing on B. Let $W = \{w_1, w_2, \ldots, w_{d-t}\}$. Then the only nonzero weights of B' lie in W. The minimum weight in A is at least $d - t$. Employing the MacWilliams equations for B' and A' (cf. Section 1.5), the number x_j of vectors of weights v_j in B' is uniquely determined in terms of n, t, q, and k by the $d - t$ equations

$$\sum_{j \in W} \binom{n - t - j}{u} x_j = q^{n-t-k-u} \binom{n - t}{u} - \binom{n - t}{u}$$

$u = 0, 1, 2, \ldots, d - t - 1$, since it is readily verified that the determinant of the coefficient matrix is nonzero. Since the weight distribution alone of the code determines the distribution for the orthogonal code (MacWilliams, 1962; cf. Section 1.5) the weight distribution for A' is determined as well, and both are independent of the particular t-subset T chosen.

We use $D(C)$ to denote the set of domains of the code C. Now since B is scalar complete to level w_0, we see that the number of domains of weight $w \leq \min(n - t, w_0)$ in $D(B'(T))$ is independent of the particular subset chosen.

Let F_w be the set of domains of weight w in $D(B)$. Let $\bar{F}_w$ be the set of complements of the members of F_w in S. To show that F_w is a t design it is sufficient, by Proposition 2.1, Corollary 3, to show that $\bar{F}_w$ is a t design. But the number of members of $\bar{F}_w$ containing a given t-subset T of S is the number of members of F_w that exclude it, which is the number of domains of weight w in $D(B'(T))$, and this number is, as stated, independent of the choice of the t-subset T. This completes the part of the theorem relating to code B. The proof of the part of the theorem involving $D(A)$ is inductive. Since $d \leq v_0$ and d is the girth of $D(A)$, the number of domains of weight d in $D(A)$ that contain a given t-subset T of S is the number of domains in $D(A'(T))$ of weight $d - t$, which is independent of the choice of T since A is scalar complete to level d. Thus the domains of weight d in $D(A)$ form a t design. Proceeding by induction, we assume that for some v satisfying $d < v \leq v_0$, if v satisfies $d \leq v' < v$, then the domains of weight v' in $D(A)$ form a t design. Now for any t-subset T of S, the number of vectors of weight $v - t$ in $A'(T)$ is independent of T. Let F_v be the family of domains of weight v in $D(A)$. We need only show that the number of domains of weight $v - t$ in $D(A'(T))$ arising from members of F_v is independent of T. But all domains of weight $v - t$ in $D(A'(T))$ arise from domains of weight $\leq v$ in $D(A)$ and by Proposition 2.1, Corollary 2, and the induction hypothesis, the number arising from domains of weight $< v$ is independent of T. Since the number of vectors of weight $v - t$ in $A'(T)$ arising from members of F_v is the total number of vectors of weight $v - t$ in $A'(T)$ less those arising from vectors of weight $< v$ in A and this number is independent of T. By the scalar completeness condition, we see that the number of domains of weight $v - t$ in $D[A'(T)]$ arising from members of F_v is independent of T, which completes the theorem. □

Theorem 6.1 can be used to create new t designs by applying the theorem to codes for which the weight distribution is known. A catalog of several of these is given by Assmus and Mattson (1969), and the automorphism groups of some of the codes and designs discussed are given. This catalog gives nontrivial 5-designs that are not associated with the Mathieu groups, as well as an infinite family of 3-designs. See also Section 2.7 for the work of Pless (1972a).

3.7 Perfect Codes and t Designs

In this section we consider, without loss of generality, that every code is a set of functions from an n-set S to a ring R, since the alphabet A may be identified with a ring of order $|A|$, one of which, of course, exists for each finite cardinal A. In most cases R will be a finite field; however, this need not always be essential to the discussion. The set of all such functions we will refer to as the *entire module*.

Definition A code C over R of minimum distance $d = 2e + 1$ is *perfect* if every member of the entire module is within distance e of some member of the code.

Definition The set of members of the entire module that lie within distance r of a fixed member f is called a sphere of radius r about the center f.

We then see that a code of distance $2e + 1$ is perfect if and only if the set of disjoint spheres of radius e about the code functions exhausts the entire module.

If C is a code consisting of functions from an n-set into a ring of order q, then each sphere of radius r contains $\sum_{j=0}^{r} \binom{n}{j}(q - 1)^j$ members of the entire module. If the code is of distance $d = 2e + 1$, we immediately have the "sphere-packing bound"

$$|C| \leq q^n \bigg/ \sum_{j=0}^{e} \binom{n}{j}(q-1)^j$$

with equality if and only if C is perfect.

If q is a prime power, it is now known (Tietäväinen, 1973) that the only perfect q-ary codes are the following, as shown in Chapter 2:

(i) The binary repetition codes.
(ii) The perfect q-ary codes with $d = 3$ and $n = (q^r - 1)/(q - 1)$ for some integer r.
(iii) The perfect Golay binary code with $d = 7$ and $n = 23$.
(iv) The ternary Golay code with $d = 5$ and $n = 11$.

For future reference we note that in any code C of distance $2e + 1$ containing zero, functions f and g of weight $2e + 1$ on a set S into a ring R cannot agree on a subset X of S where $|X| \geq e + 1$, since $d(f, g) \leq w(f) + w(g) - 2|X|$, whenever f and g agree on X.

Theorem 7.1 (*Assmus and Mattson, 1967a*) If a scalar complete code C on a set S over $GF(q)$ and of minimum distance $d = 2e + 1$ is perfect and contains the zero chain, then the set D of weight d domains of C is a $(q - 1)^e$-$(e + 1, 2e + 1, n)$ configuration, where $|S| = n$.

Proof Let V be the set of all chains on S over $GF(q)$. Let X be any $(e + 1)$-subset of S. Then X is the domain of $(q - 1)^{e+1}$ chains of V. Let $V(X)$ denote the set of all such chains.

Now every chain f of $V(X)$ is within distance e of some uniquely determined code chain $x(f)$. Since the code C contains the zero chain, each nonzero code chain is of weight $2e + 1$ or greater, and we see that in order to be within distance e of $x(f)$, we must have dom $f \subset$ dom $x(f)$ and further f and $x(f)$ must agree on dom f. Thus it is evident that x is a one-to-one map of $V(X)$ onto the set of code chains of weight $2e + 1$. Let $D(X) = \{\text{dom } x(f): f \in V(X)\}$. Clearly, for any $Y \in D$, $X \subseteq Y$ if and only if $Y \in D(X)$. But since the code is scalar complete,

$$|D(X)| = \frac{|V(X)|}{q-1} = (q-1)^e$$

independent of the choice of $(e + 1)$-set X, which establishes the theorem. □

Corollary If a binary perfect code of distance $d = 2e + 1$ exists, then $\binom{n-i}{e+1-i}/\binom{2e+1-i}{e+1-i}$ is an integer for $i = 0, 1, 2, \ldots, e$. In particular, $(n + 1)/(e + 1)$ is an integer.

3.7 Perfect Codes and t Designs

Proof The result is immediate from Corollary 1 of Proposition 2.1, noting especially the case $i = e$. □

As mentioned in the previous chapter, this result rules out the existence of a binary (90, 78) perfect code with $d = 5$.

The reader might observe that if the code C is defined over an arbitrary ring, an appropriate condition can be chosen to replace scalar completeness. At present such generality seems to be unwarranted in any practical sense.

The theorem also follows directly from Theorem 6.1 in the case of linear codes, after appropriate analysis of the weight distribution in a perfect linear code or, equivalently (and in fact), that of its dual.

When the above theorem is applied to the perfect Hamming–Golay codes of distance three, the associated $(q - 1)$-$(2, 3, (q^m - 1)/(q - 1))$ configuration consists of the set of all collinear triples in the projective geometry $PG(m - 1, q)$ (see Chapter 2). When applied to the nonlinear binary Vasil'ev codes (see Chapter 2) the results yield an abundance of nonisomorphic Steiner triple systems (Assmus and Mattson, 1966c). The perfect Golay codes with $d = 7$ and 5, respectively, yield the Mathieu systems, $S(4, 7, 23)$ and $S(4, 5, 11)$ (see Section 3.2).

Theorem 7.1 has a partial converse. This is given in the following theorem.

Theorem 7.2 (*Assmus and Mattson, 1967a*) Let C be linear code of weight $2e + 1$ on a set S over $GF(q)$. If D, the set of belts of $M(C)$, forms a $(q - 1)^e$-$(e + 1, 2e + 1, n)$ design, then C is perfect.

Proof Let us assume that the above hypotheses are valid, but that C is not perfect. Let v be a chain of least weight which is not within distance e of any code chain. Since the zero chain belongs to C, we have $w(v) \geq e + 1$. Let X be an $(e + 1)$-subset of dom v. Then X is in $(q - 1)^e$ members of D, which will constitute a subset $D(X)$. Since C is scalar complete, there are $(q - 1)^{e+1}$ code vectors whose domains are in $D(X)$. By linearity, these vectors must all be distinct on X. Thus one of these chains, say a, agrees with v on X. Thus $0 < w(a - v) \leq w(v) - 1$. Thus by choice of v, $a - v$ is within distance e of some code chain b. But $d(v - a, b) \leq e$ implies that $w(v - (a + b)) \leq e$. But $a + b \in C$, which contradicts the existence of v and establishes the theorem. □

Corollary A perfect linear code is spanned by its minimum weight vectors.

Proof Let V be the subspace spanned by the minimum weight vectors of C. Then the belts of $M(V)$ are the belts of $M(C)$. Thus V is perfect. In order to exhaust the entire module, we see that $|V| = |C|$; hence $V = C$. □

The interested reader can also verify that if one makes a simple parity check extension of a perfect binary code that contains the zero chain, then the set of minimal weight domains of the code yields a Steiner system $S(e + 2, d + 1, n + 1)$.

3.8 Nearly Perfect Codes and t Designs

It has been seen that one can obtain t designs from perfect codes, that is, codes that satisfy the equality of the Hamming sphere-packing bound. It is possible to obtain t designs from those codes called *nearly perfect* [by Goethals and Snover (1972), whose work is presented in this section] that satisfy a specialized version of the Johnson bound (see Section 1.11). The specialized version of the bound is proved here, using the following notation. Given the vector space $V_n(q)$, let $S_r(v)$ denote the sphere of radius r (in the Hamming metric) with center v. Given a nonperfect code C of distance $2e + 1$, there is at least one vector of the space that is at distance greater than e from every code vector. Let $T(v)$ denote the set of vectors at distance $e + 1$ from the particular code vector v. The set $T(v)$ can be partitioned into two classes, namely

$$T_\alpha(v) = \{x \in T(v) \mid \exists u \in C, x \in S_e(u)\}$$

and

$$T_\beta(v) = T(v) - T_\alpha(v)$$

Lemma 8.1 Let C be a binary code of distance $d = 2e + 1$. Then for each $v \in C$,

$$|T_\alpha(v)| \leq \left\lfloor \frac{n-e}{e+1} \right\rfloor \binom{n}{e}$$

Proof Choose $x \in T_\alpha(v)$. Then $x \in S_e(u)$ for some $u \in C$, and $d(x, v) = e + 1$, $d(x, u) \leq e$, and $d(u, v) \geq 2e + 1$, which, together with the triangle inequality, yields $d(u, v) = 2e + 1$. This implies

$$|T_\alpha(v) \cap S_e(u)| = \binom{2e+1}{e+1}$$

since each vector y of this set must agree with v in all coordinate positions except for those $2e + 1$ in which u and v differ; further, y must agree with v in all but $e + 1$ of these. Consequently,

$$|T_\alpha(v)| = \binom{2e+1}{e+1} |N_{2e+1}(v)|$$

3.8 Nearly Perfect Codes and t Designs

where
$$N_{2e+1}(v) = \{x \mid x \in C, d(v, x) = 2e + 1\}$$

Since any two vectors of $N_{2e+1}(v)$ are at least distance $2e + 1$ from one another, the $2e + 1$ sets of coordinate places in which both differ from v can agree on at most an e-subset. Since there are at most $[(n - e)/(e + 1)]\,(2e + 1)$-subsets of an n-set intersecting in a given e-subset, then

$$|N_{2e+1}(v)| \leq \lfloor(n - e)/(e + 1)\rfloor \binom{n}{e} \Big/ \binom{2e+1}{e+1}$$

and

$$|T_\alpha(v)| \leq \lfloor(n - e)/(e + 1)\rfloor \binom{n}{e} \quad \square$$

Corollary For each $v \in C$
$$|T_\beta(v)| \geq \binom{n}{e+1} - \lfloor(n - e)/(e + 1)\rfloor \binom{n}{e}$$

Proof This is immediate from the fact that for given v
$$|T_\alpha(v)| + |T_\beta(v)| = \binom{n}{e+1} \quad \square$$

Theorem 8.1 (*Specialized Johnson Bound*) For any binary code C of length n and distance $d = 2e + 1$

$$|C|\left\{\left(\frac{1}{[n/(e+1)]}\right)\binom{n}{e}\left(\frac{n-e}{e+1} - \left\lfloor\frac{n-e}{e+1}\right\rfloor\right) + \sum_{j=0}^{e}\binom{n}{j}\right\} \leq 2^n$$

Proof There are at least $|\bigcup_{v \in C} T_\beta(v)|$ vectors of the space not contained in any $S_e(v)$, $v \in C$. A given vector can belong to at most $[n/(e + 1)]$ distinct sets $T_\beta(v)$, $v \in C$, and therefore by the preceding corollary

$$\left|\bigcup_{v \in C} T_\beta(v)\right| \geq \frac{|C|}{[n/(e+1)]}\left(\binom{n}{e+1} - \left\lfloor\frac{n-e}{e+1}\right\rfloor\binom{n}{e}\right)$$

which establishes the theorem. $\square$

Notice that this is a version of Theorem 11.3 of Chapter 1 in the form we require.

Definition A binary code that satisfies the equality of the specialized Johnson inequality is called *nearly perfect*.

Note that in a nearly e-error-correcting perfect code any vector is at most a distance $e + 1$ from a codeword, that is, nearly perfect codes are also quasi-perfect (see Section 1.5). Examining the requirements for equality in the overall proof of the specialized inequality shows that in any nearly perfect e-error-correcting binary code of length n any vector at a distance greater than e from every code vector is at distance $e + 1$ from exactly $\lfloor n/(e + 1) \rfloor$ code vectors; any vector at distance e from a given code vector is at distance $e + 1$ from exactly $\lfloor (n - e)/(e + 1) \rfloor$ other code vectors.

Theorem 8.2 In any nearly perfect e-error-correcting code C of length n the code vectors at distance $d = 2e + 1$ from a given code vector v determine a t design with parameters $\lfloor (n - e)/(e + 1) \rfloor$-$(e, d, n)$.

Proof Let S be the set of coordinate positions of the vector space of the code. Let $N_d(v)$ denote the set of code vectors at a distance d from v. Any vector u of $N_d(v)$ determines a d-subset of S, namely those coordinate positions in which u and v differ. But it has been noted that any e-subset of S is contained in precisely $\lfloor (n - e)/(e + 1) \rfloor$ such d subsets. □

Theorem 8.3 In any nearly perfect e-error-correcting code of length n the set $T_\beta(v)$ of vectors at a distance $e + 1$ from a code vector v and at a distance greater than e from every other code vector determine a t design.

Proof Let A denote the t design of Theorem 8.2. Let D denote the t design obtained from A by replacing each block β of A by the set of all $(e + 1)$-subsets of β. The design D has parameters λ^*-$(e, e + 1, n)$, where $\lambda^* = (e + 1) \lfloor (n - e)/(e + 1) \rfloor$.

Let C denote the trivial e design of all $(e + 1)$-subsets of S, the set of coordinate positions. It has parameters $(n - e)$-$(e, e + 1, n)$.

Now consider the set B of all $(e + 1)$-subsets of coordinate positions for which there exists an element of $T_\beta(v)$ that differs from v in this set of coordinate positions. These $(e + 1)$-subsets are precisely those that are not contained in the d-subsets of the design A; hence this set B of $(e + 1)$-subsets consists of the set of blocks of C with the blocks of D removed. Thus every e-subset occurs $(n - e) - (e + 1) \lfloor (n - e)/(e + 1) \rfloor$ times in the blocks of B, and B is the required design. □

Note that if $n - e \equiv 0 \mod(e + 1)$ in the preceding theorem, the specialized Johnson bound becomes the sphere-packing bound, and the nearly perfect code is perfect. If a is the least positive residue of $(n + 1) \mod(e + 1)$, any nearly perfect code with $n \equiv -1 \mod(e + 1)$ is perfect. However, it was shown in the corollary to Theorem 7.1 that in a perfect code n must be congruent to $-1 \mod(e + 1)$, so that this result is not surprising.

3.9 Balanced Codes and t Designs

Moreover, for $a = 0$ the design of Theorem 8.2 is an $e + 1$ design with $\lambda_{e+1} = 1$, namely the Steiner system of Theorem 7.2.

Again a simple parity check extension of a nearly perfect e-error-correcting code of length n gives rise to an $e + 1$ design on an $(n + 1)$-set with parameters $[(n - e)/(e + 1)]$-$(e + 1, 2e + 2, n + 1)$. (Consider the vectors at a distance $2e + 2$ from a given vector v, and the coordinates in which they differ.)

This in turn implies (cf. Proposition 2.2 and corollary) that in any nearly perfect e-error-correcting code of length n the vectors at a distance $2e + 2$ from a given vector v determine a t design with parameters λ^*-$(e, 2e + 2, n)$ with

$$\lambda^* = \left\lfloor \frac{n-e}{e+1} \right\rfloor \frac{n - 2e - 1}{e + 2}$$

Nearly perfect codes were also studied by Semakov et al. (1971) (under the name uniformly packed codes) from a different point of view, but similar results were obtained.

3.9 Balanced Codes and t Designs

A binary code is *balanced* if each word has the same weight r, where $r < n$, the dimension of the code space. It is clear that the incidence matrix (see Chapter 2) of a t design can be interpreted as a balanced code in two ways: One can use either the rows or columns of the matrix as code vectors. If the incidence vectors corresponding to varieties are taken as codewords, then there is the advantage that the distance between codewords is constant and easily calculated. Indeed, given any two binary vectors of length n, then

$$d(u, v) = w(u) + w(v) - 2(u, v) \tag{9.1}$$

where (u, v) denotes the real dot product or *correlation* of u and v. If u and v are vectors of an incidence matrix corresponding to two distinct varieties of a t design, then $w(u) = w(v) = \lambda_1$ and $(u, v) = \lambda_2$ and $d(u, v) = 2(\lambda_1 - \lambda_2)$. Following standard practice, λ_1 will henceforth be denoted by r and λ_2 by λ, if there is no danger of ambiguity. A code is *equidistant* if the distance between any pair of distinct code vectors is constant. Thus the vectors corresponding to the varieties of the incidence matrix of a t design form a balanced equidistant code. This section examines certain conditions that guarantee that the code vectors form the incidence matrix of a t design.

Let $C(n, r, \lambda)$ denote a balanced binary code of length n in which every word has length r and the maximum correlation λ. The following theorems are given by Semakov and Zinov'ev (1969a, b).

Theorem 9.1 (*Semakov and Zinov'ev, 1969a, b*) The number a of words in the balanced code $C = C(n, r, \lambda)$ with $\lambda < r^2/n$ satisfies $a \leq n(r - \lambda)/(r^2 - n\lambda)$ with equality if and only if the vectors of C correspond to the incidence vectors of the varieties of a balanced incomplete block design (cf. Section 2.3) with parameters

$$\left(\frac{n(r - \lambda)}{r^2 - n\lambda}, n, r, \frac{r(r - \lambda)}{r^2 - n\lambda}, \lambda\right)$$

Proof Assume that such a balanced incomplete block design exists. (*Note:* the parameters satisfy basic relations on the parameters of a balanced incomplete block design.) Clearly the incidence vectors of its varieties form the required code.

Conversely, assume that such a code exists.

Let the code vectors be denoted by $v_1, v_2, \ldots, v_a$. By hypothesis $(v_i, v_j) \leq \lambda$ for $i \neq j$. Hence

$$\sum_{i \neq j} (v_i, v_j) \leq \lambda a(a - 1) \tag{9.2}$$

Now consider the matrix

$$M = \begin{pmatrix} v_1 \\ v_2 \\ \vdots \\ v_a \end{pmatrix}$$

The sum $S = \sum_{i \neq j} (v_i, v_j)$ over the rows of M can in turn be computed along the columns. Let k_s denote the number of ones in the sth column of M. Then

$$S = \sum_{s=1}^{n} k_s(k_s - 1) = \sum_{s=1}^{n} k_s^2 - \sum_{s=1}^{n} k_s = \sum_{s=1}^{n} k_s^2 - ar$$

However,

$$\sum_{s=1}^{n} k_s^2 \geq \frac{1}{n}\left(\sum_{s=1}^{n} k_s\right)^2 = \frac{(ar)^2}{n}$$

with equality only if all k_s are equal. This yields

$$S \geq ar(ar - n)/n$$

This, with (9.2) and the fact that $\lambda < r^2/n$, yields

$$a \leq n(r - \lambda)/(r^2 - n\lambda)$$

But if $a = n(r - \lambda)/(r^2 - n\lambda)$, then all k_s must be equal and v_i, v_j must equal λ for all distinct subscripts i, j, $i, j = 1, 2, \ldots, a$. Thus the matrix M can be

3.9 Balanced Codes and t Designs

considered as an incidence matrix for a balanced incomplete block design with the parameters given. □

The case of $C(n, r, \lambda)$ codes with $\lambda = r^2/n$ is also of interest. Before dealing with this case a lemma is required.

Lemma 9.1 If in a balanced code $\lambda = r^2/n$, then $\lambda - 1 < (r - 1)^2/(n - 1)$.

Proof Consider $(n - \lambda)^2$. Since in a balanced code $n > \lambda$, we have

$$0 < (n - \lambda)^2 = n^2 + \lambda^2 - 2n\lambda = n^2 + \lambda^2 - 2r^2$$

then

$$4r^2 < n^2 + \lambda^2 + 2n\lambda$$

and thus

$$2r < n + \lambda$$

Hence

$$0 = r^2 - n\lambda > 2r - n - \lambda$$

and

$$(n - 1)(\lambda - 1) < (r - 1)^2 \quad \square$$

Theorem 9.2 (*Semakov and Zinov'ev, 1969a, b*) The number a of words in the balanced code $C = C(n, r, \lambda)$ with $\lambda = r^2/n$ satisfies $a \leq n(n - 1)/(n - r)$ with equality if and only if the code vectors of C correspond to the blocks of an affine-resolvable, balanced, incomplete block design with parameters

$$\left(n, \frac{n(n - 1)}{n - r}, \frac{r(n - 1)}{n - r}, r, \frac{r(r - 1)}{n - r}\right)$$

Proof Assume that such a balanced incomplete block design exists. Since the design is resolvable and any pair of blocks intersects in either 0 or $\lambda = r^2/n$ varieties (see Chapter 2), the incidence vectors corresponding to the blocks of the design therefore have correlation either 0 or λ, and each such vector has weight r. This guarantees the existence of a code C with the required number of words. Conversely, let C be any code $C(n, r, \lambda)$ with a words. Let the codewords be $v_1, v_2, \ldots, v_a$. Consider the matrix

$$M = \begin{pmatrix} v_1 \\ v_2 \\ \vdots \\ v_a \end{pmatrix}$$

Then M has the property that no column of M contains more than $R = r(n-1)/(n-r)$ ones. Indeed, suppose that the sth column contains R_s ones. Consider the set S of all codewords with a one in the sth column. Deleting this coordinate position from S, we obtain a code $C_s(n-1, r-1, \lambda-1)$. But by Lemma 9.1, $\lambda - 1 < (r-1)^2/(n-1)$. Hence by Theorem 9.1, this code has at most $(n-1)(r-\lambda)/[(r-1)^2 - (n-1)(\lambda-1)]$ code vectors. Simplification yields

$$R_s \leq (n-1)(r-\lambda)/(n+\lambda-2r) = (n-1)r/(n-r)$$

Now since the code is balanced,

$$a \leq nR/r = n(n-1)/(n-r)$$

with equality if and only if each column of M contains exactly $(n-1)r/(n-r)$ ones.

Now consider the case of equality. In this case the code $C_s(n-1, r-1, \lambda-1)$ for each column s has vectors corresponding to the incidence vectors of varieties of a balanced incomplete block design, with blocks of size $r(r-1)/(n-r)$. In terms of M this states that if we consider the set of rows with ones in the sth column, then each of the remaining columns of this portion on M contains exactly $r(r-1)/(n-r)$ ones, that is, the scalar product of any two rows is $r(r-1)/(n-r)$. If M is viewed as the incidence matrix of a configuration whose blocks correspond to rows and whose varieties correspond to columns, every pair of varieties occurs in exactly $r(r-1)/(n-r)$ blocks.

Now since every row contains exactly r ones, the configuration is a balanced incomplete block design with the given parameters. To see that the design is affine resolvable, we note that by Theorem 9.1 the rows of $C_s(n-1, r-1, \lambda-1)$ have correlation $\lambda - 1$. Hence any two blocks containing s intersect in exactly $\lambda = r^2/n$ varieties. Hence any pair of blocks is either disjoint or has intersection r^2/n, and by the results of Section 2.3 the design is affine resolvable. $\square$

3.10 Equidistant Codes

As mentioned earlier, an equidistant code is a binary code in which the distance between any pair of distinct codewords is independent of the pair chosen. Equidistant codes are closely related to balanced codes, as the following lemma shows.

Lemma 10.1 Let A be an equidistant code of dimension n and distance d with at least two words. Then there exists a balanced equidistant code $C(n, d, d/2)$ with $|A| - 1$ words.

3.10 Equidistant Codes

Proof Let the code vectors be $v_1, v_2, \ldots, v_a$, where $|A| = a$. Consider the matrix

$$M = \begin{pmatrix} v_1 \\ v_2 \\ \vdots \\ v_a \end{pmatrix}$$

If in any column the roles of zero and one are interchanged, the distance between codewords remains unchanged. Thus by such an operation the matrix M can be transformed into another $a \times n$ matrix M^* whose first row is zero and which is such that the distance between any two rows of M^* is d. Let M_1^* be the matrix derived from M^* by removing the first row. Then every row of M_1^* has d ones. Moreover, if u and v are any pair of distinct columns of M_1^*, then since $d(u, v) = w(u) + w(v) - 2(u, v)$, we see that the correlation $\lambda = (u, v) = d/2$. □

Corollary In an equidistant code with two or more words d is always even.

Theorem 10.1 (*Bose and Shrikhande, 1959b*) Let A be a binary code of dimension $4t - 1$ and distance $2t$. Then $|A| \leq 4t$, with equality if and only if there exists a balanced incomplete block design with parameters

$$(4t - 1, 4t - 1, 2t, 2t, t).$$

Proof Let $\bar{d}$ denote the average distance between distinct words of A. Let $|A| = a$, and let the codewords be $v_1, v_2, \ldots, v_a$. Consider the matrix

$$M = \begin{pmatrix} v_1 \\ v_2 \\ \vdots \\ v_a \end{pmatrix}$$

Then

$$\bar{d} = 2 \sum_{i>j} d(v_i, v_j)/a(a-1)$$

But $\bar{d} \geq d$ with equality if and only if the code is equidistant. Moreover, if x_s denotes the number of ones in the sth column of M,

$$\sum_{i>j} d(v_i, v_j) = \sum_{s=1}^{4t-1} x_s(a - x_s)$$

But $x_s(a - x_s) \leq a^2/4$, with equality if and only if $x_s = a/2$. Hence

$$2t = d \leq \bar{d} \leq 2\left(\sum_{s=1}^{4t-1} x_s(a - x_s)\right)/a(a-1) \leq (4t-1)a/2(a-1)$$

whence $a \leq 4t$.

Now consider the case of equality. Since $\bar{d} = d$, the code is equidistant. Moreover, since each $x_s = a/2$, every column of M contains $2t$ ones. Applying the construction of Lemma 10.1 to the matrix M, one obtains a matrix M_1^* from M in which each of the columns contains exactly $2t$ ones. Moreover, the matrix M_1^* has correlation of t between its columns. Thus the columns of M_1^* are the incidence vectors of a balanced incomplete block design with the given parameters. □

Let C be an equidistant code of dimension n. Let $M = M(C)$ denote the $|C| \times n$ matrix whose rows are the vectors of C. Throughout this section let x_s, $s = 1, 2, \ldots, n$, denote the number of ones in the sth column of M. The code C is said to be trivial if, for $s = 1, 2, \ldots, n$, $x_s \in \{0, 1, |C|, |C| - 1\}$. Otherwise C is nontrivial.

Deza (1973) has shown that in a nontrivial equidistant code C of distance $2t$, $|C| \leq t^2 + t + 2$, with equality only if there exists a projective plane of order t. The following lemmas will be employed in the proof.

Lemma 10.2 Let α, β, α_1, and β_1 be any four real numbers, where $\alpha\beta \geq 0$. Then

$$F = \alpha_1(\beta - \beta_1) + \beta_1(\alpha - \alpha_1) - \frac{\alpha\beta_1(\beta - \beta_1)}{\beta} - \frac{\beta\alpha_1(\alpha - \alpha_1)}{\alpha} \geq 0$$

Proof The expression F simplifies to $\alpha\beta[(\alpha_1/\alpha) - (\beta_1/\beta)]^2$, which is clearly nonnegative. □

Lemma 10.3 Let C be a binary code of length n. Let E and F be nonempty subsets of C such that $E \cup F = X$ and $E \cap F = \emptyset$. Then

$$\sum_{\substack{x \in E \\ y \in F}} d(x, y) \geq \frac{|F|}{2|E|} \sum_{x, y \in E} d(x, y) + \frac{|E|}{2|F|} \sum_{x, y \in F} d(x, y)$$

Proof Consider the matrix $M = M(C)$. Since the Hamming distance is additive on the columns of M, it is sufficient to prove the result for a column of M. Let us assume without loss of generality that the vectors of E are the first e rows of M, and those of F the remaining $m - e = f$. Let us assume that column c contains e_1 ones in the first e rows and f_1 ones in the remaining f columns. Let $d^c(x, y)$ denote the Hamming distance between x and y in column c.

3.10 Equidistant Codes

Now

$$\sum_{\substack{x \in E \\ y \in F}} d^c(x, y) - \frac{|F|}{2|E|} \sum_{x, y \in F} d^c(x, y) - \frac{|E|}{2|F|} \sum_{x, y \in F} d^c(x, y)$$

$$= e_1(f - f_1) + f_1(e - e_1) - \frac{fe_1(e - e_1)}{e} - \frac{ef_1(f - f_1)}{f}$$

which by Lemma 10.2 is nonnegative. This is sufficient to establish the lemma. □

The following notation will be used in the next lemma. If C is a binary code of length n containing m words and $M = M(C)$, let x_s denote the number of ones in the sth column of M, and let $t_s = \min(x_s, m - x_s)$. Further, let $a = \min d(x, y)$ and $b = \max d(x, y)$, where (x, y) ranges over all pairs of distinct codewords, and T_s denotes the number of columns of M identical to the sth column.

Lemma 10.4 Let C be a binary code of length n containing m codewords. If the sth column of $M(C)$ is such that $t_s \neq 0$, then

$$T_s \leq b - a + \frac{am}{2t_s(m - t_s)}$$

Proof Let M^* denote the matrix derived from M by removing the T_s columns that are identical with the sth. Let E be the set of rows of M^* in which the sth column contains a one, and F the complementary set. Let v^* denote the row of M^* corresponding to the row v of M, and let d^* denote the Hamming metric defined on the rows of M^*. Note that if $x^* \in E$ and $y^* \in F$, then

$$d^*(x^*, y^*) = d(x, y) - T_s \leq b - T_s$$

Now if x^* and y^* are distinct members of E, then $d^*(x^*, y^*) = d(x, y) \geq a$. A similar statement holds in F. Applying Lemma 10.3, one obtains

$$t_s(m - t_s)(b - T_s) \geq \sum_{\substack{x^* \in E \\ y^* \in F}} d^*(x^*, y^*)$$

$$\geq \frac{|E|}{2|F|} \sum_{x^*, y^* \in F} d^*(x^*, y^*) + \frac{|F|}{2|E|} \sum_{x^*, y^* \in E} d^*(x^*, y^*)$$

$$\geq \tfrac{1}{2}|E|(|F| - 1)a + \tfrac{1}{2}|F|(|E| - 1)a$$

$$= t_s(m - t_s)a - \tfrac{1}{2}ma$$

which yields the lemma. □

Corollary 1 If $T_s > (b - a) + (a/2t_s)$ and $t_s \neq 0$, then

$$m \leq \frac{t_s^2}{[(t_s - a)/(2(T_s - b + a))]}$$

Proof By the theorem

$$2t_s(m - t_s)(T_s - b + a) \leq am$$

which reduces to the stated inequality. □

Corollary 2 Let C be an equidistant code of length n of distance $2k$ containing m words. Then $t_s(m - t_s) \leq mk$.

Proof The corollary is clearly true if $t_s = 0$. If $t_s > 0$, the result follows from the lemma since $T_s \geq 1$. □

Lemma 10.5 Let C be an equidistant code of length n of distance $2k$ with m words. If $t_s \leq [(m + 2k - 1)/(k + 1)] - 1$ for all s satisfying $1 \leq s \leq n$, then $t_s \leq 1$ for all such s.

Proof If $m = 1$, the result is trivial. If $m > 1$, apply Lemma 10.1 to obtain a new equidistant code C^* which contains the zero codeword but for which t_s remains unchanged. The remaining codewords form a balanced code $C(n, 2k, k)$. Let $[(m + 2k - 1)/(k + 1)] = t$. Then (i) $t_s < t$, for $s = 1, 2, \ldots, n$, and (ii) $m \geq k(t - 2) + t + 1$. Partition the columns of $M = M(C^*)$ into two types, those of type I containing $t - 1$ or fewer ones and those of type II containing $m - t + 1$ or more ones. Since $t_s \leq t$, every column is of type I or type II. Since $m \geq 2t - 1$, the two sets are disjoint. To prove the lemma, it is sufficient to demonstrate that (a) the number of columns of type II does not exceed k, and that every word other than the zero word has at least k ones in the columns of type II. (Recall that the correlation in the balanced part of the code is k.) Assume that there are at least $k + 1$ columns of type II. Each such column contains $t - 2$ or fewer zeros in the nonzero rows. By (ii) $(m - 1) - (k + 1)(t - 2) \geq 2$. Therefore there are at least two (nonzero) codewords that have no zeros in their columns of type II, and these two words have a correlation of at least $k + 1$, a contradiction; hence there are at most k columns of type II. Now let u be a nonzero codeword of C^*. Let q denote the number of ones of u that lie in the columns of type II. Now $2k - q$ columns of type I contain ones in u. Consider the submatrix M' of M obtained from M by removing the row u and all columns of M that contain a zero in the row u.

3.10 Equidistant Codes

The matrix M' contains $q(m-2)$ ones in the columns that were of type II and no more than $(2k-q)(t-2)$ ones in the columns that were of type I. Since the correlation between nonzero codewords is k, the matrix M' contains $k(m-2)$ ones. Hence

$$(2k-q)(t-2) + q(m-2) \geq k(m-2), \qquad q \geq k - \frac{k(t-2)}{m-t}$$

But by (ii), $k(t-2) \leq m-t$, and since q is an integer, $q \geq k$. □

Recall that an equidistant code is trivial if and only if $t_s \leq 1$, for $s = 1, 2, \ldots, n$. Such trivial codes are easily constructed. Let J_{ms} denote the $m \times s$ matrix of ones and 0_{ms} the $m \times s$ matrix of all zeros. Let I_m denote the identity matrix of order M_1 and $K_m = J_{mm} - I_m$. Then the rows of the matrix

$$J_{ms} I_m \cdots I_m K_m \cdots K_m 0_{mt}$$

will always form a trivial equidistant code. In particular given any pair of parameters m and $2k$, one can always form a trivial equidistant code with distance $2k$ and m words by forming the matrix

$$I_m I_m \cdots I_m$$

where I_m occurs k times.

Theorem 10.2 Let C be an equidistant code of distance $2k$ which contains m words. If $m \geq k^2 + k + 3$, then the code is trivial.

Proof Note that if $m \geq k^2 + k + 3$, then $t_s \leq k+1$ for $s = 1, 2, \ldots, n$. Indeed, if $t_s \geq k+2$ and $m \geq k^2 + k + 3$, then

$$t_s(m - t_s) - mk \geq (k+2)(k^2+1) - k(k^2+k+3) > 0$$

a contradiction of Corollary 2 of Lemma 10.4. Also

$$[(m+2k-1)/(k+1)] - 1 \geq k+1 \qquad \text{for} \qquad m \geq k^2 + k + 3$$

Hence $t_s \leq 1$, and the code is trivial. □

Clearly, if there is a projective plane π of order k (cf. Chapter 2), then there exists an equidistant code of distance $2k$ with $k^2 + k + 2$ words. To construct such a code, one can proceed as follows. Let P be the incidence matrix of π. Form a matrix M by adding $k-1$ columns of ones to P and adjoin a row of zeros to the resulting matrix. The rows of this matrix produce the required code.

Van Lint (1973) has proved the converse theorem, namely, if there exists an equidistant code of distance $2k$ with $k^2 + k + 2$ codewords ($k > 1$), then there is a projective plane of order k. Before proving this result, let us dispose of the case $k = 1$. The code consisting of the rows of

$$\begin{pmatrix} 0 & 0 & 0 \\ 1 & 1 & 0 \\ 1 & 0 & 1 \\ 0 & 1 & 1 \end{pmatrix}$$

forms a nontrivial code of distance two and with four words. Having disposed of this situation, henceforth it will be assumed that $k > 1$.

Theorem 10.3 For $k > 1$ there exists a nontrivial equidistant code of distance $2k$ with $k^2 + k + 2$ words if and only if there exists a projective plane of order k.

Proof In view of the preceding comments, it is only necessary to show that the existence of such a code guarantees the existence of a projective plane of order k. Let $M' = M(C)$. Without loss of generality we may assume that M' has no columns of zeros only or ones only since such columns contribute nothing to distances between words. Applying the transformation of Lemma 10.1 to M', one obtains a matrix M whose first row is zero. Considering the sth column of M, by Lemma 10.4, Corollary 2 one obtains

$$t_s(m - t_s) \leq mk \tag{10.1}$$

There are two cases to consider. If $k = 2$, then there exists the projective plane $PG(2, 2)$. Let A denote the incidence matrix of its complementary design, a block design with parameters $(7, 7, 4, 4, 2)$. By adjoining a row of zeros to A we obtain a matrix whose rows are an equidistant code of distance four with eight code vectors. Hence the theorem is true for $k = 2$.

For $k > 2$ note that since the left side of $t_s(m - t_s)$ is quadratic, and if $t_s \geq k + 2$, recalling that $t_s \leq m/2$, it follows that $t_s(m - t_s) \geq k^3 + 2k^2$. But $mk = k^3 + k^2 + 2k$. Since $k > 2$, then (10.1) is violated. Hence for $k > 2$ the inequality $t_s \leq k + 1 < k^2 + 1$ is satisfied.

Now split the columns of M into two classes, namely those columns with at most $k + 1$ ones (type I) and those with at least $k^2 + 1$ ones (type II). These two classes are disjoint. Assume that the columns of type II precede the columns of type I.

Case (a) Suppose there are at least $k + 1$ columns of type II. Since all rows of C except the first one have weight $2k$ and mutual distance $2k$, there can be at most one row which has ones in $k + 1$ fixed columns of type II. However, these $k + 1$ columns of type II contain a total of at least

3.10 Equidistant Codes

$(k + 1)(k^2 + 1)$ ones and this is possible only if, besides the zero row, there is indeed one row with $k + 1$ ones in these $k + 1$ columns of type II and furthermore all other rows have k ones in these $k + 1$ columns. This means that after a suitable permutation M has the following form:

$$M = \begin{pmatrix} 0 & 0 & \cdots & 0 & 0 & 0 & \cdots & 0 & 0 & 0 & \cdots & 0 \\ & E^{(1)}_{k,\,k+1} & & & F^{(1)}_{k,\,k} & F^{(2)}_{k,\,k} & \cdots & F^{(k)}_{k,\,k} & & & & \\ & E^{(2)}_{k,\,k+1} & & & & & & & & & & \\ & \vdots & & & & M^* & & & & 0_{k^2+k,\,k-1} & & \\ & E^{(k+1)}_{k,\,k+1} & & & & & & & & & & \\ 1 & 1 & \cdots & 1 & 0 & 0 & \cdots & 0 & 1 & 1 & \cdots & 1 \end{pmatrix}$$

where 0_{pq} and J_{pq} are as defined previously, $E^{(i)}_{pq}$ is the $p \times q$ matrix with zeros in column i and ones elsewhere, and $F^{(i)}_{pq}$ is the $p \times q$ matrix with ones in row i and zeros elsewhere.

We have the following information about the rows of M^*:

(i) Each row of M^* has k ones, one below each of the matrices $F^{(i)}_{k,\,k}$, $i = 1, \ldots, k$.

(ii) If two rows of M^* follow $E^{(i)}_{k,\,k+1}$, resp. $E^{(j)}_{k,\,k+1}$, $i \neq j$, then these rows have one one in common; if they follow the same $E^{(i)}_{k,\,k+1}$, then they have no one in common.

From (i) and (ii) we see that M^* is the incidence matrix of a *transversal system* $T_k(k, k)$ [cf. Hall (1967a, p. 224)]. The existence of such a transversal system is equivalent to the existence of an *orthogonal array* $OA(k, k + 1)$ (Hall, 1967a, p. 224), which implies the existence of a $PG(2, k)$ [cf. Hall 1967a, p. 191)].

Case (b) In the remaining part of the proof we may assume that in M there are at most k columns of type II. Now, consider any nonzero row of M without loss of generality the second row, and let this row have q ones in the columns of type II. Then M has the following form:

$$\begin{pmatrix} \overbrace{\begin{matrix} 0 & 0 & \cdots & 0 \\ 0 & 0 & \cdots & 0 \end{matrix}}^{\text{type II}} & \overbrace{\begin{matrix} 0 & 0 & \cdots & 0 \\ 1 & 1 & \cdots & 1 \end{matrix}} & \overbrace{\begin{matrix} 0 & 0 & \cdots & 0 \\ 1 & 1 & \cdots & 1 \end{matrix}}^{\text{type I}} & \overbrace{\begin{matrix} 0 & 0 & \cdots & 0 \\ 0 & 0 & \cdots & 0 \end{matrix}} \\ C_1 & C_2 & C_3 & C_4 \end{pmatrix}$$

where C_2 is $k^2 + k$ by q and C_3 is $k^2 + k$ by $2k - q$. We count the total number of ones in C_2 and C_3 in two ways, using the fact that mutual distances are $2k$ and the definitions of types I and II. This yields

$$k(k^2 + k) \leq q(k^2 + k) + (2k - q)k$$

i.e.,
$$q \geq k - 1 \tag{10.2}$$

Now assume there are k columns of type II. Now C must have the following form:

$$\begin{pmatrix} 0 & 0 & \cdots & 0 & 0 & 0 & \cdots & 0 & 0 & 0 & \cdots & 0 \\ \hline & E^{(1)}_{m_1, k} & & & & & & & & & & \\ & E^{(2)}_{m_2, k} & & & & & & & & & & \\ & \vdots & & & & & & & & & & \\ & E^{(k)}_{m_k, k} & & & & & & & & & & \\ \hline & J_{\bar{m}, k} & & & F^{(1)}_{\bar{m}, k} & F^{(2)}_{\bar{m}, k} & \cdots & F^{(\bar{m})}_{\bar{m}, k} & & & & \end{pmatrix}$$

Clearly $m_i \leq k$, $i = 1, \ldots, k$. If $m_1 = m_2 = \cdots = m_k = 0$, then the code is trivial. Hence $m_i \neq 0$ for some i. Each of the $m_1 + m_2 + \cdots + m_k$ rows that has a zero in a type II column must have exactly one one above each of the matrices $F^{(i)}_{\bar{m}, k}$, $i = 1, \ldots, \bar{m}$. Hence $\bar{m} \leq k + 1$. This implies $m_1 = m_2 = \cdots = m_k = k$, $\bar{m} = k + 1$. We now count the number of ones in M that are below the ones of the second row in two ways and find

$$k(k^2 + k) \leq (k - 1)^2 + (k + 1)k$$

i.e., $k \leq 1$, a contradiction.

This contradiction combined with (10.2) shows that it remains to consider the possibility that M has the form

$$C = \begin{pmatrix} 0 & 0 & \cdots & 0 & 0 & 0 & \cdots & 0 \\ J_{k^2 + k + 1, k - 1} & & & & C_1 & & & \end{pmatrix}$$

where the columns of C_1 are all of type I. We have the following information about C_1:

(i) Each row of C_1 has exactly $k + 1$ ones.
(ii) Two distinct rows of C_1 have exactly one one in common.
(iii) Every column of C_1 has at most $k + 1$ ones, hence exactly $k + 1$ ones.

These are precisely the conditions required for C to be the incidence matrix of a projective plane. □

Vanstone (1974) has shown that any nontrivial balanced equidistant code of weight $k + 1$ and mutual distance $2k$ has at most $k^2 + k + 1$ words. He has also found nontrivial equidistant codes for $k = 2$ with eight words, which do not have the form of those given for $k > 2$ in the preceding theorem.

3.11 Comments

From the results of this chapter it is obvious that the ties between various combinatorial structures and coding theory are very close and deep. The arguments leading to the construction of certain designs often parallel those used to construct good codes. The matroidal approach to coding taken here was in keeping with our aim to provide as broad a mathematical background as was reasonable. Most of the results contained in this chapter were not originally derived with the use of matroids. However, their application to problems of coding is interesting and provides an alternative point of view.

Exercises

1 A regular t-wise balanced design is a generalization of a t design. Such a design consists of a finite set V of elements called varieties together with a family of subsets of V called blocks, which satisfies the following axiom:

For every $i \leq t$ every i-subset of V occurs in precisely λ_i blocks of the design where λ_i is independent of the i-subset chosen.

Show that if D is a regular $2t$-wise balanced design, then the design D' formed by taking the blocks of D together with their complements is a regular $(2t + 1)$-wise balanced design.

2 If there exists a λ_1-(t, x, v) design and a λ_2-(t, k, x) design, show that there exists a $\lambda_1 \lambda_2$-(t, k, v) design.

3 Show that $(m - 1)$-$(3, 2m, 4m)$ designs are invariant under complementation.

4 Let M be a matroid defined on a set E. A *separator* of M is a subset S of E such that each circuit of M is contained either in S or $E - S$.

(i) Show that the set of separators of a matroid M is closed with respect to union, intersection, and complementation.

(ii) Show that S is a separator of M if and only if $M \cdot S = M \times S$.

(iii) Let N be a chain group over the integers. A chain is *primitive* if it is elementary and its coefficients are restricted to the values 1, -1, and 0. N is regular if every elementary chain is an (integral) multiple of a primitive chain. Show that if N is regular, then a subset S of E is a separator of $M(N)$ if and only if each chain f of N can be written in the form $g + h$, where g and h are chains of N such that dom $g \subseteq S$ and dom $h \subseteq E - S$.

5   A kth-order Steiner system can be defined as follows: Let D_i be a collection of nonempty i-sets of S, $i = 3, 4, \ldots, k$; a free t-set of S is one not containing any member of $D_3 \cup D_4 \cup \cdots \cup D_t$; for $t = 2, 3, \ldots, k - 1$, every free t-set is contained in precisely one member of D_{t+1} and every proper subset of every element of D_{t+1} is free. The system is closed if and only if there are no free k-sets.

Show that if $n = 2^{k-1} - 1$, then there is a closed kth-order Steiner system on n points. Can this be generalized?

6   Let D be an affine-resolvable, balanced, incomplete block design (ARBIBD) (Shrikhande (1953)).

(i) Show that the parameters can always be written in the form
$$v = nk = n^2[(n-t)t + 1], \qquad b = nr = n[n^2t + n + 1], \qquad \lambda = nt + 1$$

(ii) Show that no ARBIBD with these parameters exists when n and t are odd and (a) k is not a perfect square or (b) k is a perfect square and $nt \equiv 1 \bmod 4$ and the square-free part of n contains a prime $\equiv 3 \bmod 4$.

(iii) Show that no ARBIBD with these parameters exists when n is odd and t is even and (a) $(n-1)t + 1$ is not a perfect square, or (b) $(n-1)t + 1$ is a perfect square and $n + t \equiv 1 \bmod 4$ and the square-free part of n contains a prime $\equiv 3 \bmod 4$.

(iv) Show that no ARBIBD with these parameters exists for any value of t if $n \equiv 2 \bmod 4$ and the square-free part of n contains a prime congruent to $3 \bmod 4$.

7   If $A_n(t)$ denotes an affine-resolvable design with parameters listed in Exercise 6, show that for n a prime power the existence of a design $A_n(t)$ implies the existence of a design $A_n(nt + 1)$.

8   A BIBD with parameters (v, b, r, k, λ) is said to be α-resolvable if the blocks can be grouped into t sets $S_1, S_2, \ldots, S_t$, each of β blocks such that in each group every variety is replicated α times. An α-resolvable design is affine α-resolvable if any pair of blocks of the same group intersects in q_1 varieties and any pair from different groups intersects in q_2 varieties. Show that in an α-resolvable design $v \mid k^2$ and $b = v + (b/\beta) - 1$.

9   Show that, in any nontrivial, balanced, equidistant code C of distance $2k$, $|C| \leq k^2 + k + 1$, with equality if and only if there exists a projective plane of order k.

10   Show that if C is a balanced equidistant code of distance $2k$ and correlation one, then if no projective plane of order k exists, and $k \neq 6$, that $|C| \leq k^2 - 2$.

Exercises

11 What is the size of the largest balanced equidistant code of correlation one and distance 12? (This problem is unsolved.)

12 Show that the binary linear code generated as the row space of the incidence matrix of a projective plane of order ten (should such exist) would have dimension 56. Generalize this result.

4 The Structure of Semisimple Rings

4.1 Introduction

The inclusion of a chapter on semisimple rings in a book on the mathematics of coding serves two purposes. We observed in the first chapter that a cyclic code is an ideal in the algebra $GF(q)[x]/(x^n - 1)$. This algebra is isomorphic to the group algebra of a cyclic group, a concept we introduce in Section 4.6. The condition that $(n, q) = 1$ ensures that $x^n - 1$ has distinct roots in any extension field of $GF(q)$. But this condition is also that required in order for the group algebra to be semisimple. Thus the study of semisimple rings provides an interesting alternative viewpoint into the structure of cyclic codes. From this viewpoint, an obvious way of extending the concept of a cyclic code is to define the code as an ideal in the group algebra of a more general group. While these more general codes have not been studied extensively, Section 4.8 contains some information on ideals in Abelian group algebras.

In Chapter 6 we will consider codes for the Gaussian channel, which for our purposes will be a collection of vectors on the unit sphere in Euclidean n-space E_n. The problem is to place the vectors on the sphere to maximize the minimum distance between any two points. An important method for the construction of such codes utilizes representations of finite groups, a subject studied in the next chapter. The study of such representations is intimately connected with the structure of semisimple rings and hence this

4.2 Rings, Ideals, and the Minimum Condition

chapter also serves as background for the study of group representations. It is interesting to note the common ground shared by the continuous and discrete coding problems.

4.2 Rings, Ideals, and the Minimum Condition

We assume some familiarity with the notion of a ring but include the definition for completeness. We shall be concerned only with associative rings, which are sets of elements with two operations, denoted addition and multiplication, such that the elements form an Abelian group under addition, the associative law of multiplication holds, and both distributive laws hold. We shall further assume throughout that R has an identity element and, unless specified otherwise, by a ring R we shall mean an associative ring with an identity in which the identity does not equal zero. Some of the results given will, of course, be valid for more general rings.

A subset S of a ring R, together with the operations inherited from R, is called a subring of R if it is a ring. A nonempty subset A of R is called a left ideal of R, if, whenever $a \in A$, then $ra \in A$ for all $r \in R$ and $a - b \in A$ for any two elements $a, b \in A$. A right ideal is defined in a similar manner and a two-sided ideal is a set that is both a left and a right ideal. Since we have assumed that R contains the identity element, it follows that $RA = A$ for a left ideal A of R, where RA denotes the set of all finite sums

$$RA = \{\sum r_i a_i, r_i \in R, a_i \in A\}$$

Of course, if A contains the unit, then $A = R$. A ring is commutative if multiplication is commutative and in such rings the concepts of left, right, and two-sided ideals coincide.

If A is a left ideal of R and $A = Ra$ for some element $a \in R$ (and hence $a \in A$), then the element a is called a generator of A and A is called a principal left ideal of R. Similarly, if A is an ideal of R such that any element of A is a sum $\sum_{i=1}^{m} r_i a_i$ for a set of elements $a_1, a_2, \ldots, a_m, a_i \in A$, then A is generated by the elements $a_1, a_2, \ldots, a_m$ and these elements form a finite basis of A. If A_1 and A_2 are left ideals, we define their sum as

$$A_1 + A_2 = \{a_1 + a_2 | a_1 \in A_1, a_2 \in A_2\}$$

and the sum is also a left ideal, the smallest ideal containing both A_1 and A_2, i.e., the ideal generated by $A_1 \cup A_2$. If each element of $A_1 + A_2$ is uniquely expressible in the form $a = a_1 + a_2$, $a_i \in A_i$, then the sum is said to be direct and is denoted $A_1 \oplus A_2$. Every element has such a unique representation if there exist no nonzero elements $a_1 \in A_1$, $a_2 \in A_2$ such that

$0 = a_1 + a_2$, i.e., the zero element has the unique representation $0 + 0$. For, suppose $a \in A_1 + A_2$ can be written in the two ways $a = a_1 + a_2 = a_1' + a_2'$; then $0 = a_1 - a_1' + a_2 - a_2'$, $a_i, a_i' \in A_i$. Then if 0 has the unique representation $0 + 0$, we have $a_i = a_i'$ and the representation of a is unique. The concepts of sums, direct sums, and unions of a finite set of left ideals carry over trivially.

The intersection of two left ideals is also a left ideal. We define the product of two left ideals as

$$A_1 A_2 = \{\sum a_1 a_2 \mid a_1 \in A_1, a_2 \in A_2\}$$

where the summation is over a finite number of terms. The product is a left ideal and similar definitions hold for right and two-sided ideals. Clearly, if A_1 and A_2 are left ideals, then $A_1 A_2 \subseteq A_2$ and if they are two-sided ideals, then $A_1 A_2 \subseteq A_1 \cap A_2$.

Let A be a left ideal of R. Then we say that A is nilpotent if there exists a positive integer n such that $A^n = AA \cdots A$ (n times) is the zero ideal. This implies that any product of n elements, distinct or not, in such an ideal is the zero element. We say that an element $a \in A$ is nilpotent if, for some positive integer m, $a^m = 0$. If every element of A is nilpotent, then we call A a nil ideal. Clearly, every nilpotent left ideal is a nil ideal since, by definition, $a^n = 0$, for all $a \in A$ if $A^n = (0)$. Under certain conditions, although not in general, the converse is also true. An element $r \in R$ is an idempotent element if $r^2 = r$. It follows that for any idempotent element, $r^n = r$, and hence that a nilpotent ideal cannot contain any idempotent elements. We will show in the next section that if the left ideal A is not nilpotent, then it must contain an idempotent element.

A left ideal A of R is called a maximal left ideal of R if there exists no left ideal A' of R such that $A \subset A' \subset R$. Similarly, A is a minimal left ideal of R if there exists no left ideal A'' of R such that $(0) \subset A'' \subset A$.

Definition If every strictly decreasing sequence of left ideals of R terminates in a finite number of steps, then we say that R satisfies the descending chain condition for left ideals, i.e., in any sequence of left ideals $A_1, A_2, \ldots$ such that

$$A_1 \supset A_2 \supset \cdots$$

there can be only a finite number of left ideals.

Analogously, we have the following definition:

Definition If every strictly increasing sequence of left ideals of R terminates in a finite number of steps, we say that R satisfies the ascending chain condition for left ideals.

Any ring R that satisfies the ascending chain condition will be called Noetherian. A ring satisfying the descending chain condition (d.c.c.) is usually called Artinian. It can be shown that for any commutative ring R the ascending chain condition is equivalent to the existence of a finite basis for R. Any field satisfies trivially both the ascending and descending chain conditions, since it has only the two trivial ideals. It can also be shown that in a ring R satisfying the d.c.c. any collection of left ideals contains an ideal that does not properly contain any other ideal, i.e., a minimal ideal in the collection.

An important structure, from a coding theory point of view, is the group algebra, which will be examined in Section 6. This is, of course, a particular type of algebra for which we have the following definition:

Definition An algebra A is a ring that is also a vector space over a field K such that the two multiplications satisfy

$$\alpha(a_1 a_2) = (\alpha a_1)a_2 = a_1(\alpha a_2), \qquad \alpha \in K, \quad a_1, a_2 \in A$$

A subalgebra of A is then a subring that is also a vector subspace. Moreover, since we have assumed all rings to be associative and to possess an identity, if A_1 is a left ideal of A, then

$$(\alpha \cdot 1)a = \alpha(1 \cdot a) = \alpha a \in A_1, \qquad \alpha \in K, \quad a \in A_1$$
$$\alpha(a_1 + a_2) = \alpha a_1 + \alpha a_2$$

and thus any left ideal of A is a vector subspace of A. The dimension of A is the dimension of A as a vector space over K. We shall only consider finite-dimensional algebras, and it is clear that such algebras satisfy both chain conditions. While our later interests will be exclusively with group algebras, we continue to investigate the more general associative ring with identity.

4.3 Nilpotent Ideals and the Radical

We begin by considering several important properties of nilpotent ideals. As before, we shall assume R is an associative ring with identity and unless explicitly mentioned otherwise, the term ideal will mean left-sided ideal.

Theorem 3.1 Let N_1 and N_2 be two nilpotent ideals of R. Then $N_1 + N_2$ and $N_1 R$ are nilpotent ideals.

Proof Suppose that $N_1^{n_1} = N_2^{n_2} = (0)$ and consider the ideal $(N_1 + N_2)^{n_1 + n_2}$. Each element of this ideal is a sum of products of elements of N_1 and N_2 and each such product contains either at least n_1 elements from N_1 or at least n_2 elements from N_2. Assume it contains at least n_2 elements from N_2 and let γ be a typical such element, where

$$\gamma = (\alpha_1 \alpha_2 \cdots \alpha_{k_1})\beta_1(\alpha_{k_1+1} \cdots \alpha_{k_2})\beta_2 \cdots, \qquad \alpha_i \in N_1, \quad \beta_i \in N_2$$

and since N_2 is a left ideal, $(\alpha_1 \alpha_2 \cdots \alpha_{k_1})\beta_1 \in N_2$, etc. Thus γ can be expressed as the product of at least n_2 elements of N_2 multiplied possibly by an element of N_1, and hence $\gamma = 0$. It follows that $N_1 + N_2$ is nilpotent.

For the last part of the theorem we observe that $N_1 R$ is a two-sided ideal and to show that it is nilpotent, consider

$$(N_1 R)^{n_1} = N_1(RN_1) \cdots (RN_1)R \subseteq N_1^{n_1} R = (0) \qquad \square$$

Corollary The sum of a finite number of nilpotent left ideals is nilpotent.

It is conventional to refer to a ring whose left ideals satisfy the descending chain condition as a ring with minimum condition and we shall adopt this notation also. For the discussion of rings with minimum condition we shall essentially follow the references of Artin *et al.* (1944) and Curtis and Reiner (1966), whose approach to the subject seems difficult to improve upon.

Theorem 3.2 Let N be a nonnilpotent left ideal of the ring R with minimum condition. Then N contains a nonzero idempotent.

Proof The proof is by the construction of an idempotent element. Recall that in any collection of left ideals of R there exists a minimal ideal. Consider the collection of all nonnilpotent left ideals of R contained in N, of which N is a member. Let N_1 be a minimal ideal of this set. Since N_1^2 is a nonnilpotent left ideal and $N_1^2 \subseteq N_1$, we must have $N_1^2 = N_1$. Denote by $\mathcal{M}$ the set of left ideals M that satisfy the conditions (a) $N_1 M \neq (0)$, (b) $M \subseteq N_1$. Since $N_1 \in \mathcal{M}$, $\mathcal{M}$ is not empty. Let M_1 be a minimal ideal of $\mathcal{M}$. By condition (a) there exists an $m_1 \in M_1$ such that $0 \neq N_1 m_1 \subseteq M_1 \subseteq N_1$. But $N_1 m_1$ is a left ideal and a member of $\mathcal{M}$, implying $N_1 m_1 = M_1$. Since $m_1 \in M_1$, there must exist $n_1 \in N_1$ such that $n_1 m_1 = m_1$, from which it follows that $n_1^k m_1 = m_1$ for all positive integers k. Since $m_1 \neq 0$, we conclude that $n_1^k \neq 0$ for any k and hence n_1 is not a nilpotent element. The remainder of the proof uses n_1 to construct the desired idempotent element. Define the set

$$N' = \{x \in N_1 \mid xm_1 = 0\}$$

from which it follows that N' is a left ideal and, since $n_1 m_1 = m_1 \neq 0$, N' is properly contained in N_1, implying that N' is nilpotent. Furthermore, since $(n_1^2 - n_1)m_1 = 0$, $(n_1^2 - n_1) \in N'$, which implies that $n_1^2 - n_1$ is nilpotent, while n_1 is not nilpotent. Now if $n_1^2 - n_1 = 0$, then n_1 is an idempotent element of N and we are finished. Suppose that $x_1 = n_1^2 - n_1 \neq 0$ and let

4.3 Nilpotent Ideals and the Radical

$n_2 = n_1 + x_1 - 2n_1x_1$. Now $n_2 \in N_1$ is a sum of powers of n_1 and hence x_1, n_2, and n_1 commute. Furthermore, n_2 cannot be nilpotent, for if it were, then $n_1 = n_2 - x_1 + 2n_1x_1$ would also be nilpotent since it is the sum of commuting nilpotent elements. So $n_2 \in N_1$ is not nilpotent and by direct calculation we find

$$n_2{}^2 - n_2 = 4x_1{}^3 - 3x_1{}^2 = x_2$$

and hence $n_2{}^2 - n_2$ is expressed as a sum of the powers of the nilpotent element x_1. We continue this process by defining successively the element

$$n_{i+1} = n_i + x_i - 2x_i n_i$$

where $x_i = n_i{}^2 - n_i$ and n_i is not nilpotent. At each stage we have

$$n_{i+1}^2 - n_{i+1} = 4x_i{}^3 - 3x_i{}^2 = x_{i+1}$$

It is readily checked that x_{i+1} can be expressed as a sum of powers of x_1, whose lowest power is 2^i. Since x_1 is nilpotent, we eventually arrive at an element n_j such that $n_j{}^2 - n_j = 0$ and thus n_j is an idempotent of N_1, which completes the theorem. □

Since by definition every element of a nilpotent ideal is nilpotent, it follows directly from the theorem that in a ring with minimum condition a left ideal is nilpotent if every element of it is nilpotent. An arbitrary sum of nilpotent left ideals is easily seen to be a left ideal. By definition, each element in such a sum is a sum of elements from a finite number of the ideals. Since a finite sum of nilpotent left ideals is, from the corollary to Theorem 3.1, a nilpotent left ideal, all such elements must be nilpotent. We conclude that an arbitrary sum of nilpotent left ideals in a ring with minimum condition is a nilpotent left ideal. This motivates the following definition.

Definition The radical of a ring R with minimum condition is the sum of all nilpotent left ideals of R, and is denoted rad R.

Theorem 3.3 Let R be a ring with minimum condition and radical rad R. Then rad R is a two-sided nilpotent ideal of R that contains every nilpotent right ideal of R.

Proof Clearly rad R is a nilpotent left ideal, and from Theorem 3.1, (rad R)R is a nilpotent two-sided ideal. Since (rad R)$R \subseteq$ rad R, rad R is a nilpotent two-sided ideal. Furthermore, if M is a nilpotent right ideal, then, arguing as in Theorem 3.1, RM is a nilpotent two-sided ideal and again using the fact that $1 \in R$, $M \subseteq RM \subset$ rad R, which completes the proof. □

Definition A ring R with minimum condition will be called semisimple if rad R is (0). It is said to be simple if its only two-sided ideals are the trivial ones (0) and R.

Since rad R is a two-sided ideal, a simple ring may have either rad $R = (0)$ or rad $R = R$. However, since we have assumed $1 \in R$ and rad R is nilpotent, $1 \notin$ rad R, implying that rad $R = 0$ for any simple ring. Thus any simple ring is also semisimple. The converse, of course, is not true and in the next section we will see that any semisimple ring can be expressed as the direct sum of simple two-sided ideals.

Lemma For a ring R with minimum condition, $R/\text{rad } R$ is semisimple.

Proof Every left ideal of $R/\text{rad } R$ is of the form $M/\text{rad } R$, where M is a left ideal of R containing rad R. Since R satisfies the minimum condition, so does $R/\text{rad } R$. The left ideal $M/\text{rad } R$ is nilpotent if some power of M is contained in rad R. However, this would imply that M is a nilpotent left ideal and hence $M \subseteq$ rad R. Thus $R/\text{rad } R$ contains no nilpotent left ideals and hence is semisimple. □

4.4 The Structure of Semisimple Rings

In the previous section it was seen that a semisimple ring contains no nilpotent ideals, either left, right, or two-sided, and that a simple ring is defined to have no nontrivial two-sided ideals. In this section we will show how a semisimple ring can be decomposed into disjoint simple components. We will prove this result by a series of theorems. In the next section we give a brief discussion of the structure of a simple ring and in Section 4.6 conclude with the application of these results to group algebras. We assume, as before, that R is a ring with identity that satisfies the minimum condition.

We first characterize left ideals in the ring R, which for the section we will assume is semisimple with identity and satisfies the minimum condition.

Theorem 4.1 Every minimal left ideal M_1 in the semisimple ring R has a generating idempotent and if M is a left ideal containing M_1, then there exists a left ideal M_2 such that $M = M_1 \oplus M_2$. Every left ideal of R can be written as a direct sum of minimal left ideals. Furthermore, every left ideal M in R has a generating idempotent e and M is minimal iff eRe is a division ring.

Proof Let M_1 be a minimal left ideal in R. Since R is semisimple, M_1 is not nilpotent and hence, by Theorem 3.2, contains a nonzero idempotent

4.4 The Structure of Semisimple Rings

e_1. It follows that $M_1 = Re_1$ and that e_1 is a right unit for M_1. Also, $r \in R$ is an element of M_1 iff $re_1 = r$. Now define the set

$$M_0 = \{r \in R \mid re_1 = 0\}$$

The set M_0 is clearly a left ideal and $r \in M_0$ iff $r(1 - e_1) = r$, i.e., $M_0 = R(1 - e_1)$. Since any element of R can be written $r = re_1 + r(1 - e_1)$ and since $M_0 \cap M_1 = (0)$, we have that $R = M_1 \oplus M_0$. If M is any ideal containing M_1, then it follows that $M = M_1 \oplus (M_0 \cap M) = M_1 \oplus M_2$, where $M_2 = M_0 \cap M$.

Now let M be any left ideal of R and suppose M_1 is a minimal left ideal of R contained in M. By the above argument there exists a left ideal M_1' such that $M = M_1 \oplus M_1'$. Similarly, if M_2 is a minimal left ideal contained in M_1', then there exists a left ideal M_2' such that $M_1' = M_2 \oplus M_2'$ and the process continues. Since $M_1' \supset M_2' \supset \cdots$ is a strictly decreasing sequence of left ideals, it must terminate as R, hence M, satisfies the minimum condition. Thus we can write

$$M = M_1 \oplus \cdots \oplus M_j$$

where each M_i, $i = 1, \ldots, j$ is a minimal left ideal of R contained in M. In particular, if we let $M = R$, we see that R can be expressed as a direct sum of minimal left ideals. In such a decomposition, which is not, in general, unique, let the representation of the identity be

$$1 = e_1 + e_2 + \cdots + e_r$$

It follows that

$$e_i = e_i e_1 + e_i e_2 + \cdots + e_i^2 + \cdots + e_i e_r$$

and, since the sum is direct, we must have that $e_i e_j = \delta_{ij} e_i$, where δ_{ij} is the usual Kronecker delta function, i.e., $\delta_{ij} = 0$ if $i \neq j$ and 1 if $i = j$. Thus, the set of elements $\{e_i\}$ is a set of orthogonal idempotents, and since the M_i are minimal left ideals, we have that $M_i = Re_i$. Assume that we have a decomposition of R of the form

$$R = Re_1 \oplus \cdots \oplus Re_j \oplus Re_{j+1} \oplus \cdots \oplus Re_s$$

where

$$M = Re_1 \oplus \cdots \oplus Re_j$$

which, by the procedure outlined, can always be accomplished. The element

$$e = e_1 + \cdots + e_j$$

is an idempotent of M. Let $r = \sum_{i=1}^{j} \alpha_i e_i$ be an arbitrary element of M; then since

$$re = \sum_{i=1}^{j} \alpha_i e_i = r$$

it follows that e is a generating idempotent of M and a right identity for M. As before,

$$M = \{r \in R \mid re = r\}$$

and if we denote $R = M \oplus M'$, then $M' = R(1 - e)$ and $1 - e$ is the generating idempotent for M'.

To prove that $M = Re$ is a minimal ideal iff eRe is a division ring, assume first that M is a minimal ideal. Notice that eRe is a subring of R and has e as a two-sided unit. It is then sufficient to show that every element of eRe has an inverse in eRe.

Let d be a nonzero element of eRe and consider the ideal Rd which, since M is minimal, is equal to M. Thus $eRd = eRe$ and since $d \in eRe$ and e is a two-sided identity, we have $d = ede = ed$ and hence that $(eRe)d = eRe$. We conclude that there must exist an element $c \in eRe$ such that $cd = e$, i.e., $d^{-1} = c$ and eRe is a division ring.

Conversely, suppose that $M = Re$ is not minimal; then we can write $M = M_1 \oplus M_2$ and $e = e_1 + e_2$, where $M = Re$ is a left ideal with the idempotent generator e and $M_i = Re_i$, $i = 1, 2$, e_1 and e_2 orthogonal idempotents. It follows that $e_1 e = e_1$ and $e_2 e = e_2$ since e is an identity on M. Since e is idempotent, $e = e^2 = ee_1 + ee_2$ and thus $ee_1 = e_1$ and $ee_2 = e_2$. It follows that $e_1 = ee_1 e$ and $e_2 = ee_2 e$. But since $e_1 e_2 = 0$ and $ee_1 e$ and $ee_2 e$ are nonzero elements of eRe, then eRe contains divisors of zero and cannot be a division ring. This completes the proof of the theorem. □

Notice that a generating idempotent of any left ideal M of R is unique and acts as an identity on M. Suppose e is a generating idempotent of $M = Re$. Every element $m \in M$ can be written as $m = re$, $r \in R$, and thus $me = re^2 = re = m$. Since M is a subring, its identity is unique.

Definition We call e a primitive idempotent if it cannot be expressed as the sum of two orthogonal idempotents.

Implicit in the proof of the preceding theorem is that the generating idempotent of a minimal left ideal is primitive.

Before going on to the key theorem on the structure of semisimple rings we prove an intermediate theorem on isomorphic left ideals.

4.4 The Structure of Semisimple Rings

Theorem 4.2 The sum of all left ideals of R isomorphic to a given minimal left ideal of R is a simple two-sided ideal S of R, where R is a semisimple ring with identity satisfying the minimum condition.

Proof Let M and M_1 be isomorphic minimal left ideals of R. If $M_1 = Ma$ for some $a \in M_1$, then the mapping $\eta: m \mapsto ma$ from M onto M_1 is a ring isomorphism from M to M_1, i.e., $\eta(m_1 + m_2) = \eta(m_1) + \eta(m_2)$ and $\eta(rm_1) = r\eta(m_1)$, $m_1, m_2 \in M$, $r \in R$. On the other hand, if η is a ring isomorphism and e a generating idempotent of M, then $\eta(re) = r\eta(e)$ for all $r \in R$. If $r \in M$, then $re = r$ and $\eta(r) = r\eta(e)$ and hence η is an isomorphism only if it is of the form $\eta: m \mapsto ma$ for some $a \in M_1$. For any $m \in M$, since M is minimal, we have that Mm is either (0) or M and a restatement of this result is that M is isomorphic to M_1 iff $MM_1 = M_1$. If M_1 is not isomorphic to M, then we must have that $MM_1 = (0)$.

Now let S be the sum of all left ideals of R isomorphic to M. Clearly $RS \subset S$ since S is a left ideal. To show that $SR \subset S$, we note that from the above discussion M is isomorphic to M_1 if $SM_1 \neq (0)$ since $M \subset S$. Also, M is isomorphic to M_1 if $M_1 \subset S$ and hence $SM_1 \subset M_1$. Considering the decomposition of R into a direct sum of minimal left ideals, we conclude that $SR \subset S$ and that S is a two-sided ideal.

To show that S is simple, let $T \subset S$ be a two-sided ideal of S containing a minimal left ideal M'. Since $M' \subset S$, we must have that M' is isomorphic to M, where M and S are as before. For any $r \in R$, $M'r \subset T$ since $M' \subset T$ and T is a two-sided ideal. However, all left ideals isomorphic to M' or, equivalently, to M are of the form $M'r$. Consequently, T contains all left ideals isomorphic to M. Thus $T = S$ and S is simple. $\square$

We shall call the two-sided ideal S of the above theorem a simple component of the ring R. Clearly R has the same number of simple components as it has nonisomorphic left ideals. The significance of simple components to the structure of semisimple rings is contained in the next theorem, for which we need a little more terminology. The set of elements of R that commute with all elements of R is called the center of R and is denoted by C. An idempotent in C is called a central idempotent.

Theorem 4.3 Let R be a semisimple ring with identity satisfying the minimum condition. Then R can be written, uniquely up to ordering, as a direct sum of simple two-sided ideals, the simple components of R. Any two-sided ideal of R is a direct sum of simple components and has a central idempotent generator. Furthermore, an ideal of R is two sided if and only if it has a central idempotent generator.

Proof Consider the decomposition of R into a direct sum of minimal left ideals M_i,

$$R = M_1 \oplus \cdots \oplus M_k$$

where, for convenience, we assume that $M_1, \ldots, M_l$ are pairwise nonisomorphic and hence M_j, $j > l$, is isomorphic to one of these. Let S_i denote the simple component containing M_i. From this decomposition we conclude that $R \subset S_1 + \cdots + S_l$ and we need only show that the sum is direct. Suppose that $S_1 \cap (S_2 + \cdots + S_l) = N$ and recall that $S_i \cdot S_j = (0)$, $i \neq j$. Then we must have that $S_1 N = (0)$ since $S_1(S_2 + \cdots + S_l) = (0)$ and $(S_2 + \cdots + S_l)N = 0$ since $(S_2 + \cdots + S_l)S_1 = (0)$ and hence $RN = (0)$. Since R contains an identity, $N = (0)$ and

$$R = S_1 \oplus \cdots \oplus S_l$$

Clearly every simple component of R is contained in the sum. Notice also that in any decomposition of R into minimal left ideals we must have that S_1 is the direct sum of those ideals in the decomposition isomorphic to M_1.

Now let J be any two-sided ideal of R and let M_1 be a minimal left ideal of R contained in J. It follows that S_1 is contained in J, where S_1 is the sum of all minimal left ideals of R isomorphic to M_1. This is true because $M_1 r \subset J$ for all $r \in R$ and all left ideals isomorphic to M_1 are generated in this manner. Thus we can write $J = S_1 \oplus L$, where L is a left ideal, and we continue the process of extracting the simple components of L. Suppose we reach the stage $J = S \oplus J'$, where S is the direct sum of all simple components of J and J' is a left ideal of R. If $J' \neq (0)$ and contains the minimal left ideal L', then, by the same argument as above, J' must contain every minimal left ideal isomorphic to L'; hence $J' = (0)$ and J is the direct sum of the simple components it contains.

In the direct sum decomposition of R into simple components let

$$1 = s_1 + s_2 + \cdots + s_l$$

where the s_i are idempotents, $S_i = Rs_i$, and s_i is a right identity for S_i. Since S_i is a two-sided ideal, we also have that $rs_i = s_i r = r$ for any $r \in S_i$. Also, s_i commutes with all $r \in R$, which follows from the uniqueness of the decomposition of R into two-sided ideals. The idempotent generators are themselves unique, which follows from the unique decomposition and the fact that the representation of the identity is unique. If J is any two-sided ideal of R, its generating central idempotent is the sum of the generating idempotents of the simple components it contains. This completes the proof of the theorem, the last sentence of the theorem being an immediate consequence of the above. ∎

4.5 The Structure of Simple Rings

To summarize, we have shown that any semisimple ring R with identity satisfying the minimum condition can be expressed as a direct sum of its unique simple components, each such component having a unique central idempotent generator. The ring R can also be written as a direct sum of minimal left ideals and the simple components appear as the direct sum of the isomorphic left ideals in this decomposition. The decomposition of a simple component into a direct sum of minimal left ideals may, in general, be accomplished many ways. Any left ideal has an idempotent generator, say e, and is minimal if and only if eRe is a division ring. It can also be shown that if e_1 and e_2 are primitive idempotents of a simple ring S, then $e_1 S e_1$ and $e_2 S e_2$ are isomorphic division rings.

Perhaps the most convenient example of a semisimple ring is a certain type of group algebra. Since we shall be discussing these at some length in Section 4.6, we postpone the example until then.

4.5 The Structure of Simple Rings

In Section 4.4 a semisimple ring with identity that satisfies the minimum condition was shown to be the direct sum of its minimal two-sided ideals, and hence the question of the structure of such rings reduces to investigating the structure of simple rings. We consider this problem here, preferring to give only a brief discussion rather than prove any of the theorems involved.

From the previous section any simple ring S is a direct sum of isomorphic left ideals, each such ideal having a generator which is a primitive idempotent. While the decomposition of S into minimal left ideals is not unique, the number of summands appearing is a constant of the ring S. It is also possible to show that if e and e' are two primitive idempotents of S, then the division rings eSe and $e'Se'$ are isomorphic. The key theorem on the structure of simple rings is due to Wedderburn, which we state without proof:

Theorem 5.1 Let S be a simple ring that satisfies the minimum condition. Then S is isomorphic to the ring of $n \times n$ matrices over a division ring D. The dimension n is the number of minimal left ideals appearing in a direct sum decomposition of S, and D is isomorphic to eSe, where e is any primitive idempotent of S. The ideal Se is a vector space over the division rings eSe, and n is also the dimension of this vector space. Conversely, the ring of $n \times n$ matrices over any division ring D is simple.

This theorem gives a detailed characterization of simple rings. To demonstrate the ideas contained in the theorem, we give the following example.

Example Let $\mathbb{R}_n$ denote the ring of $n \times n$ matrices over the real field $\mathbb{R}$. Every two-sided ideal of $\mathbb{R}_n$ is of the form J_n, where J is a two-sided ideal of $\mathbb{R}$. Since $\mathbb{R}$ has only the trivial ideals (0) and $\mathbb{R}$, it follows that $\mathbb{R}_n$ is simple. Let e_1 denote the $n \times n$ matrix with the unit element in the $(1, 1)$ position and zeros elsewhere. Let $A = (\alpha_{ij})$; then e_1 is an idempotent and

$$Ae_1 = A_1 = \begin{bmatrix} \alpha_{11} & & \\ \alpha_{21} & & 0 \\ \vdots & & \\ \alpha_{n1} & & \end{bmatrix}, \quad e_1 A e_1 = \begin{bmatrix} \alpha_{11} & \vline & 0 \\ \hline 0 & \vline & 0 \end{bmatrix} \simeq \mathbb{R}$$

The dimension of $\mathbb{R}_n e_1$ over $e_1 \mathbb{R}_n e_1$ is n, implying that $\mathbb{R}_n$ can be written as the direct sum of n minimal left ideals. Define e_i to be the $n \times n$ matrix with the unit element in the (i, i) position and zero elsewhere. Then it is easily seen that e_i is an idempotent and

$$\mathbb{R}_n = \mathbb{R}_n e_1 \oplus \mathbb{R}_n e_2 \oplus \cdots \oplus \mathbb{R}_n e_n$$

and

$$1 = e_1 + e_2 + \cdots + e_n, \quad 1 \in \mathbb{R}_n$$

That this decomposition is not unique is also easily seen by replacing e_1 and e_2 by

$$e_1' = \begin{bmatrix} 1 & 1 & \vline & 0 & \cdots & 0 \\ 0 & 0 & \vline & & & \\ \vdots & \vdots & \vline & & 0 & \\ 0 & 0 & \vline & & & \end{bmatrix}, \quad e_2' = \begin{bmatrix} 0 & -1 & \vline & 0 & \cdots & 0 \\ 0 & 1 & \vline & 0 & \cdots & 0 \\ \hline 0 & 0 & \vline & & & \\ \vdots & \vdots & \vline & & 0 & \\ 0 & 0 & \vline & & & \end{bmatrix}$$

Then the set $e_1', e_2', e_3, \ldots, e_n$ is a set of primitive orthogonal idempotents satisfying

$$1 = e_1' + e_2' + e_3 + \cdots + e_n$$

and again we have

$$\mathbb{R}_n = \mathbb{R}_n e_1' \oplus \mathbb{R}_n e_2' \oplus \mathbb{R}_n e_3 \oplus \cdots \oplus \mathbb{R}_n e_n$$

Thus the decomposition of a simple two-sided ideal into its minimal left ideals is not unique. From Theorem 5.1 (Wedderburn's theorem) a prototype for all semisimple rings satisfying the minimum condition would be the direct sum of complete matrix rings

$$R \simeq M_{n_1}(D_1) \oplus \cdots \oplus M_{n_k}(D_k)$$

where D_i, $i = 1, \ldots, k$, are division rings and n_i, $i = 1, \ldots, k$, are the appropriate dimensions.

4.6 The Group Algebra and Group Characters

The structure that we are principally interested in, and that much of the remainder of the book is concerned with, either implicitly or explicitly, is the group algebra, which we now define.

Definition Let G be a finite group and K an arbitrary field. The group algebra KG is then the set of formal sums

$$KG = \left\{ \sum_{g \in G} \alpha_g g, \quad \alpha_g \in K \right\}$$

with an addition and multiplication defined, respectively, by

$$\sum_{g \in G} \alpha_g g + \sum_{g \in G} \beta_g g = \sum_{g \in G} (\alpha_g + \beta_g) g$$

$$\left(\sum_{g \in G} \alpha_g g \right) \left(\sum_{h \in G} \beta_h h \right) = \sum_{g, h \in G} \alpha_g \beta_h gh = \sum_{f \in G} \gamma_f f$$

where

$$\gamma_f = \sum_{g \in G} \alpha_g \beta_{g^{-1} f}$$

Scalar multiplication on the left is defined in the usual manner.

The group algebra is a vector space over K, a ring with identity, and, of course, an algebra over K. As a vector space over K, the group elements themselves form a convenient basis. The following theorem specifies precisely the conditions for application of the results of Section 4.4.

Theorem 6.1 Let G be a finite group of order n and let K be an arbitrary field. Then the group algebra KG is semisimple if and only if either char $K = 0$ or char $K \nmid n$.

Proof We first verify that if char $K = p$, a prime, and $p \mid n$, then KG cannot be semisimple. Consider the element

$$\eta = \sum_{g \in G} g$$

and notice that $g\eta = \eta$, $g \in G$, since multiplication by g merely permutes the elements of the sum. Also, g commutes with η for any $g \in G$ and hence $KG\eta$ is a nontrivial two-sided ideal. However, since

$$\eta^2 = \left(\sum_{g \in G} g \right) \left(\sum_{h \in G} h \right) = n\eta = 0$$

it follows that $(KG\eta)^2 = (0)$ and $KG\eta$ is a nontrivial nilpotent ideal, implying that KG is not semisimple.

For convenience we denote the elements of G by g_i, $i = 1, \ldots, n$, $g_1 = 1$, and assume that either char $K = 0$ or char $K = p$ and $p \nmid n$. Let

$$r = \sum_{i=1}^{n} \alpha_i g_i$$

be an arbitrary element of rad KG. Consider the action of r on the elements $g_1, g_2, \ldots, g_n$, which form a basis of KG over K. Denote

$$rg_i = \left(\sum_{j=1}^{n} \alpha_j g_j\right) g_i = \sum_{j=1}^{n} \rho_{ij} g_j, \qquad i = 1, \ldots, n$$

and notice that left multiplication by r on KG is a linear transformation on KG that we can represent by the matrix $\boldsymbol{\rho} = (\rho_{ij})$ with respect to the basis $g_1, \ldots, g_n$. Similarly, the linear transformation effected by r^k has the representation $\boldsymbol{\rho}^k$. Since r is nilpotent, $r^l = 0$ for some integer l, and it follows that $\boldsymbol{\rho}^l = \mathbf{0}$ and hence the eigenvalues of $\boldsymbol{\rho}^l$ are all zero. However, the eigenvalues of $\boldsymbol{\rho}^l$ are the lth powers of the eigenvalues of $\boldsymbol{\rho}$ and hence all eigenvalues of $\boldsymbol{\rho}$ are zero. This implies that the trace of $\boldsymbol{\rho}$ is zero. From the defining equation of $\boldsymbol{\rho}$ the coefficient of g_i in rg_i is α_1 and the trace of $\boldsymbol{\rho}$ is $n\alpha_1$. It follows that $0 = n\alpha_1$ and since, by assumption, $n \neq 0$, $\alpha_1 = 0$. Since rad KG is a two-sided ideal, the element

$$rg_i^{-1} = \alpha_i g_1 + \sum_{\substack{j=1 \\ j \neq i}}^{n} \alpha_j g_j g_i^{-1} = \sum_{i=1}^{n} \beta_l g_i$$

belongs to rad KG and we repeat the argument to show that $\beta_1 = \alpha_i = 0$, $i = 1, \ldots, n$, and hence rad $KG = (0)$ and KG is semisimple. $\square$

The following information on the center of the group algebra, which we shall denote by $C(KG)$, will be useful.

Theorem 6.2 Let G be a finite group and K an arbitrary field. Denote by $C_1, C_2, \ldots, C_s$ the conjugacy classes of G and define

$$\eta_i = \sum_{g \in C_i} g$$

Then the elements η_i, $i = 1, \ldots, s$, span the center of the group algebra $C(KG)$ as a vector space over K.

Proof The elements η_i, $i = 1, \ldots, s$, are clearly linearly independent in the group algebra. Also, we have $g\eta_i g^{-1} = \eta_i$, since such action by g is an automorphism of G preserving conjugacy classes, and $\eta_i \in C(KG)$. Now let

$$\gamma = \sum_{i=1}^{n} \alpha_i g_i = h\gamma h^{-1} = \sum_{i=1}^{n} \alpha_i h g_i h^{-1} = \sum_{i=1}^{n} \alpha_i' g_i$$

4.6 The Group Algebra and Group Characters

for any $h \in G$, where γ is an arbitary element of $C(KG)$. It follows that $\alpha_i = \alpha_j$ if $g_i, g_j \in C_k$, i.e., the coefficients in γ are class functions, implying that any element of $C(KG)$ is a linear sum over K of the η_i and these elements span $C(KG)$. ☐

Example As a trivial example of a semisimple ring, consider the group algebra of the cyclic group of order three, C_3, over $GF(2)$. The elements are

$$GF(2)C_3 = \{0, 1, g, g^2, 1+g, 1+g^2, g+g^2, 1+g+g^2\}$$

and, since C_3 is commutative, all ideals are two sided. This implies that all ideals are pairwise nonisomorphic and each contains a unique (central) generating idempotent. There are two nontrivial ideals in this algebra,

$$A_1 = \{0, 1+g+g^2\}, \qquad A_2 = \{0, g+g^2, 1+g^2, 1+g\}$$

and they have the generating idempotents

$$e_1 = 1 + g + g^2, \qquad e_2 = g + g^2$$

respectively. It is easily seen that

$$GF(2)C_3 = A_1 \oplus A_2$$

Unfortunately, the two-sided ideals A_i contain no nontrivial left ideals and with this example we are unable to demonstrate that the decomposition of such an ideal into a direct sum of minimal left ideals is not, in general, unique. The simplest example of a noncommutative group algebra seemed too cumbersome to discuss.

In the proof of Theorem 6.1 the idea of a group representation was implicitly introduced and now we shall develop the concept further. Our immediate goal is to arrive at the character theory for Abelian groups, which we shall need in Section 4.8. as quickly as possible. After considering the relationship between semisimple group algebras and group representations, we shall merely state some properties of these representations that we need for Abelian group characters and prove these properties in the next chapter.

Definition A representation of a finite group G over a field K is a homomorphism

$$\rho : G \longrightarrow GL(V)$$

from G to the general linear group on a vector space V over a field K. If V has dimension n, by fixing a basis in V, we can also consider the map ρ as a map from G to $GL(n, K)$, the group of nonsingular $n \times n$ matrices over K.

Now let KG be a semisimple group algebra and define a transformation $\rho(g)$ of the left ideal L of KG by

$$\rho(g)a = ga, \qquad g \in G, \quad a \in L$$

The ideal L is a vector space over K and a vector subspace of KG. The transformation $\rho(g)$ is a linear transformation of L into L. As a vector space over K, L has a finite dimension, say m, since KG is finite dimensional. By fixing a basis of L, $\rho(g)$ can be interpreted as an $m \times m$ matrix. Since

$$\rho(g_1 g_2) = \rho(g_1)\rho(g_2), \qquad g_1, g_2 \in G$$

the above transformation is clearly a representation of G. The interesting property of group algebras is that the converse is also true, in a certain sense, i.e., given a $k \times k$ representation of G over K, this representation is derivable from representations found by the transformations $\rho_i(g)$ defined on a certain sequence of left ideals $L_1, L_2, \ldots, L_s$ such that

$$\rho_i(g)a = ga, \qquad a \in L_i, \quad i = 1, \ldots, s$$

This direct correspondence between left ideals of KG and K representations of G is of fundamental importance to group representation theory.

Let $\{\rho(g), g \in G\}$ be a matrix representation over the field K of the group G. The representation is said to be faithful if the homomorphism ρ has a kernel consisting of the identity only. If U is any nonsingular matrix over K, then $\{U\rho(g)U^{-1}, g \in G\}$ is also a representation of G and we call any two representations of G related in this manner equivalent representations. We shall show in a later chapter that if K is either $\mathbb{R}$ or $\mathbb{C}$, then every representation of G is equivalent to a representation by orthogonal or unitary matrices, respectively.

If there exists a matrix U over K such that for a K representation $\{\rho(g)\}$

$$U\rho(g)U^{-1} = \begin{bmatrix} \rho_1(g) & \tau(g) \\ \hline 0 & \rho_2(g) \end{bmatrix}, \qquad g \in G$$

where the dimension of $\rho_1(g)$ is independent of g, then we say that the representation $\rho(g)$ is reducible and $\{\rho_1(g)\}$ and $\{\rho_2(g)\}$ are also representations of G. If no such matrix exists, then we say that $\{\rho(g)\}$ is an irreducible representation of G. If, whenever a representation $\{\rho(g)\}$ is reducible as above there exists a matrix over K, U_1, such that

$$U_1 \rho(g) U_1^{-1} = \begin{bmatrix} \rho_1'(g) & 0 \\ \hline 0 & \rho_2'(g) \end{bmatrix}$$

4.6 The Group Algebra and Group Characters

then we say that the representation $\{\rho(g)\}$ is completely reducible. We will show later that when KG is semisimple every K representation of G is completely reducible.

When KG is semisimple, every K representation of G can be written as a direct sum of irreducible representations of G up to equivalence. For the remainder of the section we require that char $K = 0$ and that K be algebraically closed and we shall assume that $K = \mathbb{C}$, the complex numbers. The number of irreducible $\mathbb{C}$ representations of a group G is the number of conjugacy classes of G. The dimensions n_i of these irreducible representations ρ_i, $i = 1, \ldots, s$, have the following properties:

$$\text{(i) } n_i \,|\, |G|, \qquad \text{(ii) } \sum_{i=1}^{s} n_i^2 = |G| = n$$

where s is the number of conjugacy classes of G.

If $\rho(g)$ is a representation of G, then the character of G is $\chi_\rho(g) = \operatorname{tr} \rho(g)$, the trace of the matrix $\rho(g)$. The character of a representation is clearly a class function, i.e., $\chi_\rho(g_1) = \chi_\rho(g_2)$ if g_1 and g_2 are in the same conjugacy class of G. Also, characters of equivalent representations are identical. Denote by $\chi_{\rho_i}(g)$ the characters of the irreducible representations ρ_i, $i = 1, \ldots, s$. The character table of the group G is the $s \times s$ array whose entry in the (i, j) position is $\chi_{\rho_i}(C_j)$, the character of the ith irreducible representation ρ_i on the jth conjugacy class C_j. It is conventional to take $C_1 = \{1\}$, the conjugacy class of the identity, and $\rho_1(g) = 1$ for all $g \in G$, which is the identity representation and is an irreducible representation of every group G.

The characters satisfy some interesting relationships. Let $|C_j| = c_j$; then we have the orthogonality relationships

$$\text{(i) } \sum_{k=1}^{s} \chi_{\rho_k}(C_i^*)\chi_{\rho_k}(C_j) = (n/c_i)\delta_{ij}, \qquad \text{(ii) } \sum_{g \in G} \chi_{\rho_i}(g)\chi_{\rho_j}(g^{-1}) = n\delta_{ij}$$

where $C_i^* = \{g^{-1} \,|\, g \in C_i\}$. It is easily shown that $\chi_{\rho_j}(g^{-1}) = \overline{\chi_{\rho_j}(g)}$, the conjugate, and so (ii) can be written

$$\text{(ii) } \sum_{k=1}^{s} c_k \chi_{\rho_i}(C_k)\overline{\chi_{\rho_j}(C_k)} = n\delta_{ij}$$

where the summation over the group elements is replaced by a summation over conjugacy classes. These equations are easily interpreted as weighted orthogonality relationships on the rows and columns of the character table.

It is readily shown that two representations of G are equivalent if and only if their characters are identical. Furthermore, a representation $\{\rho(g)\}$ is irreducible if and only if

$$\sum_{i=1}^{s} c_i \chi_\rho(C_i)\overline{\chi_\rho(C_i)} = n = |G|$$

where $\chi_\rho(g) = \text{tr } \rho(g)$. These are two useful criteria in group representation theory.

The foregoing has been a very brief introduction to group representations and all the statements made will be proven in the next chapter. We turn now to our purpose of determining the characters of Abelian groups that we shall need in Section 4.8. In an Abelian group every element is in a conjugacy class by itself. Thus, if $|G| = n$, there are n conjugacy classes and hence n irreducible representations ρ_i, $i = 1, \ldots, n$. It follows that the degree n_i of ρ_i is unity for $i = 1, \ldots, n$, since $\sum_{i=1}^{n} n_i^2 = n$, $n_i \geq 1$.

Suppose that G is cyclic of order n and that α is a primitive nth root of unity in $\mathbb{C}$. If g is a generator of G, then the representations

$$\rho_i(g) = \alpha^i, \quad i = 0, 1, \ldots, n-1$$

are all the irreducible representations of G. Notice that $\rho_i(g^j) = \alpha^{ij}$ and the character table follows readily from this equation.

Now let G be an arbitrary Abelian group of order n. Then it is known that we can write this as the direct product of cyclic groups A_i, $i = 1, \ldots, k$,

$$G = A_1 \times \cdots \times A_k$$

and if $|A_i| = m_i$, then $m_i | n$. Now let a_i be a generator for A_i and let α_i be a primitive m_ith root of unity. Label the n irreducible representations of G by $\rho_{(j_1, \ldots, j_k)}$, where the components of the k-tuple $(j_1, \ldots, j_k)$ may take values $1 \leq j_i \leq m_i$, and recall that $n = m_1 m_2 \cdots m_k$. Then the representations defined by

$$\rho_{(j_1, \ldots, j_k)}(a_i) = \alpha_i^{j_i}, \quad i = 1, \ldots, k$$

give all n nonequivalent irreducible representations as $(j_1, \ldots, j_k)$ run over all possible n k-tuples, as can be checked by using the criteria on the characters for nonequivalence and irreducibility.

4.7 The Structure of Cyclic Codes

We recall some facts on the structure of $A_n = GF(q)[x]/(x^n - 1)$ from Chapter 1, where $(n, q) = 1$. Suppose $x^n - 1$ factors into the polynomials $g_i(x)$, $i = 1, \ldots, s$. Then the ideal N_i generated by $g_i(x)$ is a maximal ideal of A_n, while that generated by $(x^n - 1)/g_i(x)$ is a minimal ideal M_i. The ideal generated by $g_i(x) \cdot g_j(x)$, $i \neq j$, is $N_i \cap N_j$. The map σ_n maps the integer i to $iq \mod n$. This map partitions the integers $\{0, 1, \ldots, n-1\}$ into the sets s_i, $i = 1, \ldots, s$. We denote by $\sum_q(n)$ the set of sets $\{s_i, i = 1, \ldots, s\}$, $|\sum_q(n)| = s$.

4.7 The Structure of Cyclic Codes

If α is a primitive nth root of unity in some extension field of $GF(q)$, then we assume that

$$g_i(x) = \prod_{i \in s_i} (x - \alpha^j)$$

If we let $C_n = \langle x \rangle$, $x^n = 1$, be the multiplicative cyclic group of order n, then there is an obvious isomorphism between the group algebra $GF(q)C_n$ and $GF(q)[x]/(x^n - 1)$, where we use the variable x for both algebras to emphasize the identification. The condition $(n, q) = 1$ ensures that $x^n - 1$ has distinct zeros, while in $GF(q)C_n$ it ensures semisimplicity and hence the decomposition into a direct sum of unique minimal ideals. The two effects are of course essentially the same. To see the effect of nonsemisimplicity, recall from Chapter 1 the example of optimal codes obtained from the algebra $GF(p)/(x^p - 1)$, or equivalently, $GF(p)C_p$. Each ideal A_i has a generator polynomial $(x - 1)^i$ and these are the only ideals. We have the inclusion relationships $A_0 \supset A_1 \supset \cdots \supset A_p$.

From the previous material in this chapter we have that

$$GF(q)C_n = M_1 \oplus M_2 \oplus \cdots \oplus M_s$$

where M_i are the unique minimal ideals of $GF(q)C_n$. Every ideal of $GF(q)C_n$ is a direct sum of certain of these minimal ideals. Let $h_i(x) = (x^n - 1)/g_i(x)$ be the generator polynomial of the minimal ideal M_i. Denote by $e_i(x) \in M_i$ the unique idempotent element such that

$$1 = e_1(x) + e_2(x) + \cdots + e_s(x), \qquad e_i(x)e_j(x) = \delta_{ij} e_i(x) \qquad (7.1)$$

Recall also that $M_i = GF(q)C_n e_i(x)$ and that $e_i(x)$ is the identity on M_i and annihilates any element in the complement of M_i i.e., if $f(x) = \sum f_i(x)$, $f_i(x) \in M_i$, then $e_i(x)f(x) = f_i(x)$. Let N_i be the maximal ideal generated by $g_i(x)$. How are the idempotents of M_i and N_i related? Observe first that $M_i \cup M_j = M_i \oplus M_j$ is generated by

$$\left(\frac{x^n - 1}{g_i(x)}, \frac{x^n - 1}{g_j(x)} \right) = \frac{x^n - 1}{g_i(x) g_j(x)}, \qquad i \neq j$$

Thus the ideal generated by $g_i(x)$ is $\oplus_{j=1, j \neq i}^{s} M_j = N_i$. From (7.1) the idempotent of N_i is

$$e_1(x) + \cdots + e_{i-1}(x) + e_{i+1}(x) + \cdots + e_s(x) = 1 - e_i(x)$$

To summarize, M_i has generator polynomial $h_i(x) = (x^n - 1)/g_i(x)$ and idempotent $e_i(x)$, while N_i has generator polynomial $g_i(x)$ and idempotent $1 - e_i(x)$. Similarly if

$$\{i_1, \ldots, i_r\} \cup \{j_1, \ldots, j_{s-r}\} = \{1, 2, \ldots, s\}$$

then the ideal generated by
$$g_{i_1}(x) \cdots g_{i_r}(x) = (x^n - 1)/g_{j_1}(x) \cdots g_{j_{s-r}}(x)$$
is just
$$\bigoplus_{k=1}^{s-r} M_{j_k}$$
and the idempotent of this ideal is
$$1 - (e_{i_1}(x) + \cdots + e_{i_r}(x)) = e_{j_1}(x) + \cdots + e_{j_{s-r}}(x)$$

Note that if an arbitrary ideal of M is generated by $g(x)$ [i.e., $g(x)$ is the unique monic polynomial of minimal degree in M], then $h(x) = (x^n - 1)/g(x)$ and $g(x)$ are relatively prime since we assumed $(n, q) = 1$. Thus, by Euclid's algorithm there exist two polynomials $a(x)$ and $b(x)$ such that

$$a(x)g(x) + b(x)h(x) = 1 \tag{7.2}$$

Let $e(x) = a(x)g(x) \in M$ and multiply (7.2) by $e(x)$ to give

$$e^2(x) + a(x)b(x)(x^n - 1) = e(x) = e^2(x)$$

giving $e(x) = a(x)g(x)$ as an idempotent. This idempotent is unique and it readily follows that it generates M and is a unit on M. Every idempotent of $GF(q)C_n$ is a sum of certain of the $e_i(x)$'s. Although we have observed this in a previous section, we can also show it in the following way. To see this, we recall that for a commutative ring R with identity, R/M is a finite field iff M is a maximal ideal of R. Thus M_i, $i = 1, \ldots, s$, are all finite fields of the appropriate size. Suppose $e(x)$ is an arbitrary idempotent of $GF(q)C_n$, which we write as $\sum_{i=1}^{s} a_i(x)$, $a_i(x) \in M_i$, and this decomposition is unique since $GF(q)C_n$ is a direct sum of the M_i. Since $e(x)$ is idempotent, then

$$e^2(x) = \sum a_i^2(x) = \sum a_i(x)$$

implying that $a_i(x) \in M_i$ is idempotent. Since $e_i(x)$ is idempotent and the identity on M_i, then

$$a_i^2(x) = a_i(x) = a_i(x)e_i(x)$$

implying that $a_i(x)(a_i(x) - e_i(x)) = 0$. This gives $a_i(x) = 0$ or $(a_i(x) - e_i(x)) = 0$, since the operation is in a finite field, which proves the statement. An arbitrary ideal M can thus have many idempotents, but they are all of the form $\sum \alpha_i e_i(x)$, $\alpha_i = 0, 1$. If $M = M_{i_1} \oplus \cdots \oplus M_{i_r}$, then the unique idempotent that acts as an identity on M (i.e., the generating idempotent) is $e_{i_1}(x) + \cdots + e_{i_r}(x)$.

Let M and M' be two arbitrary ideals of $GF(q)C_n$ with generating idempotents $e(x)$ and $e'(x)$, respectively. Then $e(x)e'(x)$ is certainly an idempotent

4.7 The Structure of Cyclic Codes

of $M \cap M'$ and since it acts as a unit on $M \cap M'$, it is the unique generating idempotent of $M \cap M'$.

Example Suppose M_i, $i = 1, 2, 3, 4$, are minimal ideals of $GF(q)C_n$ with idempotents $e_i(x)$. Let $M = M_1 \oplus M_2 \oplus M_3$ and $M' = M_2 \oplus M_3 \oplus M_4$. M has the idempotent

$$e(x) = e_1(x) + e_2(x) + e_3(x)$$

while M' has the idempotent

$$e'(x) = e_2(x) + e_3(x) + e_4(x)$$

We have $M \cap M' = M_2 \oplus M_3$, which has idempotent $e_2(x) + e_3(x)$, which, by a simple calculation, is also

$$e(x)e'(x) = (e_1(x) + e_2(x) + e_3(x))(e_2(x) + e_3(x) + e_4(x))$$
$$= e_2(x) + e_3(x)$$

With the same notation the ideal $M \cup M'$ certainly contains the idempotent $e(x) + e'(x) - e(x)e'(x)$, which acts as a unit on it. Also, $M \cup M' = M_1 \oplus M_2 \oplus M_3 \oplus M_4$, which has idempotent

$$e_1(x) + e_2(x) + e_3(x) + e_4(x)$$
$$= (e_1(x) + e_2(x) + e_3(x)) + (e_2(x) + e_3(x)$$
$$+ e_4(x)) - (e_2(x) + e_3(x))$$

which verifies the calculation.

In general, of course, if $M + M' = GF(q)C_n$ and $M \cap M' = \{0\}$, then $e'(x) = 1 - e(x)$. If $M \cap M' = \{0\}$, then the product of any two elements $f(x) \in M$, $f'(x) \in M'$ yields zero. This leads us to consider again the concept of a dual code. If C is a cyclic (n, k) code with generator polynomial $g(x)$, then the dual code $C^\perp$ is generated by the reciprocal polynomial $h^*(x)$, where $h(x) = (x^n - 1)/g(x)$. If $C'^\perp$ is the $(n, n - k)$ code generated by $h(x)$, then $C^\perp$ and $C'^\perp$ are equivalent codes, and viewing the codes as ideals in $GF(q)C_n$, the mapping $\eta \colon x \mapsto x^{-1}$ maps $C^\perp$ into $C'^\perp$. In terms of multiplication in $GF(q)C_n$, the ideal $C'^\perp$ is a natural dual of C since $a(x)b(x) = 0$ if $a(x) \in C$ and $b(x) \in C'^\perp$. In terms of the more conventional inner product $C^\perp$ is the natural dual of C. If $e(x)$ is the generating idempotent of C, then $1 - e(x)$ is the generating idempotent of $C'^\perp$ and $e(x)[1 - e(x)] = 0$.

Consider the mapping

$$\eta_r \colon GF(q)C_n \longrightarrow GF(q)C_n$$
$$x^i \longmapsto x^{ri}, \quad i = 1, \ldots, n - 1$$

which is readily shown to be an algebra isomorphism iff $(r, n) = 1$. The set of mappings $\{\eta_r : (r, n) = 1\}$ forms a group of automorphisms under composition since $\eta_{r_1}\eta_{r_2} = \eta_{r_1 r_2}$. Clearly any such map η_r maps a minimal ideal into another minimal ideal of the same dimension. Equivalently, η_r maps a set $s_i \in \Sigma_q(n)$ into a set s_j where $|s_i| = |s_j|$, and a similar equivalence can be stated in terms of the irreducible generating polynomials of minimal ideals. If r is in the set s_i which contains unity, then η_r is an automorphism of M_i, the minimal ideal corresponding to s_i. This follows since $r = q^i$ mod n for some i and it is readily seen that if $r_1, r_2 \in s_j$, then $\eta_{r_1}(M_j) = \eta_{r_2}(M_j)$.

Example Let $q = 3^2$ and $n = 11$. The cycles of $\Sigma_q(11)$ are

$$s_1 = \{0\}, \qquad s_2 = \{1, 9, 4, 3, 5\}, \qquad s_3 = \{2, 7, 8, 6, 10\}$$

In this case, since n is prime, the condition that $r_1, r_2 \in s_i$ is the same as $r_1 r_2^{-1}$ being in the same cycle as unity.

Any of the automorphisms η_r fixes the identity of the algebra. Suppose η_r maps an arbitrary ideal M to M'. If $e(x)$ is the generating idempotent of M, then $\eta_r(e(x))$ is the generating idempotent of M'. Recall that in general an ideal will contain many idempotents but the generating idempotent is the unique idempotent that acts as an identity on that ideal.

Let N_1 be the maximal ideal in $GF(2)[x]/(x^n + 1)$ generated by the polynomial $g_1(x) = \prod_{i=0}^{m-1}(x - \alpha^{2^i})$ which is associated with the cycle $\{1, 2, 2^2, \ldots, 2^{m-1}\}$. Suppose v is an integer prime to n and let $uv \equiv 1$ mod n, where u also is prime to n. We want to show that $\eta_u(N_1)$ is the maximal ideal associated with the cycle $\{v, 2v, \ldots, 2^{m-1}v\}$. The irreducible polynomial associated with this cycle also has exponent n as $(v, n) = 1$. Since $uv \equiv 1$ mod n, we can write

$$\eta_u(g_1(x)) = g_1(x^u) = \prod_{i=0}^{m-1}(x^u - \alpha^{2^i}) = \prod_{i=0}^{m-1}(x^u - \alpha^{2^i uv})$$

Since $(x - \alpha^{2^i v}) | (x^u - \alpha^{2^i uv})$, we have that $g_2(x) | \eta_u(g_1(x))$, where $g_2(x)$ corresponds to the cycle $\{v, 2v, \ldots, 2^{m-1}v\}$. It follows that $\eta_u(g_1(x))$ is contained in the maximal ideal N_2 generated by $g_2(x)$. Since η_u is an automorphism, $\eta_u(N_1)$ is a maximal ideal and we have that $\eta_u(N_1) = N_2$. It follows that if M_1 and M_2 are the minimal ideals associated with the cycles $\{1, 2, \ldots, 2^{m-1}\}$ and $\{v, 2v, \ldots, 2^{m-1}v\}$, respectively, then $\eta_u(M_1) = M_2$. In each case η_u maps an idempotent to an idempotent.

The cycles used in the above are maximum exponent. Suppose M_1 and M_2 are two minimal ideals of exponent $e = n/r$ with the same dimension. Their associated cycles, say $\{a, 2a, \ldots, 2^{l-1}a\}$ and $\{b, 2b, \ldots, 2^{l-1}b\}$, are of the same length and there exists an integer s, $(s, n) = 1$, such that $a = sb$. As with the above proof, we have $\eta_s(M_1) = M_2$. Thus if M_1 and M_2 are of the same dimension and exponent, then there exists an automorphism η_u such that $\eta_u(M_1) = M_2$.

Conversely, suppose that M_1 and M_2 are minimal ideals and that there exists an automorphism η_u such that $\eta_u(M_1) = M_2$. Let M_1 be associated with the cycle $s_1 = \{a, 2a, \ldots, 2^{l-1}a\}$ and M_2 with the cycle $s_2 = \{b, 2b, \ldots, 2^{l-1}b\}$. These cycles must have the same length since η_u is an isomorphism, which implies M_1 and M_2 have the same dimension. Using precisely the same argument as above we must have that η_u maps one set into the other. We assume that $a = ub$. Suppose that s_1 has exponent e_1 and s_2 has exponent e_2. We then have that $\alpha^{ae_1} = 1 = (\alpha^{ub})^{e_1}$. If $uv = 1 \bmod n$, then $va = b \bmod n$ and

$$\alpha^{vae_1} = 1 = \alpha^{uvbe_1} = (\alpha^b)^{e_1}$$

which implies that e_2, the order of α^b, divides e_1. Similarly, it can be shown that $e_1 | e_2$ and hence $e_1 = e_2$ and we conclude that if $\eta_u(M_1) = M_2$ for some automorphism η_u, then M_1 and M_2 must have the same exponent. We have demonstrated the following:

Theorem 7.1 Let M_1 and M_2 be minimal ideals of $GF(q)C_n$. Then M_1 and M_2 have the same dimension and exponent iff there exists an automorphism η_u such that $\eta_u(M_1) = M_2$.

It is also possible to show that two minimal ideals are isomorphic iff they have the same weight enumerator, but this is left as a problem for the reader.

4.8 Abelian Codes

Since we have observed that a cyclic code can be viewed as an ideal in the group algebra of a cyclic group over a finite field, it is natural to consider the structure of ideals in more general group algebras. A class of groups of particular interest because of their simplicity is the class of Abelian groups. While the structure of all such groups is well known, they still represent a significant extension of cyclic groups. Furthermore, Camion (1970) has recently shown that there exist ideals in Abelian group algebras (i.e., Abelian codes) that are not obtainable by taking direct products of cyclic codes. Berman (1967b) has also proven the existence of asymptotically good Abelian codes. Apart from these considerations, it is interesting to study Abelian codes to gain a better understanding of the cyclic assumption.

In keeping with the aim of this book, we shall present the mathematical basis required for an investigation of Abelian codes and only survey results of a specific nature. We shall mainly confine our interest to the binary field in this section. Some of the results presented will carry over easily to more general fields, while others will not.

Let G be an arbitrary, finite multiplicative Abelian group of order n with the unique (up to ordering) factorization into cyclic groups

$$G = C_{p_1^{s_1}} \times C_{p_2^{s_2}} \times \cdots \times C_{p_k^{s_k}}$$

where the p_i are primes that are not necessarily distinct and where $C_{p_i^{s_i}} = \langle a_i \rangle$ is the cyclic group of order $p_i^{s_i}$ generated by the element a_i. We now need some of the results stated in Section 4.6. As mentioned there, all of these results are proven in Chapter 5. For an arbitrary finite group the number of irreducible representations is equal to the number of its conjugacy classes. Since an Abelian group of order n has n conjugacy classes, it has n irreducible representations and they are all of degree one over a field which contains the nth roots of unity and whose characteristic does not divide n. Let G be an arbitrary group of order n for the moment and let $\{\chi_i, i = 1, \ldots, s\}$ be a complete set of its irreducible characters (over $\mathbb{C}$ say). Then we have the orthogonality relationships

(i) $\quad \sum_{i=1}^{s} c_i \bar{\chi}_j(C_i) \chi_k(C_i) = n \, \delta_{jk}, \qquad c_i = |C_i|, \quad C_i = i\text{th conjugacy class}$

(ii) $\quad \sum_{i=1}^{s} \bar{\chi}_i(C_j) \chi_i(C_k) = \dfrac{n}{c_k} \delta_{jk}$

and a character χ of G is irreducible iff

$$\sum_{i=1}^{s} c_i \bar{\chi}(C_i) \chi(C_i) = n$$

We return now to the case where G is Abelian, and for this case there is no difference between an irreducible representation and an irreducible character. In particular, let χ_i and χ_j be two irreducible representations of G and consider the representation $\chi_i \chi_j$ which is defined by

$$(\chi_i \chi_j)(g) = \chi_i(g) \chi_j(g)$$

That it is a representation follows from the commutativity of the group and since it is of dimension one, it is an irreducible representation and hence an irreducible character. We conclude that the irreducible characters of an Abelian group form a group under multiplication, where the group identity is the principal character. We shall denote this group G^*. It is noted that the above observations are not valid for a noncommutative group.

We will show that the group G^* is isomorphic to G, which will allow us to label the characters with group elements rather than the integers. If the least common multiple of the orders of all the group elements of G is m, then an irreducible character is a homomorphism from G into the mth roots of unity

4.8 Abelian Codes

contained in some extension field of $GF(2)$. Define the character χ_{a_i} on the generators of the cyclic groups in the decomposition of G by

$$\chi_{a_i}(a_i) = \varkappa_i, \qquad \chi_{a_i}(a_j) = 1, \qquad j \neq i$$

where $\varkappa_i$ is a primitive $p_i^{s_i}$th root of unity. Applying the above criterion for irreducibility, we observe that this is indeed an irreducible character of G. Henceforth the term character will automatically imply irreducible since we are only concerned with the multiplicative group G^*, i.e., strictly speaking, if χ_i and χ_j are two irreducible characters (representations) of G, then $\chi_i + \chi_j$ is the character of the direct sum representation $\chi_i \oplus \chi_j$, but we will not consider such characters.

Now suppose that χ is an arbitrary element of G^* and that

$$\chi(a_i) = \alpha_i, \qquad i = 1, \ldots, s$$

Then since a_i is of order $p_i^{s_i}$ we have that

$$\chi(a_i)^{p_i^{s_i}} = \chi(a_i^{p_i^{s_i}}) = \chi(1) = 1 = \alpha_i^{p_i^{s_i}}$$

which implies α_i is a $p_i^{s_i}$th root of unity. Thus we can write

$$\chi(a_i) = \varkappa_i^{k_i}$$

for some positive integer k_i. For an arbitrary element $g \in G$, where

$$g = a_1^{l_1} a_2^{l_2} \cdots a_s^{l_s}$$

we have that

$$\chi(g) = \prod_{i=1}^{s} \chi(a_i^{l_i}) = \prod_{i=1}^{s} \varkappa_i^{k_i l_i} = \prod_{i=1}^{s} \chi_{a_i}(a_i^{l_i})^{k_i}$$

from which we conclude that an arbitrary character in G^* can be expressed as a product of the "basic" characters χ_{a_i}, $i = 1, \ldots, s$.

This fact leads to the mapping

$$f: G \longrightarrow G^*$$
$$g = a_1^{l_1} a_2^{l_2} \cdots a_s^{l_s} \longmapsto \chi_{a_1}^{l_1} \chi_{a_2}^{l_2} \cdots \chi_{a_s}^{l_s} = \chi_g$$

That the map is a homomorphism follows from commutativity of the group. The injectivity and surjectivity of f is clear from the definition. We can thus label the elements of G^* with elements of G in such a way that

$$\chi_g \chi_h = \chi_{gh}$$

The orthogonality relationships for characters of a group reduce in the Abelian case to

$$\sum_{\chi_g \in G^*} \chi_g(h) = \begin{cases} |G| = n & \text{if } h = 1 \\ 0 & \text{otherwise} \end{cases}$$

and
$$\sum_{h \in G} \chi_g(h) = \begin{cases} |G| = n & \text{if } g = 1 \quad (\text{i.e., } \chi_1, \text{ the principal character}) \\ 0 & \text{otherwise} \end{cases}$$

It is also relatively straightforward to show that
$$\chi_g(h) = \chi_h(g) \quad \text{and} \quad \chi_{g^{-1}}(h) = \chi_h(g^{-1}), \quad h, g \in G$$
which are properties we shall need later.

It will be convenient at times to assume that the elements of G are ordered in some fashion, say $g_1 = 1, g_2, \ldots, g_n$. We can then define the character matrix Λ whose (i, j) element is $\chi_{g_j}(g_i)$. From the orthogonality relationships we conclude that
$$\Lambda \Lambda^T = nI_n$$
and hence that Λ is nonsingular.

It is convenient for later work to extend the notion of a group character to a character on the algebra in the following manner. Suppose $\chi \in G^*$ is some character of G; then define $\chi(\alpha)$, $\alpha \in GF(2)G$, $\alpha = \sum_G \alpha_g g$, by
$$\chi(\alpha) = \chi\left(\sum_G \alpha_g g\right) = \sum_G \alpha_g \chi(g)$$

Since G is assumed Abelian, it is a simple matter to check that χ is actually a representation of the algebra $GF(2)G$ and thus
$$\chi(\alpha\beta) = \chi\left(\left(\sum_G \alpha_g g\right)\left(\sum_G \beta_g g\right)\right) = \chi\left(\sum_G \alpha_g g\right)\chi\left(\sum_G \beta_g g\right), \quad \chi \in G^*$$
and in particular
$$\chi(\alpha^2) = \chi^2(\alpha), \quad \chi \in G^*, \quad \alpha \in GF(2)G$$

While for $\chi_{g_i}, \chi_{g_j} \in G^*$ it is true that
$$\chi_{g_i}(g)\chi_{g_i}(h) = \chi_{g_i}(gh), \quad g, h \in G$$
and
$$\chi_{g_i}(g)\chi_{g_j}(g) = \chi_{g_i g_j}(g)$$
it is not true in general that
$$\chi_{g_i}(g)\chi_{g_j}(h) = \chi_{g_i g_j}(gh)$$
Thus for two distinct characters $\chi_{g_i}, \chi_{g_j} \in G^*$ then, in general,
$$\chi_{g_i}\left(\sum_G \alpha_g g\right)\chi_{g_j}\left(\sum_G \beta_g g\right) \neq \chi_{g_i g_j}\left(\left(\sum_G \alpha_g g\right)\left(\sum_G \beta_g g\right)\right)$$

4.8 Abelian Codes

We are now in a position to use this information on characters to consider the structure of ideals in the group algebra, i.e., the Abelian codes. We first recall some basic facts on semisimple group algebras. We shall confine ourselves to the semisimple case, although the results of Berman (1967a) dealing with a particular nonsemisimple case will be briefly discussed later. Since the field characteristic assumed is two, we shall assume that G is of odd order n unless otherwise specified.

For this semisimple case we can express the group algebra as

$$GF(2)G = \bigoplus_{i=1}^{r} M_i$$

where M_i, $i = 1, \ldots, r$, is a (unique) minimal ideal. Furthermore, there exists a set of mutually orthogonal idempotents e_i, $i = 1, \ldots, r$, such that

$$1 = e_1 + \cdots + e_r, \qquad e_i e_j = \delta_{ij} e_j$$

and

$$M_i = GF(2)G e_i, \qquad i = 1, \ldots, r$$

The idempotent e_i is a unit for M_i and annihilates M_j, $j \neq i$. It is primitive and a generating idempotent for M_i.

Suppose we extend the field $GF(2)$ to contain all the characters of the group G, say $GF(2^e)$. Then, as is proven in the next chapter [Eq. (8.1)], there exists a simple formula for the generating idempotents $\{e_i'\}$ of the minimal ideals $\{M_i'\}$ of $GF(2^e)G$, namely

$$e_i' = \frac{1}{n} \sum_{g \in G} \bar{\chi}_{g_i}(g) g, \qquad i = 1, \ldots, n$$

and the number of minimal ideals is precisely the same as the order of the group. It is possible to relate these minimal ideals with the minimal ideals of the algebra $GF(2)G$. Specifically, it is possible to express the minimal ideal $M_i \subset GF(2)G$ as the direct sum

$$M_i = M'_{i_1} \oplus M'_{i_2} \oplus \cdots \oplus M'_{i_{m_i}}, \qquad M'_{i_j} \subset GF(2^e)G$$

and the generating idempotent of M_i is

$$e_i = e'_{i_1} + e'_{i_2} + \cdots + e'_{i_{m_i}}$$

There is another method for obtaining expressions for the orthogonal generating idempotents $\{e_i\}$ that requires some computation. Since it is an informative exercise, we consider this method now. We shall denote arbitrary elements of the algebra by lowercase Greek letters $\alpha, \beta, \ldots$. Notice that $\alpha = \sum_G \alpha_g g$, $\alpha \in GF(2)G$, is an idempotent if

$$\alpha^2 = \left(\sum_G \alpha_g g\right)^2 = \sum \alpha_g g^2 = \alpha = \sum \alpha_g g$$

which implies that $\alpha_g = \alpha_{g^2}$, $g \in G$. With this observation we can generate a trivial set of idempotents, which we can use to determine the primitive orthogonal generating idempotents. We first decompose the group G into sets S_i which we form in the following manner. Let $S_1 = \{g_1 = 1\}$, and choose $g_2 \notin S_1$. Then

$$S_2 = \{g_2, g_2^2, g_2^{2^2}, \ldots, g_2^{2^{k-1}}\}$$

where $g_2^{2^k} = g_2$, which, since G is finite, must eventually happen. Choosing $g_3 \notin S_1 \cup S_2$, we continue in this manner until we have decomposed G and $G = S_1 \cup S_2 \cup \cdots \cup S_{r'}$. It is then clear that the elements

$$\sigma_i = \sum_{g \in S_i} g, \qquad i = 1, \ldots, r'$$

are idempotents in $GF(2)G$. Viewing $GF(2)G$ as a vector space over $GF(2)$, these elements are linearly independent. If $g \in G$ appears with a nonzero coefficient in the expression for a generating idempotent e_i, then g^2 must also appear with a nonzero coefficient. It follows that every generating idempotent may be expressed as a sum of trivial idempotents. Now the set of all idempotents in $GF(2)G$ is a vector subspace of the algebra and any idempotent is the sum over $GF(2)$ of generating idempotents. To see this, suppose that $\{e_i\}$, $i = 1, \ldots, r$, is a set of orthogonal generating idempotents of the ideals and let $\alpha \in GF(2)G$ be an idempotent. Then we can write

$$\alpha = \alpha_1 + \alpha_2 + \cdots + \alpha_r, \qquad \alpha_i \in GF(2)Ge_i$$

where α_i is idempotent in $GF(2)Ge_i$. Using the same argument as in the previous section, since e is a unit of $GF(2)Ge_i$, we have that

$$\alpha_i e_i = \alpha_i \qquad \text{or} \qquad \alpha_i(e_i - \alpha_i) = 0$$

However, since $GF(2)Ge_i$ is a minimal ideal, it has no nontrivial subideals and hence can have no divisors of zero. [Actually, since $GF(2)Ge_i$ is thus a finite division ring it is, by Wedderburn's theorem, a finite field.] We conclude that $\alpha_i = e_i$. Thus the number of trivial idempotents is the same as the number of orthogonal generating idempotents, i.e., $r = r'$. Since the generating idempotents generate disjoint subspaces of the algebra, they are independent and thus there exists an invertible matrix $T = (t_{ij})$ such that

$$e_i = \sum_{j=1}^r t_{ij} \sigma_j, \qquad i = 1, \ldots, r$$

4.8 Abelian Codes

Before establishing the formula we require, we need the fact that an element of the group algebra α is uniquely determined by the set of values $\{\chi_g(\alpha), g \in G\}$ as expressed by the following lemma.

Lemma If $\alpha = \sum_G \alpha_g g$, then

$$\alpha_g = \frac{1}{n} \sum_{f \in G} \chi_f(\alpha) \chi_f(g^{-1})$$

Conversely if

$$\alpha_g = \frac{1}{n} \sum_{h \in G} \beta_h \chi_h(g^{-1})$$

then $\chi_h(\alpha) = \beta_h$. Thus the set of values $\chi_h(\alpha)$ as h runs through G uniquely determines α.

Proof Suppose that $\alpha = \sum_G \alpha_g g$ and consider the sum

$$\sum_{h \in G} \chi_h(\alpha) \chi_h(g^{-1}) = \sum_{h \in G} \left[\sum_{f \in G} \alpha_f \chi_h(f) \right] \chi_h(g^{-1})$$

$$= \sum_{f \in G} \alpha_f \left[\sum_{h \in G} \chi_h(f) \chi_h(g^{-1}) \right]$$

$$= \sum_{f \in G} \alpha_f (n \, \delta_{fg}) = n \alpha_g$$

Conversely, suppose that

$$\alpha_g = \frac{1}{n} \sum_{h \in G} \beta_h \chi_h(g^{-1})$$

and consider the expression

$$\chi_h(\alpha) = \sum_{g \in G} \alpha_g \chi_h(g) = \sum_{g \in G} \left[\frac{1}{n} \sum_{f \in G} \beta_f \chi_f(g^{-1}) \right] \chi_h(g)$$

$$= \frac{1}{n} \sum_{f \in G} \beta_f \sum_{g \in G} \chi_f(g^{-1}) \chi_h(g)$$

$$= \frac{1}{n} \sum_{f \in G} \beta_f \sum_{g \in G} \chi_g(f^{-1}) \chi_g(h)$$

$$= \frac{1}{n} \sum_{f \in G} \beta_f (n \, \delta_{fh}) = \beta_h$$

as required. Thus the values of the characters on the element uniquely determine the element. □

The minimal ideals M_i, $i = 1, \ldots, r$, of the group algebra are unique and every ideal is a direct sum of minimal ideals. Thus there are $2^r - 1$ distinct nonzero ideals. The generating idempotent of each such ideal is the sum of the generating idempotents of its constituent minimal ideals and the generating idempotent is a unit for the ideal. If we can find a set of r orthogonal idempotents e_i'', $i = 1, \ldots, r$, such that $e_i'' e_j'' = \delta_{ij} e_j''$ and

$$1 = e_1'' + \cdots + e_r''$$

then $GF(2)Ge_i''$ is a nonzero minimal ideal of $GF(2)G$ with unit e_i''. Since the unit of an ideal is unique, we must have a one-to-one correspondence between the set of idempotents $\{e_i''\}$ and the desired set $\{e_i\}$.

In order to define such a set of idempotents, we first decompose the group G^* into subsets Σ_i such that

$$G^* = \Sigma_1 \cup \Sigma_2 \cup \cdots \cup \Sigma_r$$

where

$$\Sigma_i = \{\chi_g \in G^* : g \in S_i\}$$

where χ_g is the character corresponding to $g \in G$ under the isomorphism between G and G^*.

Now define the elements e_i, $i = 1, \ldots, r$, by the property that

$$\chi_g(e_i) = \begin{cases} 1 & \chi_g \in \Sigma_i \\ 0 & \text{otherwise} \end{cases}$$

From the above lemma then we have that

$$e_i = \sum_{g \in G} \left[\frac{1}{n} \sum_{\chi_h \in \Sigma_i} \chi_h(g^{-1}) \right] g$$

and it remains to show that this defines a set of orthogonal idempotents whose sum is unity. For e_i to be idempotent, it is only necessary to check that $\alpha_g = \alpha_{g^2}$ or, in this case,

$$\frac{1}{n} \sum_{\chi_h \in \Sigma_i} \chi_h(g^{-1}) = \left[\frac{1}{n} \sum_{\chi_h \in \Sigma_i} \chi_h(g^{-1}) \right]^2 = \frac{1}{n} \sum_{\chi_h \in \Sigma_i} \chi_h^2(g^{-1})$$

But since by definition Σ_i is closed under the squaring operation, equality follows. To show that the idempotents sum to the unit element, consider the expression

$$\sum_{i=1}^r e_i = \sum_{g \in G} \left\{ \sum_{i=1}^r \left[\frac{1}{n} \sum_{\chi_h \in \Sigma_i} \chi_h(g^{-1}) \right] \right\} g = \sum_{g \in G} \left[\frac{1}{n} \sum_{\chi_h \in G^*} \chi_h(g^{-1}) \right] g = 1$$

since the inner sum of the last equation is zero for any $g \neq 1$. To show the orthogonality of the idempotents, suppose

$$e_i = \sum \alpha_g g \quad \text{and} \quad e_j = \sum \beta_g g, \quad i \neq j$$

4.8 Abelian Codes

Then we have

$$e_i e_j = (\sum \alpha_g g) \sum \beta_g g = \sum_{f \in G} \left(\sum_{g \in G} \alpha_g \alpha_{g^{-1}f} \right) f = \sum_{f \in G} \gamma_f f$$

From the definition of the coefficients we can write

$$\gamma_f = \sum_{g \in G} \left[\frac{1}{n} \sum_{\chi_h \in \Sigma_i} \chi_h(g^{-1}) \right] \left[\frac{1}{n} \sum_{\chi_{h'} \in \Sigma_j} \chi_{h'}((g^{-1}f)^{-1}) \right]$$

$$= \frac{1}{n^2} \sum_{\chi_h \in \Sigma_i} \sum_{\chi_{h'} \in \Sigma_j} \left[\sum_{g \in G} \chi_h(g^{-1}) \chi_{h'}(f^{-1}g) \right]$$

$$= \frac{1}{n^2} \sum_{\chi_h \in \Sigma_i} \sum_{\chi_{h'} \in \Sigma_j} \left[\sum_{g \in G} \chi_h(g^{-1}) \chi_{h'}(g) \right] \chi_{h'}(f^{-1})$$

$$= 0$$

since the summation in brackets is always zero since $h \neq h'$ by assumption. Thus the equation

$$e_i = \sum_{g \in G} \left[\frac{1}{n} \sum_{\chi_h \in \Sigma_i} \chi_h(g^{-1}) \right] g, \quad i = 1, \ldots, r$$

defines the set of orthogonal generating idempotents of the minimal ideals M_i, $i = 1, \ldots, r$.

There are other methods of specifying ideals which are more convenient for such problems as determining the dimensions of an ideal as a vector space. We consider some of these methods now. If we form the ideal

$$M = M_{i_1} \oplus \cdots \oplus M_{i_k}$$

then there is an ideal

$$M' = M_{j_1} \oplus \cdots \oplus M_{j_{r-k}}$$

where $j_l \neq i_m$, $l = 1, \ldots, r - k$, $m = 1, \ldots, k$, and $GF(2)G = M \oplus M'$. Since the idempotent generators of the ideals M_{j_l}, $l = 1, \ldots, r - k$, are annihilators of the ideals M_{i_m}, $m = 1, \ldots, k$, we can as well define the ideal M by

$$M = \{\alpha \in GF(2)G : \alpha e_{j_l} = 0, l = 1, \ldots, r - k\}$$

or equivalently

$$M = \{\alpha \in GF(2)G : \alpha e_{i_m} \neq 0, m = 1, \ldots, k\}$$

In the case of a cyclic code, however, we had a very convenient criterion for determining the dimension of the code in terms of the roots of the codeword polynomials. We consider the cyclic case for a moment with a view to extending those methods to the general Abelian case. Suppose then that $G = \langle g \rangle$ is a cyclic group of order n and let z be a primitive nth root of unity. The characters of G are then defined by

$$\chi_{g^i}(g) = z^i, \qquad i = 1, \ldots, n$$

Every cyclic code is generated by a codeword polynomial $\sum_{i=0}^{n-1} \alpha_{g^i} g^i$, which is completely specified by its roots in an extension field of $GF(2)$. The dimension of the code is n minus the degree of the generator polynomial. If we let S be that subset of nth roots of unity that are roots of the generator polynomial, then the code or ideal M can be defined by

$$M = \left\{ \left(\sum_i \alpha_{g^i} g^i \right) \in GF(2)G : \sum_i \alpha_{g^i}(z^k)^i = 0, \; z^k \in S \right\}$$

and dim $M = n - |S|$, assuming S is the maximal set of nth roots of unity that are also roots of every codeword polynomial. Since $z^k = \chi_{g^k}(g)$, we can define a subset Σ of G^* such that

$$M = \left\{ \left(\sum_i \alpha_{g^i} g^i \right) \in GF(2)G : \chi_{g^k}\left(\sum_i \alpha_{g^i} g^i \right) = 0, \; \chi_{g^k} \in \Sigma \right\}$$

or, conversely,

$$M = \left\{ \left(\sum_i \alpha_{g^i} g^i \right) \in GF(2)G : \chi_{g^k}\left(\sum \alpha_{g^i} g^i \right) \neq 0 \text{ for at least one } \chi_{g^k} \in \bar{\Sigma} \right\}$$

where $\bar{\Sigma} = G^* \backslash \Sigma$ and $\dim(M) = |\bar{\Sigma}|$. In the polynomial description of ideals if (z^i) is a root of the generating polynomial, then $(z^i)^2$, $(z^i)^{2^2}$, ... are also roots and thus S is a union of chains of powers of z, as mentioned in Chapter 1. Similarly, if $\chi_{g^k}(\alpha) = 0$ for $\alpha \in M$, then $\chi_{g^k}^2(\alpha) = 0$ for $\alpha \in M$ and thus Σ is a union of the Σ_i defined previously for the general Abelian case. In the cyclic case a polynomial and roots of polynomials approach to the problem simplifies matters. A natural extension to the Abelian case would be to consider multivariable polynomials and their roots. This appears to be cumbersome, however, and it seems advantageous in this case to use characters.

Every ideal M in the general Abelian group G has an idempotent generator $e = \sum_G \varepsilon_g g$, which is also a unit for the ideal. The analog to the above situation for the general Abelian case is then contained in the following theorem.

Theorem 8.1 (*MacWilliams*, 1970) The dimension of the ideal $GF(2)Ge$, where e is the idempotent generator, is the number of characters $\chi_g \in G^*$ such that $\chi_g(e) \neq 0$.

4.8 Abelian Codes

Proof As before, we assume some ordering $\{g_i, i = 1, \ldots, n\}$ of the elements of G, and recall that $\Lambda = (\lambda_{ij})$, $\lambda_{ij} = \chi_{g_j}(g_i)$. If the generator e is given by $\sum_G \varepsilon_{g_i} g_i$, then define the $n \times n$ matrix $\varepsilon = (\varepsilon_{ij})$, $\varepsilon_{ij} = \varepsilon_{g_i^{-1} g_j}$. Since Λ is nonsingular, we have that $\overline{\Lambda}^T \varepsilon \Lambda$ has the same rank as ε. But a straightforward computation shows that

$$(\overline{\Lambda}^T \varepsilon \Lambda)_{ij} = \sum_{k,l} \bar{\chi}_{g_k}(g_i) \varepsilon_{g_k^{-1} g_l} \chi_{g_j}(g_l) = \sum_k \bar{\chi}_{g_k}(g_i) \sum_l \varepsilon_{g_k^{-1} g_l} \chi_{g_j}(g_l)$$

$$= \sum_k \bar{\chi}_{g_k}(g_i) \sum_h \varepsilon_{g_h} \chi_{g_j}(g_k g_h) = \sum_k \bar{\chi}_{g_k}(g_i) \chi_{g_j}(e) \chi_{g_j}(g_k)$$

$$= \chi_{g_j}(e) \sum_k \bar{\chi}_{g_k}(g_i) \chi_{g_k}(g_j) = n \chi_{g_j}(e) \delta_{ij}$$

Thus

$$\overline{\Lambda}^T \varepsilon \Lambda = \mathrm{diag}(n \chi_{g_1}(e), n \chi_{g_2}(e), \ldots, n \chi_{g_n}(e))$$

and the rank of ε is the number of characters that do not vanish on e. Now the elements $(g_j e, j = 1, \ldots, n)$ span $GF(2)Ge$ as a vector space and the jth element in the ith row of ε is the coefficient of g_j in $g_i e$, i.e., the ith row corresponds to the codeword $g_j e$. Thus the rank of ε is the dimension of $GF(2)Ge$. □

The above theorem remains true for any generator for the ideal, the idempotent generator being chosen as a matter of convenience. Applying this theorem to minimum ideals, we observe that for the minimum ideal M_i generated by the idempotent e_i defined by

$$\chi_g(e_i) = \begin{cases} 1, & \chi_g \in \Sigma_i \\ 0, & \text{otherwise} \end{cases}$$

it follows trivially that $\dim M_i = |\Sigma_i|$. Similarly, for any ideal M

$$M = M_{i_1} \oplus \cdots \oplus M_{i_k}$$

and $\dim M = \sum_{j=1}^{k} |\Sigma_{i_j}|$. It also follows that if M and N are two ideals corresponding to subsets Σ and Σ' of G^* (Σ, Σ' are unions of certain of the Σ_i) and if $\Sigma \subset \Sigma'$, then $M \subset N$. Thus we have a method of determining the dimension of an ideal which is a direct extension of a method used for the case of cyclic codes.

Finally, we mention how to obtain the dual code of an ideal. To each element $\alpha = \sum \alpha_g g$ of M, M an ideal of $GF(2)G$, we can associate an n-tuple $(\alpha_1, \alpha_2, \ldots, \alpha_n)$ of the coefficients of α. Two n-tuples associated with $\alpha, \beta \in GF(2)G$ are orthogonal if

$$\sum_{i=1}^n \alpha_i \beta_i = 0$$

and the dual code or ideal of M is defined by

$$M^\perp = \{\alpha \in GF(2)G : \sum_{i=1}^{n} \alpha_i \beta_i = 0 \quad \text{for all} \quad \beta \in M\}$$

Now suppose M is generated by the idempotent e. Then $1 + e$ generates an ideal of dimension $n - \dim M$, since, if

$$e = e_{i_1} + \cdots + e_{i_k}$$

then

$$1 + e = e_{j_1} + \cdots + e_{j_{r-k}}$$

where $i_l \neq j_m$, $l = 1, \ldots, k$, $m = 1, \ldots, r - k$, and

$$GF(2)Ge \oplus GF(2)G(1 + e) = GF(2)G$$

If $\alpha = \sum_G \alpha_g g$, then we define α^* as $\sum_G \alpha_{g^{-1}} g$. It is readily checked that $(\alpha\beta)^* = \alpha^*\beta^*$. Suppose that

$$\beta = \sum \beta_g g \in GF(2)G(1 + e)^*$$

Then

$$\beta^* = \sum \beta_{g^{-1}} g \in GF(2)G(1 + e)$$

However, since $e(1 + e) = 0$, we have that $\alpha\beta^* = 0$ for any element $\alpha \in GF(2)Ge$ and $\beta^* \in GF(2)G(1 + e)$. If $\alpha = \sum \alpha_g g$, then

$$\alpha\beta^* = 0 = \sum_g \sum_h \alpha_g \beta_{h^{-1}} gh = \sum_f \left(\sum_g \alpha_g \beta_{gf^{-1}}\right) f$$

The coefficient of $f = 1$ is thus zero or

$$\sum_{i=1}^{n} \alpha_{g_i} \beta_{g_i} = 0$$

implying that $GF(2)G(1 + e)^* \subseteq (GF(2)Ge)^\perp$. If the dimension of $GF(2)Ge$ is k, then, from Chapter 1, $\dim(GF(2)Ge)^\perp = n - k$. Since the operation * preserves independence of elements as vectors

$$\dim(GF(2)G(1 + e)^*) = \dim(GF(2)G(1 + e)) = \dim((GF(2)Ge)^\perp)$$

and we conclude that $(GF(2)Ge)^\perp = GF(2)G(1 + e)^*$, giving us the dual of the code.

The above has been a rather detailed account of the structure of ideals in Abelian group algebras over $GF(2)$. The remainder of the section surveys the results that have been obtained in the literature of Abelian codes. We begin with the work of MacWilliams (1970), from which much of the foregoing material was adapted.

4.8 Abelian Codes

Consider the case when $G = H \times K$, the direct product of subgroups H and K of Abelian group G, and let e be an idempotent of $GF(2)H$ and e' an idempotent of $GF(2)K$. If M and M' are the corresponding codes of e and e', then $M \otimes M'$, where $\otimes$ indicates the usual tensor or Kronecker product, corresponds to the ideal generated by ee' in $GF(2)G$. Furthermore, if Σ and Σ' are the sets of nonzeros of M and M', respectively, then the set of nonzeros of $M \otimes M'$ is given by

$$\{\chi_{1h_i}\chi_{2k_j}, \chi_{1h_i} \in \Sigma, \chi_{2k_j} \in \Sigma'\}$$

Conversely, now suppose that Σ'' is the set of nonzeros of characters of the ideal M'' in $GF(2)G$. Then if we can find subgroups H and K such that $G = H \times K$ and

$$\Sigma'' = \{\chi_{1h_i}\chi_{2k_j}, \chi_{1h_i} \in \Sigma, \chi_{2k_j} \in \Sigma'\}$$

then M can be expressed as the Kronecker product of subcodes of $GF(2)H$ and $GF(2)K$. As mentioned previously, however, not all Abelian codes are such simple direct product codes (Camion, 1970). A necessary and sufficient condition to determine those that are, however, seems to be unavailable at the moment.

On observing the duality between the definitions of an orthogonal generating idempotent

$$\chi_{g_j}(e_i) = \begin{cases} 1 & \chi_{g_j} \in \Sigma_i \\ 0 & \text{otherwise} \end{cases}$$

and the trivial idempotent $\sigma_i = \sum_G \alpha_g g$, where

$$\alpha_g = \begin{cases} 1 & g \in S_i \\ 0 & \text{otherwise} \end{cases}$$

MacWilliams (1970) was able to show that if

$$e_i = \sum_{j=1}^{r} \lambda_j \sigma_j$$

then

$$\sigma_i^* = \sum_{j=1}^{r} \lambda_j e_i$$

As a final interesting property of idempotents, Macwilliams (1970) proved that if T is the $r \times r$ matrix which maps σ_i into e_i, $i = 1, \ldots, r$, then

$$T^2 = P$$

where P is the permutation matrix which takes σ_i into σ_i^*, $i = 1, \ldots, r$. These properties were used in examples of Abelian codes and yielded interesting results even in the cyclic case.

Some very significant work on Abelian codes appears in the papers of the Soviet mathematician Berman (1967a, b). The first paper (Berman, 1967a) is concerned with the nonsemisimple case. Considering the semisimple case first, the contents of his Theorems 2.1 and 2.2 can be stated as follows. Let $p_1, p_2, \ldots, p_s$ be fixed odd prime numbers and M_n a cyclic code of length n over $GF(2)$. Denote the dimension and minimum distance of M_n by $k(M_n)$ and $d(M_n)$ respectively. Consider a sequence of codes $\{M_n, n \in I\}$ where I is an index set containing integers of the form $n = p_1^{\alpha_1} p_2^{\alpha_2} \cdots p_s^{\alpha_s}$, α_i a positive integer, $i = 1, \ldots, s$. Then if $k(M_n)/n \geq \beta > 0$ for each code in the set, then there exists a constant c such that $d(M_n) \leq c$ for each $n \in I$. Thus, if I is arranged as a sequence of increasing integers, then $\lim[d(M_n)/n] \to 0$. On the other hand, there exists a class of Abelian codes $\{M_n, n \in I\}$, M_n of length $n = p_1^{\alpha_1} p_2^{\alpha_2} \cdots p_s^{\alpha_s}$ such that

$$\lim_{(\alpha_1 + \cdots + \alpha_s) \to \infty} \frac{k(M_n)}{n} = 1$$

while $d(M_n) \to \infty$. Thus, in this sense, for a constant rate code of length n, n a composite number with a fixed set of primes, there exist Abelian codes having better error-correcting capabilities than cyclic codes.

Almost as a side issue, and apparently to compare the error-correcting capabilities of Abelian and cyclic codes, Berman (1976a, b) finds the minimum distance of any cyclic code of length p^n for both the semisimple and non-semisimple cases. Because his results for this case are so interesting, we shall consider them briefly here. Again considering the semisimple case first, we let $G = \langle g \rangle$ be a cyclic group of order p^n for some prime p and $GF(q)$ be a finite field, where q is a prime such that q is a primitive root mod p and mod p^2. This of course implies that q is a primitive root with respect to the moduli p^i for any positive integer i. In this situation the group algebra has precisely $n + 1$ minimal ideals and a complete set of orthogonal generating idempotents is given by the equations

$$e_0 = \frac{1}{p^n} \sum_{i=0}^{p^n - 1} g^i$$

$$e_i = \frac{1}{p^{n-i+1}} (1 + g^{p^i} + g^{p^i 2} + g^{p^i 3} + \cdots + g^{p^i (p^{n-i} - 1)})$$

$$\times [(p-1) - g^{p^{i-1}} - g^{2p^{i-1}} - \cdots - g^{(p-1)p^{i-1}}], \quad i = 1, \ldots, n$$

When $GF(q) = GF(2)$ then these equations, up to a relabeling, are identical to the ones given previously. Verification of the conditions required for idempotents is tedious in this general case. The dimension of the ideal

4.8 Abelian Codes

$M_i = GF(q)Ge_i$ is $\varphi(p^i)$, $i = 0, 1, \ldots, n$, where φ is Euler's phi function. By considering forms of elements in the ideal M_i it can be shown that

$$d(M_0) = p^n \quad \text{and} \quad d(M_i) = 2p^{n-i}, \quad i = 1, \ldots, n$$

where, as before, $d(M)$ is the minimum distance of the ideal M. Now if M is an arbitrary ideal

$$M = M_{i_1} \oplus M_{i_2} \oplus \cdots \oplus M_{i_r}$$

where $i_1 > i_2 > \cdots > i_r$, then by similar arguments it can be shown that

$$d(M) = \begin{cases} 2p^{n-i_1} & \text{if } i_r > 0 \\ p^{n-i_1} & \text{if } i_r = 0 \end{cases}$$

As mentioned previously, these results were used to compare Abelian codes with cyclic codes. Specifically, the minimum distance of ideals in the group algebras $GF(2)C_9$, $GF(2)(C_3 \times C_3)$, $GF(2)C_{25}$, and $GF(2)(C_5 \times C_5)$ are compared and the results appear encouraging for continued investigation of Abelian codes.

As a general rule, the nonsemisimple case is more difficult to deal with since the ideal structure of these algebras is not so simple. Berman (1967a) restricts his attention to Abelian p-groups, i.e., one that can be expressed as

$$G = C_{p^{\alpha_1}} \times C_{p^{\alpha_2}} \cdots \times C_{p^{\alpha_s}}, \quad |G| = n$$

where $C_{p^{\alpha_i}}$ is a cyclic group of order p^{α_i}, with generator g_i, and we assume that $\alpha_1 \geq \alpha_2 \geq \cdots \alpha_s \geq 0$. We consider the group algebra of G over a field $GF(q)$, characteristic p, and in this nonsemisimple case the radical R of the group algebra is an important ideal. It is easily shown that R has dimension $n - 1$ over $GF(q)$ and that the elements

$$(g_1 - 1)^{k_1}(g_2 - 1)^{k_2} \cdots (g_s - 1)^{k_s}, \quad 0 \leq k_i \leq p^i - 1$$

(where not all the powers are identically zero) form a basis for R over $GF(q)$. If we define the index of nilpotency of R by

$$t = (p^{\alpha_1} - 1) + \cdots + (p^{\alpha_s} - 1)$$

then no two powers R, R^2, R^3, $\ldots$, R^{t-1} are equal and in fact when $G = C_{p^m} = \langle g \rangle$, a cyclic group of order p^m, then these powers of the radical form *all* the nontrivial ideals of the group algebra $GF(q)G$. In the general case a basis for R^j can be taken of the form

$$(g_1 - 1)^{k_1}(g_2 - 1)^{k_2} \cdots (g_s - 1)^{k_s}, \quad k_1 + k_2 + \cdots + k_s \geq j$$

while for the cyclic case R^j is the ideal generated by the element $(g - 1)^j$.

Theorem 8.2 (Berman, 1967a) Let $G = C_{p^m} = \langle g \rangle$ be a cyclic group of order p^m and $GF(q)$ a field of characteristic p. Consider the ideal R^i generated by $(g-1)^i$ and let the expansion to the base p of the integer i be

$$i = \beta_1 p^{m-1} + \beta_2 p^{m-2} + \cdots + \beta_m, \qquad 0 \leq \beta_j \leq p-1$$

If $\beta_1 = \cdots = \beta_m = p-1$, then the corresponding ideal R^{p^m-1} is generated by the element $1 + g + g^2 + \cdots + g^{p^m-1}$, is one dimensional, and has distance p^m. Let r be the smallest integer for which $\beta_r \neq p-1$ in the expansion of the given integer i. Then we have that

$$d(R^i) = \begin{cases} (\beta_r + 2)p^{r-1}, & (\beta_{r+1}, \ldots, \beta_m) \neq (0, \ldots, 0) \\ (\beta_r + 1)p^{r-1}, & (\beta_{r+1}, \ldots, \beta_m) = (0, \ldots, 0) \end{cases}$$

where $i \neq p^m - 1$.

A subclass of the codes described in the above theorem, namely those of prime length p over $GF(p^k)$, were discussed as optimal codes by Assmus and Mattson (1969).

For a more general Abelian group than the cyclic case above, the situation becomes considerably more intricate. Let G be the Abelian group

$$G = \underbrace{C_{p^{\alpha_1}} \times \cdots \times C_{p^{\alpha_1}}}_{s_1} \times \underbrace{C_{p^{\alpha_2}} \times \cdots \times C_{p^{\alpha_2}}}_{s_2}$$
$$\times C_{p^{\alpha_3}} \times \cdots \times C_{p^{\alpha_{m-1}}} \times \underbrace{C_{p^{\alpha_m}} \times \cdots \times C_{p^{\alpha_m}}}_{s_m}$$

where

$$|G| = p^{s_1 \alpha_1 + \cdots + s_m \alpha_m} \quad \text{and} \quad \alpha_1 > \alpha_2 > \cdots > \alpha_m$$

Consider the integers $m_i = l_i(p-1)p^i$, $i = 0, 1, \ldots, \alpha_1$, defined by the equations

$$l_{\alpha_1} = m_{\alpha_1} = 0$$
$$l_{\alpha_1 - 1} = l_{\alpha_1 - 2} = \cdots = l_{\alpha_2} = s_1$$
$$l_{\alpha_2 - 1} = l_{\alpha_2 - 2} = \cdots = l_{\alpha_3} = s_1 + s_2$$
$$\vdots$$
$$l_{\alpha_m - 1} = l_{\alpha_m - 2} = \cdots = l_0 = s_1 + s_2 + \cdots + s_m$$

It is then easy to check that

$$\sum_{i=0}^{\alpha_1} m_i = s_1(p^{\alpha_1} - 1) + \cdots + s_m(p^{\alpha_m} - 1) = t$$

Now for any integer $1 \leq j < t$, we can always find an integer r such that

$$m_{\alpha_1} + m_{\alpha_1 - 1} + \cdots + m_{\alpha_1 - r} \leq j < m_{\alpha_1} + \cdots m_{\alpha_1 - r} + m_{\alpha_1 - r - 1}$$

4.8 Abelian Codes

which implies that we can write

$$j = m_{\alpha_1} + \cdots + m_{\alpha_1 - r} + \lambda$$

where, since $\lambda < m_{\alpha_1 - r - 1}$, we have that

$$0 \leqslant \lambda < m_{\alpha_1 - r - 1} = l_{\alpha_1 - r - 1}(p - 1)p^{\alpha_1 - r - 1}$$

Thus for some integer k, $0 \leqslant k < l_{\alpha_1 - r - 1}$, we have that

$$k(p - 1)p^{\alpha_1 - r - 1} \leqslant \lambda < (k + 1)(p - 1)p^{\alpha_1 - r - 1}$$

and hence that

$$\lambda_1 = \lambda - k(p - 1)p^{\alpha_1 - r - 1} < (p - 1)p^{\alpha_1 - r - 1}$$

and thus we can consider the p expansion of λ_1 as

$$\lambda_1 = \delta_{\alpha_1 - r - 1} p^{\alpha_1 - r - 1} + \cdots + \delta_0$$

$$0 \leq \delta_j \leq p - 1, \quad j = 0, 1, \ldots, \alpha_1 - r - 2, \quad \delta_{\alpha_1 - r - 1} < p - 1$$

We finally conclude that for any integer j, $1 \leqslant j < t$, we have the unique expansion

$$j = m_{\alpha_1} + \cdots + m_{\alpha_1 - r} + k(p - 1)p^{\alpha_1 - r - 1} + \delta_{\alpha_1 - r - 1} p^{\alpha_1 - r - 1} + \cdots + \delta_0$$

Using this expansion, we define the function $d(j)$ as

$$d(j) = \begin{cases} p^{l_{\alpha_1} + \cdots + l_{\alpha_1 - r} + t(\delta_{\alpha_1 - r - 1} + 2)}, & (\delta_{\alpha_1 - r - 2}, \ldots, \delta_0) \neq (0, \ldots, 0) \\ p^{l_{\alpha_1} + \cdots + l_{\alpha_1 - r} + t(\delta_{\alpha_1 - r - 1} + 1)}, & (\delta_{\alpha_1 - r - 2}, \ldots, \delta_0) = (0, \ldots, 0) \end{cases}$$

We are now in a position to state the following theorem of Berman:

Theorem 8.3 (Berman, 1967a) Let G be an Abelian p-group of the form

$$G = \underbrace{C_{p^{\alpha_1}} \times \cdots \times C_{p^{\alpha_1}}}_{s_1} \times C_{p^{\alpha_2}} \times \cdots \times C_{p^{\alpha_{m-1}}} \times \underbrace{C_{p^{\alpha_m}} \times \cdots \times C_{p^{\alpha_m}}}_{s_m}$$

and let R be the radical of the group algebra $GF(q)G$, char $GF(q) = p$. Then the distance of the code R^j,

$$1 \leqslant j \leqslant s_1(p^{\alpha_1} - 1) + \cdots + s_m(p^{\alpha_m} - 1)$$

is given by the formula

$$d(R^j) = \begin{cases} |G| & \text{when } j = s_1(p^{\alpha_1} - 1) + \cdots + s_m(p^{\alpha_m} - 1) \\ d(j) & \text{when } 1 \leqslant j < s_1(p^{\alpha_1} - 1) + \cdots + s_m(p^{\alpha_m} - 1) \end{cases}$$

For the special case when $G = C_{p^m}$ the formulas for distances of ideals in the group algebra contained in this and the previous theorem coincide, with the appropriate identification of parameters.

Another interesting special case of this theorem occurs when
$$G = C_p \times C_p \times \cdots \times C_p \quad (s \text{ times})$$
From the above theorem we have
$$d(R^j) = p^k(r+1), \quad 1 \leq j \leq (p-1)s$$
when $j = (p-1)k + r$, $0 \leq r < p-1$, $0 \leq k \leq s$. When $p = 2 = q$, then this formula becomes
$$d(R^j) = 2^l, \quad 1 \leq l \leq s$$
and it turns out that these codes are simply the Reed–Muller codes.

The work of Delsarte (1970b, and 1971b) on Abelian codes is also of interest but a little too detailed to consider in depth here. Delsarte (1970b) examines automorphisms and equivalence of Abelian codes and gives particular attention to the elementary Abelian group algebras. A weight enumeration formula of McEliece for cyclic codes over $GF(p)$ is extended to Abelian codes over $GF(p)$ by Delsarte (1971b).

From the results described in this section it would appear that Abelian codes merit further investigation.

4.9 Comments

The basic theory of semisimple rings and their relationship to group representations has been discussed. The books by Artin *et al.* (1944) and Curtis and Reiner (1966) are excellent references and our presentation tended to follow these, at least in spirit. The work of Burrow (1965) also contains much interesting material. The two books by McCoy (1948, 1964) are very readable accounts of ring theory and the work of Lang (1971) is a comprehensive reference. The approach to coding contained in Sections 4.7 and 4.8 is the work of MacWilliams (1965, 1969, 1970).

Exercises

1 Let e be a primitive idempotent of a semisimple algebra A. Show that eAe is a division algebra.

2 Show that if a semisimple algebra has an identity and no other idempotent element, then it is a division algebra.

3 Show that the only division algebras over the real field are $\mathbb{R}$, $\mathbb{C}$, and $\mathbb{H}$.

Exercises

4 Show that if R is a simple ring with identity, then the complete matrix ring R_n of $n \times n$ matrices over R is also simple. Conclude that if R is a division ring, then R_n is simple.

5 Show that the d.c.c. holds in a ring R with identity iff in every nonempty set of left ideals of R there is a minimal ideal that does not properly contain any other ideal in the set.

6 Show that if I is a two-sided ideal of R, then I_n is a two-sided ideal of R_n, where I_n and R_n are the corresponding complete matrix rings.

7 If e_1 and e_2 are two idempotents in a ring R, show that $u \in e_1 R e_2$ iff it satisfies the equation $u = e_1 u e_2$.

8 For an arbitrary ring R with identity (not equal to zero) suppose that its only left ideals are $\{0\}$ and R. Show that R must be a division ring. Show that any commutative simple ring must be a field.

9 Show that the set of all nilpotent elements in a commutative ring is the radical. What is the radical of the ring of integers mod n?

10 Consider the group algebra $GF(2)C_n$. Show that the elements

$$\beta_i(x) = \sum_{j \in s_i} x^j, \quad i = 1, \ldots, s$$

are the idempotents, $s_i \in \Sigma_2(n)$, $(n, 2) = 1$. Show that any idempotent can be written as a sum of the $\beta_i(x)$ and give an alternative proof to the one in the text that the number of primitive idempotents equals $s = |\Sigma_2(n)|$. Generalize this situation to $GF(q)C_n$.

11 (MacWilliams, 1965) Let M and M' be minimal ideals of $GF(2)C_n$, $(n, 2) = 1$. Show that M and M' have the same weight enumerator iff they have the same dimension and exponent.

5 Group Representations

5.1 Introduction

In this chapter we will give a brief introduction to the theory of group representations. Most of the standard results on group representations are proved, although we avoid some of the theorems that require lengthy and specialized methods for proof.

One of the main applications of group representations is in quantum mechanics, and consequently there are many excellent books on the subject written by chemists and physicists. By and large, these books take a linear algebra point of view and express much of the work in terms of matrices. Mathematicians, on the other hand, tend to use the module-theoretic approach. For this chapter we have chosen the former approach since it is more suitable to the application for which we shall need this theory. The relationships between the two approaches are explored in Section 5.8.

5.2 Representation of Groups

Throughout this chapter the terms group, vector space, representation, etc. will mean finite group, finite-dimensional vector space, finite-dimensional representation, etc. Let V be a vector space over the field $\mathbb{C}$ of complex

5.2 Representation of Groups

numbers and denote its general linear group by $GL(V)$, i.e., the multiplicative group of invertible linear transformations or isomorphisms of V. Any element of $GL(V)$ can be expressed as a matrix by fixing a basis in V. Unless stated otherwise, the underlying field will always be taken as the complex numbers.

Definition A representation of the finite group G is a group homomorphism

$$\rho: \quad G \to GL(V), \qquad \rho(ab) = \rho(a)\rho(b), \qquad a, b \in G$$

It follows from this definition that

$$\rho(1) = I, \qquad \rho(a^{-1}) = (\rho(a))^{-1}, \qquad a \in G$$

As a matter of notation we shall denote representations by lowercase Greek letters, the corresponding matrices by boldface lowercase Greek letters, and when more than one representation is under consideration we differentiate between them either by subscripts or different Greek letters. We call V the representation space. The set of elements of $GL(V)$ that form the representation will be simply denoted by the homomorphism ρ.

If ρ and γ are representations of the group G with representation spaces V_1 and V_2, respectively, then we call the representations equivalent or isomorphic if there exists a vector space isomorphism T

$$T: \quad V_1 \to V_2$$

such that the following composition mappings are identical:

$$T\rho(a) = \gamma(a)T, \qquad \forall a \in G$$

Again, by fixing bases in V_1 and V_2 we can see that the matrices of the equivalent representations are related by the equation

$$\mathbf{T}\boldsymbol{\rho}(a) = \boldsymbol{\gamma}(a)\mathbf{T}, \qquad \forall \in G, \qquad \det \mathbf{T} \neq 0$$

The concept of equivalent representations will play an important part in the theory to follow, as will the following two examples of particular types of representations. If ρ and γ are equivalent representations, we shall denote this by $\rho \cong \gamma$.

Example. *The Natural or Permutation Representation of a Permutation Group* If G is a permutation group on n symbols, which for convenience we assume to be the first n positive integers, then it has a very natural representation in terms of permutation matrices (a matrix of zeros and ones with exactly one one in each row and each column). As a specific example, consider the alternating group of four symbols A_4, which is the group of rotations of a regular tetrahedron. It is generated by the permutations (12)(34) and (123) where in the multiplication of such permutations we agree to work from left to right. If we let V be a four-dimensional vector space with a basis v_1, v_2, v_3, and v_4, then we define a linear transformation on V by

$$\rho_N(a)(v_i) = v_{a(i)}, \qquad a \in A_4$$

where $a(i)$ is the integer to which i is taken by the permutation $a \in A_4$. In terms of matrices, if $\rho_N(a)$ is the matrix corresponding to $\rho(a)$, then we can obtain $\rho_N(a)$ by permuting the *columns* of the 4×4 identity matrix according to the permutation $a \in A_4$. This clearly yields a representation which is generated by the matrices

$$\rho_N((12)(34)) = \begin{bmatrix} 0 & 1 & 0 & 0 \\ 1 & 0 & 0 & 0 \\ 0 & 0 & 0 & 1 \\ 0 & 0 & 1 & 0 \end{bmatrix}, \qquad \rho_N((123)) = \begin{bmatrix} 0 & 1 & 0 & 0 \\ 0 & 0 & 1 & 0 \\ 1 & 0 & 0 & 0 \\ 0 & 0 & 0 & 1 \end{bmatrix}$$

More generally, if G is a permutation group on n symbols, i.e., of degree n, then it has a natural or permutation representation by permutation matrices of dimension n generated in the above manner. We shall denote the natural representation of any permutation group by ρ_N.

Example. The Regular Representation Let G be an arbitrary group of order n and V a vector space of dimension n with a basis (v_g), $g \in G$, indexed by the elements of G. We define a linear transformation of V by

$$\rho_R(g)v_h = v_{gh}, \qquad g, h \in G$$

This is the regular representation of G. In terms of matrices, it is convenient to order the elements g_i, $i = 1, \ldots, n$. Then it is not difficult to show that

$$\rho_R(g_k)_{ij} = \begin{cases} 1 & \text{if } g_i = g_k g_j \\ 0 & \text{otherwise} \end{cases}$$

and this yields a matrix representation of G by permutation matrices. Notice that $\rho_R(g_k)$ has a one in positions (i, j) whenever $g_k = g_i g_j^{-1}$. If we consider the multiplication table given by $g_i g_j^{-1}$, we can read the matrices off directly. If $G = A_4$ as in the previous example, then, letting

$g_1 = 1$ $\quad g_2 = (12)(24)$ $\quad g_3 = (14)(23)$ $\quad g_4 = (12)(34)$
$g_5 = (234)$ $\quad g_6 = (243)$ $\quad g_7 = (134)$ $\quad g_8 = (143)$
$g_9 = (124)$ $\quad g_{10} = (142)$ $\quad g_{11} = (123)$ $\quad g_{12} = (132)$

5.2 Representation of Groups

and forming the table gives the matrices

$$\rho_R((123)) = \rho_R(g_{11}) = \begin{bmatrix} 0 & 0 & 0 & 0 & 0 & 0 & 0 & 0 & 0 & 0 & 0 & 1 \\ 0 & 0 & 0 & 0 & 1 & 0 & 0 & 0 & 0 & 0 & 0 & 0 \\ 0 & 0 & 0 & 0 & 0 & 0 & 0 & 0 & 1 & 0 & 0 & 0 \\ 0 & 0 & 0 & 0 & 0 & 0 & 0 & 1 & 0 & 0 & 0 & 0 \\ 0 & 0 & 0 & 0 & 0 & 0 & 0 & 0 & 0 & 1 & 0 & 0 \\ 0 & 0 & 0 & 1 & 0 & 0 & 0 & 0 & 0 & 0 & 0 & 0 \\ 0 & 0 & 1 & 0 & 0 & 0 & 0 & 0 & 0 & 0 & 0 & 0 \\ 0 & 0 & 0 & 0 & 0 & 1 & 0 & 0 & 0 & 0 & 0 & 0 \\ 0 & 0 & 0 & 0 & 0 & 0 & 1 & 0 & 0 & 0 & 0 & 0 \\ 0 & 1 & 0 & 0 & 0 & 0 & 0 & 0 & 0 & 0 & 0 & 0 \\ 1 & 0 & 0 & 0 & 0 & 0 & 0 & 0 & 0 & 0 & 0 & 0 \\ 0 & 0 & 0 & 0 & 0 & 0 & 0 & 0 & 0 & 0 & 1 & 0 \end{bmatrix}$$

$$\rho_R((12)(34)) = \rho_R(g_4) = \begin{bmatrix} 0 & 0 & 0 & 1 & 0 & 0 & 0 & 0 & 0 & 0 & 0 & 0 \\ 0 & 0 & 1 & 0 & 0 & 0 & 0 & 0 & 0 & 0 & 0 & 0 \\ 0 & 1 & 0 & 0 & 0 & 0 & 0 & 0 & 0 & 0 & 0 & 0 \\ 1 & 0 & 0 & 0 & 0 & 0 & 0 & 0 & 0 & 0 & 0 & 0 \\ 0 & 0 & 0 & 0 & 0 & 0 & 0 & 0 & 0 & 0 & 0 & 1 \\ 0 & 0 & 0 & 0 & 0 & 0 & 0 & 0 & 1 & 0 & 0 & 0 \\ 0 & 0 & 0 & 0 & 0 & 0 & 0 & 0 & 0 & 1 & 0 & 0 \\ 0 & 0 & 0 & 0 & 0 & 0 & 0 & 1 & 0 & 0 & 0 & 0 \\ 0 & 0 & 0 & 0 & 0 & 0 & 1 & 0 & 0 & 0 & 0 & 0 \\ 0 & 0 & 0 & 0 & 0 & 1 & 0 & 0 & 0 & 0 & 0 & 0 \\ 0 & 0 & 0 & 0 & 0 & 0 & 1 & 0 & 0 & 0 & 0 & 0 \\ 0 & 0 & 0 & 0 & 1 & 0 & 0 & 0 & 0 & 0 & 0 & 0 \end{bmatrix}$$

from which the rest of the representation can be generated.

The notion of the irreducibility of a representation, which we now consider, is crucial to the study. We will see later in the chapter that the concepts of equivalence and irreducibility of representations will allow the complete classification of all the representations of a finite group G (over the complex numbers).

Let ρ be a representation of the group G and V its representation space. We say that V_1 is invariant under ρ if

$$\rho(g)x \in V_1, \qquad \forall x \in V_1, \quad \forall g \in G$$

From the properties of ρ it follows that the restriction of ρ to V_1 is also a representation of G with representation space V_1. If $\dim(V) = n$ and $\dim(V_1)$

$= k$ and we assume a nontrivial case where $0 < k < n$, then it is clear that the matrices corresponding to this representation can be written

$$\rho(g) = \begin{bmatrix} \gamma(g) & \alpha(g) \\ \hline 0 & \beta(g) \end{bmatrix}$$

where $\gamma(g)$ is a k-dimensional representation with representation space V_1. Such a matrix form will result by first choosing a basis for V_1 and extending it to a basis for V. It is a simple matter to show that $\beta(g)$ is also a representation of G. Recall that if V_1 and V_2 are subspaces of V such that for any $x \in V$ we can write $x = x_1 + x_2$ uniquely, $x_1 \in V_1$, $x_2 \in V_2$, then we say that V is the direct sum of V_1 and V_2 and write $V = V_1 \oplus V_2$. We shall refer to V_2 as a complement of V_1 in V.

It is helpful in the proof of the following fact to recall the definition of a projection mapping. If V is expressed as a direct sum $V_1 \oplus V_2$ then a linear map

$$\pi: V \to V_1$$

which maps $x = x_1 + x_2$, $x_1 \in V_1$, $x_2 \in V_2$, into $\pi(x) = x_1$ for any $x \in V$, is called a projection mapping.

Conversely, if π is an (injective homomorphism) monomorphism of V with image V_1 and with the property that

$$\pi(x) = x \quad \text{if} \quad x \in V_1$$

then it is not difficult to show that

$$V = V_1 \oplus V_1', \qquad V_1' = \ker \pi$$

To see this, we recall that $V_1 \cong V/\ker \pi$ and hence the sum of the dimensions of V_1 and $\ker \pi$ is equal to the dimension of V.

That the sum is direct follows from the fact that $V_1 \cap V_1' = \{0\}$, for if $x \in V_1 \cap V_1'$, then $\pi(x) = 0$. Since $x \in V_1$, $\pi(x) = x$ and hence $x = 0$.

The following theorem is really a special case of Maschke's theorem, as will be discussed after the proof.

Theorem 2.1 (*Maschke*) Let ρ be a representation of G with representation space V. Let V_1 be a subspace of V invariant under ρ. Then there exists a complement of V_1 in V, V_1', such that $V = V_1 \oplus V_1'$ and V_1' is also invariant under ρ.

5.2 Representation of Groups

Proof Let V_2 be a complement of V_1 in V so that $V = V_1 \oplus V_2$ and let π be the projection map defined by this direct sum. Using this, map, we define a second projection π' by

$$\pi' = \frac{1}{|G|} \sum_{g \in G} \rho(g) \pi \rho(g)^{-1}$$

where $\rho(g)^{-1}$ is the inverse isomorphism of $\rho(g)$ in $GL(V)$. Since $\rho(g)^{-1}$ is an isomorphism of V, π a projection of V onto V_1, and V_1 is invariant under ρ, we conclude that π' maps V onto V_1. Similarly the restriction of π' to V_1 is an identity map and hence π' is a projection map of V onto V_1. As mentioned above the complement of V_1 with respect to π' is V_1' where

$$V_1' = \{x \in V \mid \pi' x = 0\}$$

To show that V_1' is also invariant under ρ consider, for $x \in V_1'$,

$$0 = \rho(h)\pi' x = \rho(h) \frac{1}{|G|} \sum_{g \in G} \rho(g)\pi\rho(g^{-1})x = \frac{1}{|G|} \sum_{g \in G} \rho(hg)\pi\rho(g^{-1})x$$

Letting $hg = f$ and $g^{-1} = f^{-1}h$ gives

$$0 = \frac{1}{|G|} \sum_{f \in G} \rho(f)\pi\rho(f^{-1})\rho(h)x = \pi\rho(h)x$$

which implies $\rho(h)x \in V_1'$, as required. $\square$

Actually, Maschke's theorem states that this result is valid for a vector space over any field whose characteristic is relatively prime to the order of the group $|G|$. The theorem has the important implication that if ρ is a representation with representation space V and a subspace V_1 which is invariant under ρ, then there exists a basis of V such that the matrices of the representation can always be expressed in the form

$$\rho(g) = \left[\begin{array}{c|c} \gamma(g) & 0 \\ \hline 0 & \beta(g) \end{array}\right]$$

and both $\gamma(g)$ and $\beta(g)$ are also representations of G. This is a refinement of the earlier form obtained where only V_1 was known to be invariant under ρ. When the representation can be so reduced we write

$$\boldsymbol{\rho}(g) = \boldsymbol{\gamma}(g) \oplus \boldsymbol{\beta}(g), \quad g \in G$$

and refer to this as the direct sum representation $\rho = \gamma \oplus \beta$.

Definition Let ρ be a representation of G with representation space V. If V contains no proper invariant subspace V_1 (i.e., $V_1 \neq V$, $V_1 \neq \{0\}$), then we say that ρ is an irreducible representation.

To show that every representation can be expressed as a direct sum of irreducible representations, we proceed by induction. Let ρ be the representation of G with representation space V, dim $V > 0$. Assume that dim $V = n$ and that all representations of G of dimension less than n can be expressed as a direct sum of irreducible representations. Clearly all representations of dimension, or degree, one are irreducible. If ρ is irreducible, then V has no proper invariant subspaces with respect to ρ and we are finished. If ρ is reducible (not irreducible), then V has an invariant subspace V_1 and we can write $V = V_1 \oplus V_1'$ and $\rho = \gamma \oplus \beta$. Since V_1 is a proper subspace, $0 < \dim V_1 < \dim V$ and hence dim $\gamma < n$ and dim $\beta < n$ and thus γ and β are two representations which may be expressed as a direct sum of irreducible representations, giving the same result for ρ. The uniqueness of this representation will be established later.

As a final important property of group representations for this section, we endow the representation space V of some representation ρ of G with an inner product (x, y) which has the property that $(\rho(g)x, \rho(g)y) = (x, y)$ for any $g \in G$, $x, y \in V$. That we can always do this follows from the observation that if $(x, y)'$ is any inner product on V, then the inner product

$$(x, y)'' = \sum_{g \in G} (\rho(g)x, \rho(g)y)'$$

has the desired property, as is readily verified. Now if an orthogonal basis $\{v_i\}$ in V is chosen, then it is clear that the matrix corresponding to $\rho(g)$, $\boldsymbol{\rho}(g)$, will be unitary since, by definition, it preserves angles and distances between vectors.

Although we will not need the concept of tensor product immediately, it is convenient to introduce it as this point. Let V_1 and V_2 be two vector spaces with bases $\{v_i^1\}$, $i = 1, \ldots, m$, and $\{v_j^2\}$, $j = 1, \ldots, n$, respectively. For any elements

$$v^1 = \sum_i a_i v_i^1 \in V_1 \quad \text{and} \quad v^2 = \sum_j b_j v_j^2 \in V_2$$

we define the product

$$v^1 \otimes v^2 = \sum_{i,j} a_i b_j (v_i^1 \otimes v_j^2) = \left(\sum_i a_i v_i^1\right) \otimes \left(\sum_j b_j v_j^2\right)$$

The expression is clearly linear in both variables v^1 and v^2 and it follows that the resulting vector space V has mn basis elements $v_i^1 \otimes v_j^2$, $i = 1, \ldots, m$, $j = 1, \ldots, n$. We denote V by $V_1 \otimes V_2$. Now suppose that V_1 is a representation space for the representation ρ_1 and that V_2 is one for ρ_2. Then we define a representation using $V_1 \otimes V_2$ by

$$\rho(g)(v_i^{\,1} \otimes v_j^{\,2}) = \rho_1(g)v_i^{\,1} \otimes \rho_2(g)v_j^{\,2}, \qquad i = 1, \ldots, m, \quad j = 1, \ldots, n$$

It is a simple matter to verify that ρ is indeed a representation.

The given bases of V_1 and V_2 yield matrix representations of G, $\boldsymbol{\rho}_1$ and $\boldsymbol{\rho}_2$. The defined basis of $V_1 \otimes V_2$, $v_i^{\,1} \otimes v_j^{\,2}$, $i = 1, \ldots, n$, yields the matrix representation

$$\boldsymbol{\rho}(g) = \boldsymbol{\rho}_1(g) \otimes \boldsymbol{\rho}_2(g)$$

i.e., the ordinary tensor or Kronecker product of matrices. The properties of the tensor product are well known but we recall in particular that

$$\text{tr}(\mathbf{A} \otimes \mathbf{B}) = \text{tr } \mathbf{A} \text{ tr } \mathbf{B}$$

where $\mathbf{A}$ and $\mathbf{B}$ are two square matrices and $\text{tr}(\cdot)$ is the trace function for matrices. Furthermore, the eigenvalues of $\mathbf{A} \otimes \mathbf{B}$ are the set of all products of eigenvalues of $\mathbf{A}$ with those of $\mathbf{B}$ and for any four matrices $\mathbf{A}$, $\mathbf{B}$, $\mathbf{C}$, and $\mathbf{D}$ satisfying the necessary compatibility conditions we have

$$(A \otimes B)(C \otimes D) = (AC \otimes BD)$$

and this property can be used to verify directly that the tensor product of two representations of a group G is also a representation of G.

As mentioned earlier in the section, the representation spaces and matrix representations in this chapter will all be over the complex numbers unless otherwise specified. The irreducibility of a representation depends, of course, on the field of the representation, i.e., a representation may be irreducible over a field K but reducible over an extension of K. Over some field it may be possible to reduce a matrix representation to the form

$$\rho(g) = \begin{bmatrix} \gamma(g) & \alpha(g) \\ \hline 0 & \beta(g) \end{bmatrix}$$

and yet not reduce it to a form where $\boldsymbol{\alpha}(g) = \mathbf{0}$, $g \in G$. By Maschke's theorem, however, this never occurs over the real or complex numbers.

5.3 Group Characters

Let ρ be a representation of G with representation space V and consider the matrix equivalent $\boldsymbol{\rho}$ by fixing a basis in V.

Definition The character of the matrix $\rho(g)$ of the representation ρ is defind by

$$\chi(g) = \operatorname{tr} \boldsymbol{\rho}(g)$$

The character of the representation ρ will be denoted simply by χ.

The definition of a character is invariant with respect to which basis of V is chosen since matrices resulting from a change of basis are equivalent and hence have the same trace. Thus the trace of a matrix is the sum of its eigenvalues. Unless explicitly mentioned otherwise, we shall assume that all matrix representations are by unitary, or, in the case of representations over $\mathbb{R}$, the real numbers, orthogonal matrices. We observed the possibility of this assumption in the last section. The eigenvalues for such matrices lie on the unit circle in the complex plane and the inverse for any such number is its conjugate.

We gather a collection of properties of characters, most of which are trivial, together in the following theorem. The overbar indicates conjugation.

Theorem 3.1 Let ρ be a representation of degree n with character χ. Then

  (i) $\chi(1) = \dim \rho$.
  (ii) χ is a class function of G.
  (iii) $\chi(g^{-1}) = \bar{\chi}(g)$.
  (iv) If χ_1 and χ_2 are the characters of the representations ρ_1 and ρ_2, respectively, then the character of $\rho_1 \oplus \rho_2$ is $\chi_1 + \chi_2$ and that of $\rho_1 \otimes \rho_2$ is $\chi_1 \chi_2$.

Proof For any matrix representation $\boldsymbol{\rho}(1) = \mathbf{I}$, the identity matrix, and (i) follows. To show (ii), we need to show that for any $g \in G$ we have

$$\chi(hgh^{-1}) = \chi(g) \qquad \text{for all} \quad h \in G$$

But from the commutativity of fields we have

$$\operatorname{tr}(\mathbf{AB}) = \operatorname{tr}(\mathbf{BA})$$

for any two square matrices of the same size. If we choose $\mathbf{A} = \boldsymbol{\rho}(hg)$ and $\mathbf{B} = \boldsymbol{\rho}(h^{-1})$, we obtain the result. If $\lambda_1, \ldots, \lambda_n$ are the eigenvalues of $\boldsymbol{\rho}(g)$, then

$$\chi(g^{-1}) = \sum_{i=1}^{n} \lambda_i^{-1} = \sum_{i=1}^{n} \bar{\lambda}_i = \bar{\chi}(g)$$

Part (iv) is a trivial consequence of previous considerations. □

Schur's lemma, which we now elevate to the status of theorem, expresses a fundamental property of vector space homomorphisms.

5.3 Group Characters

Theorem 3.2 (*Schur's lemma*) Let ρ_1 and ρ_2 be irreducible representations of G over $\mathbb{C}$ with representation spaces V_1 and V_2, respectively. Let T be a linear transformation of V_1 into V_2 with the property that $\rho_2(g)T = T\rho_1(g)$ for all $g \in G$. Then we have the two cases:

(i) If ρ_1 and ρ_2 are not equivalent representations, then $T = 0$, the null map.

(ii) If $\rho_1 = \rho_2$ and $V_1 = V_2$, then T is a scalar multiple of the identity map.

In terms of matrices we have that

(i) $\mathbf{T} = \mathbf{0}$ if $\boldsymbol{\rho}_1$ and $\boldsymbol{\rho}_2$ are not equivalent.
(ii) $\mathbf{T} = a\mathbf{I}$, $a \in \mathbb{C}$, if $\boldsymbol{\rho}_1 = \boldsymbol{\rho}_2$.

Proof Suppose that $T \neq 0$ and let $N = \ker T \subseteq V_1$. N is invariant with respect to ρ_1 since, if $x \in N$, $\rho_2(g)T(x) = T\rho_1(g)x = 0$. However, V_1 and V_2 were assumed irreducible, implying that $N = 0$ or V_1. If $N = V_1$, then $T = 0$, contrary to assumption, and $N = 0$. Let $M = \operatorname{Im} T \subseteq V_2$. Again, since $\operatorname{Im} T$ is a subspace of V_2, which was assumed irreducible, and since

$$\rho_2(g)T(x) = T(\rho_1(g)(x)) \in \operatorname{Im} T, \quad x \in V_1$$

we have that $\operatorname{Im} T$ is invariant with respect to ρ_2. We conclude that $V_2 = \operatorname{Im} T$. These two conditions imply that T is an isomorphism of V_1 onto V_2 satisfying $\rho_2 T = T\rho_1$, i.e., ρ_1 and ρ_2 are equivalent representations, contrary to the assumptions. We conclude that $T = 0$.

If $V_1 = V_2$ and $\rho_1 = \rho_2$, then we select an eigenvalue λ_1 of T and consider the map $T' = T - \lambda_1 i$, where i is the identity map. Since the kernel of T' is not empty and is invariant with respect to ρ_1, we have $\ker T' = V_1$ and that $T' = 0 = T - \lambda_1 i$ and hence T is a scalar multiple of the identity mapping. The matrix formulation of the problem follows immediately. □

The following, more general and weaker statement is often called Schur's lemma.

Theorem 3.2' Let ρ_1 and ρ_2 be irreducible representations of G over some field K. Let $\mathbf{T}$ be a nonzero matrix over K such that

$$\mathbf{T}\boldsymbol{\rho}_1(g) = \boldsymbol{\rho}_2(g)\mathbf{T}$$

for all $g \in G$. Then $\mathbf{T}$ is nonsingular and ρ_1 and ρ_2 are equivalent.

We shall use this form in the proof of the following theorem.

Lemma Let ρ be a representation of G over K where (char K, $|G|$) = 1. Then ρ is irreducible over K iff the set of matrices that commute with $\boldsymbol{\rho}(g)$, $g \in G$, forms a division ring.

Proof Let S be a nonzero matrix that commutes with all the matrices $\boldsymbol{\rho}(g)$, $g \in G$, which we assume to be K-irreducible. Then, from the above form of Schur's lemma, every such matrix is nonsingular and hence possesses an inverse. If $C(K)$ is the commuting ring of matrices of ρ, then if $S \in C(K)$, $S^{-1} \in C(K)$. If $S_1, S_2 \in C(K)$, then trivially $S_1 + S_2 \in C(K)$ and $C(K)$ is a division ring. Conversely, now suppose that ρ is reducible over K and let the representation space of ρ be V. From previous work we can write $V = V_1 \oplus V_2$ where V_1 and V_2 are invariant subspaces. Considering the matrix S which is the projection of V into V_1, i.e., S is the identity on V_1 and the zero map on V_2, we have that $S \in C(K)$, but S is clearly singular, which completes the theorem. □

We will later have occasion to consider the case where $K = \mathbb{R}$, the real numbers, in some detail. It is convenient to prove at this point the following lemma concerning this case.

Lemma The only real, symmetric matrices that commute with all the matrices of a real, irreducible representation of G are scalar multiples of the identity matrix.

Proof We can actually prove something stronger. Let S be a real matrix with a real eigenvalue λ, which commutes with all the matrices of a real, irreducible representation of a group G. Since $S - \lambda I$ is a real matrix that commutes with all the representation matrices, we conclude from the lemma that if $S - \lambda I \neq 0$, then, since $S - \lambda I$ is an element of a division ring, it must be nonsingular or the zero matrix. Since it is singular, we must have that $S = \lambda I$. Since all real scalar multiples of the identity are trivially in the commuting ring, the lemma follows, since the eigenvalues of a real, symmetric matrix are all real. □

The result of this lemma will follow from some stronger results to be proven in Section 5.7.

The relationships of the following theorem are commonly referred to as the orthogonality relationships of group representations. They provide some remarkably simple criteria for establishing the types of irreducible representations contained in a given representaton and reveal many important properties of group representations. For a given matrix representation $\boldsymbol{\rho}$ of G, we denote the (i, j)th element of $\boldsymbol{\rho}(g)$ as $\rho(g)_{ij}$.

5.3 Group Characters

Theorem 3.3 Let ρ_1 and ρ_2 be two irreducible matrix representations of G. Then

(i) If ρ_1 and ρ_2 are nonequivalent,

$$\frac{1}{|G|} \sum_{g \in G} \rho_2(g^{-1})_{ij} \rho_1(g)_{i_1 j_1} = 0$$

for any set of indices i, j, i_1, j_1.

(ii) If ρ_1 and ρ_2 are equal and of dimension n, then

$$\frac{1}{|G|} \sum_{g \in G} \rho_1(g^{-1})_{ij} \rho_1(g)_{i_1 j_1} = \begin{cases} 1/n & \text{if } i = j_1, j = i_1 \\ 0 & \text{otherwise} \end{cases}$$

(This last equation is not true if one of the representations is replaced by an equivalent version—they must be equal.)

Proof We give a matrix-theoretic proof of the theorem and let $\mathbf{A}$ represent a linear transformation from V_1 to V_2 ($\mathbf{A}$ is not necessarily square, of course) where V_1 and V_2 are representation spaces of ρ_1 and ρ_2, respectively. Define the mapping from V_1 to V_2

$$\mathbf{T} = \frac{1}{|G|} \sum_{h \in G} \rho_2(g^{-1}) \mathbf{A} \rho_1(g)$$

From the multiplicative property of group representations we have easily that $\rho_2(h)\mathbf{T} = \mathbf{T}\rho_1(h)$ for any $h \in G$. From this, and using Schur's lemma, we have two cases. If ρ_1 is not equivalent to ρ_2, then $\mathbf{T} = \mathbf{0}$, the zero matrix. Considering the (i, j_1)th element of $\mathbf{T}$ then yields

$$0 = \frac{1}{|G|} \sum_{g \in G} \sum_{k,l} \rho_2(g^{-1})_{ik} A_{kl} \rho_1(g)_{lj_1}$$

from which result (i) of the theorem results by choosing $A_{kl} = \delta_{kj} \delta_{li_1}$. If $\rho_1 = \rho_2$ then $\mathbf{T}$ is a scalar multiple of the identity matrix, say $\lambda \mathbf{I}$. Since $\text{tr}(\mathbf{T}) = \text{tr}(\mathbf{A}) = n\lambda$, $\dim(\rho_1) = n$, then $\lambda = (1/n)\text{tr}(\mathbf{A})$, where $\mathbf{A}$ is now necessarily a square matrix of dimension n. The result follows by considering the equation

$$\frac{1}{n} \text{tr}(\mathbf{A}) \delta_{kl} = \frac{1}{n} \left(\sum_{i,j} \delta_{ij} A_{ij} \right) \delta_{kl} = \frac{1}{|G|} \sum_{g \in G} \sum_{i,j} \rho_1(g^{-1})_{ki} A_{ij} \rho_1(g)_{jl}$$

Then equating coefficients of A_{ij} gives the result of part (ii) of the theorem. □

As a matter of convenience, we shall denote irreducible representations and their characters by ρ_i and χ_i, $i = 1, 2, \ldots$. We assume that no two such representations are equivalent. We also define an inner product on two arbitrary characters χ and χ' of G by

$$(\chi, \chi') = \frac{1}{|G|} \sum_{g \in G} \chi(g)\bar{\chi}'(g) = \frac{1}{|G|} \sum_{g \in G} \chi(g)\chi'(g^{-1})$$

The following theorem indicates why characters play such an important role in group representation theory. We use the notation introduced above.

Theorem 3.4

(i) An arbitrary representation ρ is irreducible if and only if its character satisfies $(\chi, \chi) = 1$.

(ii) For any two irreducible representations ρ_i and ρ_j we have $(\chi_i, \chi_j) = \delta_{ij}$.

(iii) In any decomposition of an arbitrary representation ρ into a direct sum of irreducible representations the number of summands equivalent to the irreducible representation ρ_i is given by (χ, χ_i), where ρ has character χ. Thus two representations with the same character are equivalent and the number of representations in the direct sum that are equivalent to ρ_i, which we call the multiplicity m_i of ρ_i in ρ, is independent of the particular decomposition.

(iv) If m_i is the multiplicity of ρ_i in ρ, then

$$(\chi, \chi) = \sum_i m_i^2$$

Proof

(i) If ρ is irreducible, then using (ii) of Theorem 3.3,

$$(\chi, \chi) = \frac{1}{|G|} \sum_{g \in G} \chi(g)\chi(g^{-1}) = \frac{1}{|G|} \sum_{g \in G} \sum_{i,j} \rho_{ii}(g)\rho_{jj}(g^{-1}) = \sum_{i,j} \frac{1}{n} \delta_{ij} = 1$$

We show the "only if" part after part (iv).

(ii) This follows directly from part (i) of Theorem 3.3.

(iii) Consider a decomposition of ρ:

$$\rho = \bigoplus_j m_j \rho_j$$

where we write $m_j \rho_j$ to denote the direct sum of m_j copies of ρ_j and where we replace representations equivalent to ρ_j by ρ_j. This can always be done by choosing the correct basis in the representation space V. The character of ρ can thus be written as

$$\chi = \sum_j m_j \chi_j$$

From the definition of the inner product we can write

$$(\chi, \chi_i) = \sum_j m_j(\chi_j, \chi_i)$$

which, from parts (i) and (ii) of this theorem, yields m_i. However, the calculation of (χ, χ_i) is independent of the given decomposition and so the multiplicity of ρ_i in ρ is independent of the decomposition. Thus if two representations have the same character, then the multiplicity of ρ_i in each of them is the same and the representations are equivalent.

(iv) From the "orthogonality" of characters of nonequivalent irreducible representations, $(\chi_i, \chi_j) = 0$, we can calculate

$$(\chi, \chi) = \left(\sum m_j \chi_j, \sum m_j \chi_j\right) = \sum_j m_j^2$$

Thus if $(\chi, \chi) = 1$, then it can contain only a single irreducible representation to multiplicity one, i.e., the representation is irreducible. This completes parts (iv) and (i) and the theorem. □

This theorem provides some simple criteria to determine the multiplicity of a given irreducible representation contained in a given representation, if the characters of the irreducible representations are known. The question remains, however, as to how many nonequivalent irreducible representations a finite group has and what their characters are. While the answer to the first part of the question is simple, that of the second part is very often extremely difficult. We investigate the problem further in the next section.

5.4 Orthogonality Relationships and Properties of Group Characters

Denote the conjugacy classes of G by $C_1 = \{1\}, C_2, \ldots, C_s$, and the number of elements in C_i by c_i. We will show that the number of nonequivalent irreducible representations of G is precisely s, the number of conjugacy classes of G. Recall the definition of the regular representation ρ_R of G, where

$$\rho_R(g)v_h = v_{gh}, \qquad g, h \in G$$

and V is a representation space with basis $\{v_g, g \in G\}$ indexed by the elements of G. Since $gh \neq h$ if $g \neq 1$, the elements of $\rho_R(g), g \neq 1$, on the main diagonal are all zero and we conclude that the character of this representation, which we denote χ_R, is given by

$$\chi_R(1) = |G|, \qquad \chi_R(g) = 0, \qquad g \in G, \quad g \neq 1$$

If χ_i is the character of an irreducible representation of dimension n_i, then

$$(\chi_R, \chi_i) = \frac{1}{|G|} \sum_{g \in G} \chi_R(g)\chi_i(g^{-1}) = \frac{1}{|G|} |G|\chi_i(1) = n_i$$

and thus any irreducible representation of dimension n_i has multiplicity n_i in the regular representation. However, this implies that there are a finite number of nonequivalent irreducible representations of G, $\rho_1, \ldots, \rho_k$, of degrees $n_1, \ldots, n_k$ such that

$$\sum_{i=1}^{k} n_i^2 = |G|$$

and that, furthermore, since $\chi_R(g) = 0$, $g \neq 1$,

$$\sum_{i=1}^{k} n_i \chi_i(g) = 0, \qquad g \neq 1$$

It can be shown that $n_i \big| |G|$, $i = 1, \ldots, k$. Actually a stronger result is possible, namely that n_i divides the index of any Abelian normal subgroup of G in G.

We now show that the number of nonequivalent irreducible representations of G, which we have tentatively labeled k, is actually equal to the number of conjugacy-classes of G, which we have assumed to be s. This result is stated as the following theorem.

Theorem 4.1 The number of nonequivalent irreducible representations of a group G is equal to the number of conjugacy classes of G.

Proof We give a vector space proof of this theorem. Consider a function which maps G into the complex numbers which is constant on the conjugacy classes of G. We call such a function (of which characters are examples) a class function. Denote the vector space (over $\mathbb{C}$) of all such functions by W. Consider the inner product

$$(f, h) = \frac{1}{|G|} \sum_{g \in G} f(g)\bar{h}(g), \qquad f, h \in W$$

This can be written as

$$(f, h) = \frac{1}{|G|} \sum_{i=1}^{s} c_i f(C_i)\bar{h}(C_i)$$

(In particular, for characters this definition agrees with the earlier one.) Thus to a class function $\{f(g), g \in G\}$ we can associate a vector

$$\left(\left(\frac{c_1}{|G|}\right)^{1/2} f(C_1), \ldots, \left(\frac{c_s}{|G|}\right)^{1/2} f(C_s) \right)$$

5.4 Orthogonality Relationships and Properties of Group Characters

and clearly the set of all such functions forms an s-dimensional vector space U over $\mathbb{C}$. Now from previous results the k irreducible characters correspond to orthogonal vectors in such a space and we conclude that $k \leq s$. We want to show that $k = s$ or, equivalently, that the characters form an orthogonal base of U. Suppose there exists a class function f such that $(f, \bar{\chi}_i) = 0, i = 1, \ldots, k$. Consider the map

$$\boldsymbol{\rho}_f = \sum_{g \in G} f(g) \boldsymbol{\rho}_i(g)$$

where ρ_i is an irreducible representation of G. Since

$$\boldsymbol{\rho}_i(h) \boldsymbol{\rho}_f \boldsymbol{\rho}_i(h^{-1}) = \sum_{g \in G} f(g) \boldsymbol{\rho}_i(h) \boldsymbol{\rho}_i(g) \boldsymbol{\rho}_i(h^{-1})$$
$$= \sum_{g \in G} f(g) \boldsymbol{\rho}_i(hgh^{-1}) = \sum_{g \in G} f(g) \boldsymbol{\rho}_i(g) = \boldsymbol{\rho}_f$$

as f is a class function, we conclude from Schur's lemma that $\boldsymbol{\rho}_f$ is a scalar multiple of the identity matrix. Taking the trace of $\boldsymbol{\rho}_f$ yields

$$\boldsymbol{\rho}_f = \left[\sum_{j=1}^{s} c_j f(C_j) \chi_i(C_j) \right] \mathbf{I} = (f, \bar{\chi}_i) \mathbf{I} = \mathbf{0}$$

since we assumed $(f, \bar{\chi}_i) = 0$, $i = 1, \ldots, k$. However, the regular representation is equivalent to $\rho_R' = \bigoplus_{i=1}^{s} n_i \rho_i$ and we can apply the same reasoning as above to conclude that

$$\boldsymbol{\rho}'_{Rf} = \sum_{g \in G} f(g) \left[\bigoplus_{i=1}^{s} n_i \boldsymbol{\rho}_i(g) \right] = \bigoplus_{i=1}^{s} n_i \left[\sum_{g \in G} f(g) \boldsymbol{\rho}_i(g) \right]$$

is the zero map. Thus we have that

$$\boldsymbol{\rho}_{Rf} = \sum_{g \in G} f(g) \boldsymbol{\rho}_R(g)$$

is also the zero map. However, from the properties of the regular representation we have

$$\boldsymbol{\rho}_{Rf} v_h = \sum_{g \in G} f(g) \boldsymbol{\rho}_R(g) v_h = \sum_{g \in G} f(g) v_{gh} = 0$$

which is impossible unless $f(g) = 0$, $g \in G$, since the $\{v_h, h \in G\}$ form a basis of V. Here, by v_h we mean a $|G|$-tuple whose coordinate positions contain zeros except that position corresponding to $h \in G$, which contains a one. Thus we have that $s = k$ and the theorem is complete. $\square$

As a simple consequence of this important theorem we have the following corollary:

Corollary Let χ_i be an irreducible character of G. Then we have

$$\sum_{i=1}^{s} \chi_i(C_j)\bar{\chi}_i(C_k) = \frac{|G|}{c_j}\delta_{jk}, \qquad i = 1,\ldots,s$$

Proof Let $\{f_i(g), g \in G\}$ be a class function that takes on the value one when $g \in C_i$ and zero elsewhere. The corresponding vector f_i in U has all coordinates zero except the ith, which is $(c_i/|G|)^{1/2}$. Denoting the vector corresponding to the ith irreducible character by v_i, we have that

$$f_i = \sum_{j=1}^{s}(f_i, v_j)v_j = \sum_{j=1}^{s}\frac{c_i}{|G|}\bar{\chi}_j(C_i)v_j$$

Thus, when $i \neq k$ we have, by considering the kth components of each side,

$$0 = \sum_{j=1}^{s} \bar{\chi}_j(C_i)\chi_j(C_k)$$

while if $i = k$,

$$\left(\frac{c_i}{|G|}\right)^{1/2} = \frac{c_i}{|G|}\sum_{j=1}^{s}\bar{\chi}_j(C_i)\left(\frac{c_i}{|G|}\right)^{1/2}\chi_j(C_i)$$

and the theorem follows. □

By a character table of a group we mean the $s \times s$ array whose (i, j)th element is $(\chi_i(C_j))$, i.e., the character of the ith irreducible representation on the jth conjugacy class. It is conventional to label the columns by conjugacy class and rows by irreducible representation according to some order and we take $C_1 = \{1\}$. Also, the first row we take to correspond to the trivial identity representation where $\rho_1(g) = 1$, $g \in G$. (Every group has such a representation, of course.) There are various orthogonality relationships between the rows and columns of the character table as expressed by the previous theorems and corollary. Indeed, if we modify the character table to form the matrix $\mathbf{U} = (u_{ij})$, $u_{ij} = (c_j/|G|)^{1/2}\chi_i(C_j)$, then these relationships imply that $\mathbf{U}$ is a unitary matrix. We apply the previous results to the tetrahedral group A_4 used in the first two examples.

Example Consider the group A_4, $|A_4| = 12$. Since A_4 has four conjugacy classes consisting of the elements $C_1 = \{g_1\}$, $C_2 = \{g_2, g_3, g_4\}$, $C_3 = \{g_6, g_7, g_{10}, g_{11}\}$, and $C_4 = \{g_5, g_8, g_9, g_{12}\}$, we know there exist precisely four irreducible representations of which one is the identity representation. Thus we have

$$1 + n_2^2 + n_3^2 + n_4^2 = 12$$

and, except for ordering, this equation has the unique solution $n_2 = n_3 = 1$, $n_4 = 3$. Since $\chi_i(1) = n_i$, the first row and column of the character table are

5.4 Orthogonality Relationships and Properties of Group Characters

known. Consider now the permutation representation of Example 1, whose character we denote by χ_N. By a simple calculation we find that $(\rho_N, \rho_1) = 1$, where ρ_N is the natural representation and ρ_1 the identity representation. Thus ρ_N is equivalent to a direct sum $\rho_1 \oplus \rho_{N'}$ and $\chi_N = 1 + \chi_{N'}$. Now $\chi_N(C_i)$ is simply equal to the number of elements fixed by the permutations in C_i. Thus, $\chi_N(C_1) = 4$, $\chi_N(C_2) = 0$, and $\chi_N(C_3) = \chi_N(C_4) = 1$, from which it follows that $\chi_{N'}(C_1) = 3$, $\chi_{N'}(C_2) = -1$, and $\chi_{N'}(C_3) = \chi_{N'}(C_4) = 0$. Now we have

$$(\chi_{N'}, \chi_{N'}) = \tfrac{1}{12}(3 \cdot 3 + 3 \cdot 1 + 0 + 0) = 1$$

and $\chi_{N'}$ is the irreducible character χ_4 and the last row of the table is known. Let $\alpha = \chi_2(C_2)$ and $\beta = \chi_3(C_2)$ and apply the corollary of Theorem 4.1 to obtain

$$\sum_{i=1}^{4} \chi_i(C_2)\bar{\chi}_i(C_2) = 12/3 = 4 = 1 + \alpha^2 + \beta^2 + 1 \tag{4.1}$$

and

$$\sum_{i=1}^{4} \chi_i(C_1)\bar{\chi}_i(C_2) = 0 = 1 + \alpha + \beta - 3 \tag{4.2}$$

From (4.2), $\alpha = 2 - \beta$ and from (4.1)

$$\alpha^2 + \beta^2 = \beta^2 + (2-\beta)^2 = 2\beta^2 + 4 - 4\beta = 2$$

or

$$\beta^2 - 2\beta + 1 = 0$$

which has the single solution $\beta = 1$. Since $\alpha = 2 - \beta$, the second column of the table is complete. To complete the table, we note that the remaining elements must all be cube roots of unity since all elements in C_3 and C_4 are of order three. It is also easily verified that none of them can be unity since if one were, from the fact that it is a representation, the remaining element of the row would also have to be. Thus if we choose $\chi_2(C_3) = \eta$, η is a primitive cube root of unity, then $\chi_2(C_4) = \eta^2$, $\chi_3(C_3) = \eta^2$, and $\chi_3(C_4) = \eta$. The situation with regard to χ_2 and χ_3 is thus symmetric, and the character table can be completed (Table VIII).

TABLE VIII

$\eta^3 = 1$, η Primitive

	C_1	C_2	C_3	C_4
χ_1	1	1	1	1
χ_2	1	1	η	η^2
χ_3	1	1	η^2	η
χ_4	3	-1	0	0

In general, the problem of finding a character table of a group is very difficult. There are many interesting applications of characters in investigating the structure of a group. However, these would take us outside the scope of the present treatment. The problem of finding the matrices of a representation is even more challenging and there exist no general techniques for doing so.

There are several properties of characters that require some results from algebraic number theory to prove. We merely state these results since little is lost for our purposes by omitting the proofs. Perhaps the principal result is that, as mentioned earlier, the dimension n_i of any irreducible representation of G divides the order of G, $|G|$. A slightly stronger result than this is that if H is any Abelian normal subgroup of G, then $n_i|(G:H)$ and in particular $n_i|(G:Z(G))$, where $Z(G)$ is the center of the group.

If χ is the character of any representation ρ of degree n, then

$$|\chi(g)| \leq \chi(1) = n$$

and if equality holds for some g, then $\rho(g) = \lambda \mathbf{I}$, where λ is a root of unity. If $\chi(g) = n$, then g is in the kernel of ρ and $\rho(g) = \mathbf{I}$, the identity matrix. Since the kernel of a group homomorphism is a normal subgroup, it is clear that we can determine some normal subgroups directly from the character table of the group. In fact, all the normal subgroups can be determined from the character table but, since they do not all appear as kernels of irreducible representations, the procedure is more involved than just finding kernels [see Burrow (1965, Section 26.B)].

In constructing some character tables, it is helpful to know the number of irreducible representations of dimension one of a group. This is not a difficult number to obtain in terms of the group properties, as we now show. We need a few results of independent interest first.

Let $g_1, g_2 \in G$ and consider an element of the form $g_1^{-1}g_2^{-1}g_1g_2$. Such an element is called a commutator of the group. We denote by G' the subgroup of G generated by the set of all commutators. To show that G' is a normal subgroup, consider

$$h^{-1}(g_1^{-1}g_2^{-1}g_1g_2)h = (h^{-1}g_1^{-1}h)(h^{-1}g_2^{-1}h)(h^{-1}g_1h)(h^{-1}g_2h)$$
$$= (g_1')^{-1}(g_2')^{-1}g_1'g_2' \in G'$$

Furthermore, we claim that the factor group G/G' is Abelian for

$$g_1 G' g_2 G' = g_1 g_2 G'$$

and since $g_1g_2 = g_2g_1(g_1^{-1}g_2^{-1}g_1g_2)$, then $g_1g_2 G' = g_2g_1 G'$.

Now suppose H is any normal subgroup of G, and let ρ be a representation of G/H. We define a representation ρ^* on G as follows:

$$\rho^*(g) = \rho(g_i H) \quad \text{if} \quad g \in g_i H$$

5.4 Orthogonality Relationships and Properties of Group Characters

We need verify only that the definition satisfies the homomorphism property. Let $g_1 = g_i h_1$ and $g_2 = g_j h_2$, $h_1, h_2 \in H$, where g_i and g_j are among a fixed set of (left) coset representatives. Then we have

$$\rho^*(g_1)\rho^*(g_2) = \rho(g_i H)\rho(g_j H) = \rho(g_i g_j H) = \rho^*(g_1 g_2)$$

since $g_1 g_2 = g_i h_1 g_j h_2$, and since $h_1 g_j h_1^{-1} = g_j h_3$, we have

$$g_1 g_2 = g_i h_1 g_j h_2 = g_i g_j h_3 h_1 h_2 \in g_i g_j H$$

Thus a representation of the factor group is also a representation of the group and the definition is independent of the coset representatives.

Finally, we observe that every irreducible representation of an Abelian group is of dimension one. This follows from the fact that if the group is of order $|G|$, then the dimensions of the representations n_i satisfy

$$\sum_{i=1}^{|G|} n_i^2 = |G|$$

since there are $|G|$ conjugacy classes. The only set of strictly positive integers satisfying this relationship are $n_i = 1$, $i = 1, \ldots, |G|$. We are now in a position to find the number of irreducible representations of a group G of dimension one.

Theorem 4.2 The number of irreducible representations of dimension one of a group G is equal to $(G:G')$, the index of G' in G.

Proof From the above comments every irreducible representation of G/G' is an irreducible representation of G of dimension one since G/G' is Abelian. Conversely, suppose that ρ is an irreducible representation of G of dimension one. In this case we have $\rho(g) = \chi(g)$, $g \in G$, and hence we can write

$$\chi(g_1^{-1} g_2^{-1} g_1 g_2) = \chi(g_1^{-1})\chi(g_2^{-1})\chi(g_1)\chi(g_2) = 1$$

or $\chi(g) = 1$, $g \in G'$. Thus ρ is also an irreducible representation of dimension one on G/G' and we conclude that the number of irreducible representations of G of dimension one is $(G:G')$.

Example Denote by Q the group of quaternions, consisting of the elements $\{\pm 1, \pm i, \pm j, \pm k\}$, with multiplications

$$i^2 = j^2 = k^2 = -1, \quad ij = k, \quad jk = i, \quad ki = j, \quad ji = -k, \quad kj = -i, \quad ik = -j$$

We want to find the character table of this group. The commutator subgroup consists of the elements $\{1, -1\}$ and thus there are four irreducible representations of dimension one. Since every element is of order two, the characters of these representations must be real. The conjugacy classes of Q are $C_1 = \{1\}$,

$C_2 = \{-1\}$, $C_3 = \{i, -i\}$, $C_4 = \{j, -j\}$, and $C_5 = \{k, -k\}$ and we observe that they are all ambivalent, i.e., the inverse of any element in a conjugacy class is in the conjugacy class. Thus the character table of Q is real since the number of real rows is equal to the number of ambivalent conjugacy classes. The one-dimensional representations are easily found from the one-dimensional representations of G/G'. The other irreducible representation must be of dimension two since there are five irreducible representations all together and

$$1^2 + 1^2 + 1^2 + 1^2 + n_5^2 = 8$$

has the solution $n_5 = 2$. We can easily fill in the last row of the character table using the orthogonality relations of characters (Table IX).

TABLE IX

	C_1	C_2	C_3	C_4	C_5
χ_1	1	1	1	1	1
χ_2	1	1	-1	-1	1
χ_3	1	1	-1	1	-1
χ_4	1	1	1	-1	-1
χ_5	2	-2	0	0	0

We wish to prove the following theorem on doubly transitive permutation groups. The proof will be somewhat lengthy and will be by a series of lemmas. The treatment is essentially similar to that of Hall (1959).

Theorem 4.3 The natural representation of a doubly transitive finite permutation group of degree n is isomorphic to the direct sum of the identity representation and an irreducible representation.

Notice that we have already seen a demonstration of the theorem in an example. Assume that G is a permutation group on the letters $X = \{x_1, \ldots, x_n\}$. We call a subset S of X a transitive constitutent of G if, for every $g \in G$ and $x_i \in S$, $g(x_i) \in S$ and, if x_1 and x_j are elements of S, then there exists an element $g' \in G$ such that $g'(x_i) = x_j$.

Lemma 4.1 If S is a transitive constituent of G containing r letters, then the subgroup H that fixes one letter of S is of index r in G.

Proof of Lemma 4.1 The set of elements of G that fixes any letter is clearly a subgroup. For convenience, we assume $x_i \in S$, $i = 1, \ldots, r$. Since S

is a transitive constituent of G, we can choose a $g_i \in G$ such that $g_i(x_1) = x_i$, $i = 1, \ldots, r$. Let H be the subgroup that fixes x_1; then we claim that

$$G = \bigcup_{i=1}^{r} g_i H$$

The $g_i H$ are left cosets of H in G and we show that they are disjoint. Notice that if $g \in g_i H$, $g = g_i h$, then $g(x_1) = g_i h(x_1) = g_i(x_1) = x_i$. Suppose the cosets were not distinct; then $g_i h = g_j h'$, $h, h' \in G$, and we have

$$g_i h(x_1) = g_i(x_1) = x_i = g_j h'(x_1) = g_j(x_1) = x_j$$

which is a contradiction. Thus the union is the coset decomposition of G with respect to H, and the result of the lemma follows. □

Lemma 4.2 Let G be a permutation group of degree n on $X = \{x_1, \ldots, x_n\}$ with k transitive constitutents $S_1, \ldots, S_k$, $|S_i| = n_i$, $i = 1, \ldots, k$. Then we have

$$\sum_{g \in G} \chi_N(g) = k|G|$$

and the natural representation ρ_N contains the identity representation k times.

Proof of Lemma 4.2 Denote by H_j the subgroup of G that fixes a particular letter $x_{i_j} \in S_j$, $j = 1, \ldots, k$. Then H_j is of index n_j in G. Thus x_{i_j} is fixed by precisely $|G|/n_j$ elements of G. If we consider the $|G|$ matrices of the natural representation ρ_N, this means that there will be $|G|/n_j$ matrices with ones on the main diagonal in position i_j. Considering the set S_j as a whole, there will be $n_j (|G|/n_j) = |G|$ ones on the main diagonal, among the $|G|$ matrices, in positions corresponding to the elements of S_j. Since there are k such transitive constitutents, we conclude that

$$\sum_{g \in G} \chi_N(g) = k|G|$$

Now the multiplicity of the identity representation in ρ_N is

$$m = (\chi_N, 1) = \frac{1}{|G|} \sum_{g \in G} \chi_N(g) \cdot 1 = k$$

and the lemma is complete. □

Lemma 4.3 Let G be a transitive permutation group of degree n on $X = \{x_1, \ldots, x_n\}$, and let H be the subgroup fixing x_1. Then

$$\sum_{g \in G} \chi_N^2(g) = t|G|$$

where t is the number of transitive constituents of H.

Proof of Lemma 4.3 From the previous lemma we have

$$\sum_{g \in H} \chi_N(g) = t|H|$$

If H_i is the subgroup fixing x_i, then, since the H_i are conjugate, we can write

$$\sum_{i=1}^{n} \sum_{g \in H_i} \chi_N(g) = t|H|n = t|G|$$

That $n|H| = |G|$ follows from the first lemma and the fact that G is transitive, i.e., has exactly one transitive constituent. Now, an element $g \in G$ may fix, say, k letters and hence will appear in k subgroups H_i. But $k = \chi_N(g)$ and thus the contribution of g to the overall sum is $\chi_N^2(g)$. Thus

$$\sum_{i=1}^{n} \sum_{g \in H_i} \chi_N(g) = \sum_{g \in G} \chi_N^2(g) = t|G| \quad \square$$

Proof of Theorem 4.3 We can now prove the theorem; let G be a doubly transitive permutation group of degree n and let H be the subgroup fixing x_1. Since G is doubly transitive, G has one transitive constituent while H has two, namely $\{x_1\}$ and $\{x_2, \ldots, x_n\}$. From Lemma 4.2 we have

$$\sum_{g \in G} \chi_N^2(g) = 2|G| = \sum_{g \in G} \chi_N(g)\bar{\chi}_N(g)$$

since $\chi_N(g)$ is always a positive integer or zero. But if ρ_N has the decomposition $\rho_N \cong \bigoplus_{i=1}^{s} m_i \rho_i$ where ρ_i is an irreducible representation with character χ_i, then

$$2|G| = \sum_{g \in G} \left[\sum_{i=1}^{s} m_i \chi_i(g) \right] \sum_{i=1}^{s} m_i \bar{\chi}_i(g) = |G| \sum_{i=1}^{s} m_i^2$$

However, from Lemma 4.2, ρ_N contains the identity representation once and so the only solution to the equation

$$\sum_{i=1}^{s} m_i^2 = 2$$

is $m_1 = 1$ and $m_i = 1$ for some i. This implies the result of the theorem and the proof is complete. $\square$

Example Consider the doubly transitive permutation group G of degree $|G| = p^n$ consisting of pairs (a, b), $a, b \in GF(q)$, $a \neq 0$, which defines a permutation on the elements of $GF(q)$:

$$x \longmapsto ax + b, \quad x \in GF(q)$$

The inverse permutation of (a, b) is $(a^{-1}, -a^{-1}b)$ and multiplication is defined by

$$(a_1, b_1)(a_2, b_2) = (a_1 a_2, a_1 b_2 + b_1)$$

5.5 Subduced and Induced Representations

Now let ρ_N be the natural representation of G and χ_N its character. If $a \neq 1$, then the transformation $y = ax + b$ can fix at most one element of $GF(q)$, namely $x = (1 - a)^{-1}b$. If $a = 1$ and $b \neq 0$, then no elements are fixed by the transformation. If $a = 1$ and $b = 0$, then every element is fixed. Thus, the characters are given by

$$\chi_N(g) = \chi_N((a, b)) = \begin{cases} 1, & a \neq 1 & [q(q-2) \text{ elements}] \\ 0, & a = 1, \ b \neq 0 & (q-1 \text{ elements}) \\ q, & a = 1, \ b = 0 \end{cases}$$

The number of times the identity representation is contained in ρ_N is given by

$$(\chi_N, \chi_1) = \frac{1}{|G|} \sum_{g \in G} \chi_N(g) \cdot 1 = \frac{1}{q(q-1)}[q(q-2) + q] = 1$$

Thus $\rho_N = \rho_1 \oplus \rho_N'$ and the character of ρ_N' is given by

$$\chi_N'(g) = \chi_N'((a, b)) = \begin{cases} 0, & a \neq 1 \\ -1, & a = 1, \ b \neq 0 \\ q-1, & a = 1, \ b = 0 \end{cases}$$

That ρ_N' is irreducible is readily verified by calculating

$$(\chi_N', \chi_N') = \frac{1}{|G|} \sum_{g \in G} \chi_N'(g)^2$$

$$= \frac{1}{q(q-1)}[(-1)(-1)(q-1) + (q-1)^2] = 1$$

and the result of the theorem is demonstrated.

We conclude the section with some observations on character tables. A character table does not, unfortunately, characterize a group in the sense that nonisomorphic groups may have the same character table. An example of this is the group of quaternions and the dihedral group D_4, both groups being of order eight. On the other hand, there exist tables which satisfy every known requirement for character tables but they are not the character table of any group.

5.5 Subduced and Induced Representations

The purpose of this section is to consider the relationships between representations of a group and those of its subgroups and in particular to prove the Frobenius reciprocity theorem. We take a concrete approach and use matrices, rather than representation spaces.

Let ρ^G be any representation of the group G. If we restrict ρ^G to the elements of a subgroup H of G, then we obtain a representation of H. We call the restriction of ρ^G to H a subduced representation of H. If ρ^G is an irreducible representation of G, it does not follow that its restriction will be an irreducible representation of H.

Conversely, suppose a representation ρ^H of dimension m of a subgroup H of G is given, where $(G:H) = n$. We will form a representation ρ^G of G from ρ^H, called the induced representation, which will be of dimension mn. Partition G into the left cosets $G = g_1 H \cup \cdots \cup g_n H$, where $\{g_1, \ldots, g_n\}$ is a set of coset representatives. We define the $mn \times mn$ matrices in the following manner. First we let $\rho^H(g) = 0$ if $g \in G \backslash H$ and let $\sigma_{ij}(g) = \rho^H(g_i^{-1} g g_j)$. We then form the $mn \times mn$ matrix whose (i,j)th block is $\sigma_{ij}(g)$, i.e.,

$$\rho^G(g) = (\sigma_{ij}(g))$$

To verify that this is indeed a representation, we first observe that

$$\rho^G(1) = \mathbf{I}_{mn}$$

since, if $g = 1$, then $g_i^{-1} g g_j = g_i^{-1} g_j \notin H$ unless $i = j$, since the g_i form a system of coset representatives. To show that

$$\rho^G(g)\rho^G(h) = \rho^G(gh), \qquad g, h \in G$$

it is necessary to verify the formula

$$\sum_{j=1}^{n} \sigma_{ij}(g)\sigma_{jk}(h) = \sigma_{ik}(gh)$$

For a fixed i the matrix $\sigma_{ij}(g)$ is nonzero iff $g_i^{-1} g g_j \in H$ or, equivalently, if $g_j H = g^{-1} g_i H$. Similarly, for a fixed k the matrix $\sigma_{jk}(h)$ is nonzero iff $g_j^{-1} h g_k \in H$ or $h g_k H = g_j H$. A nonzero product results only if this is the same j, in which case we obtain

$$\sigma_{ij}(g)\sigma_{jk}(h) = \rho^H(g_i^{-1} g g_j)\rho^H(g_j^{-1} h g_k) = \rho^H(g_i^{-1} g h g_k) = \sigma_{ik}(gh)$$

which verifies the equation.

The character of the induced representation is easily deduced from the construction. Define the character of ρ^H by χ and let

$$\chi'(g) = \begin{cases} \chi(g), & g \in H \\ 0, & g \notin H \end{cases}$$

5.5 Subduced and Induced Representations

If χ^G is the character of the induced representation ρ^G, then

$$\chi^G(g) = \sum_{i=1}^{n} \chi'(g_i^{-1} g g_i)$$

However, a more convenient form that is independent of coset representatives can also be obtained by observing that $g_i^{-1} g g_i \in H$ iff $h^{-1} g_i^{-1} g g_i h \in H$, $h \in H$, and since

$$\chi(g_i^{-1} g g_i) = \chi(h^{-1} g_i^{-1} g g_i h)$$

we can rewrite the above summation as

$$\chi^G(g) = \frac{1}{|H|} \sum_{s \in G} \chi'(s^{-1} g s)$$

which is independent of the chosen coset representatives, as desired.

The following theorem is one of the most important results on induced characters.

Theorem 5.1 (*Frobenius Reciprocity Theorem*) Let α and β be two irreducible representations of a group G and a subgroup H, respectively. Denote by α_0 and β^0 the representation α subduced onto H and the representation β induced onto G, respectively. Then β^0 contains α to the same multiplicity that α_0 contains β.

Proof Denote the characters of the various representations α, β, α_0, and β^0 by χ_α, χ_β, χ_{α_0}, χ_{β^0}, respectively. We wish to show that

$$m_1 = (\chi_\beta, \chi_{\alpha_0}) = (\chi_\alpha, \chi_{\beta^0}) = m_2$$

where m_1 is the multiplicity of β in α_0 and m_2 is that of α in β^0. From the previous section we have

$$m_2 = (\chi_\alpha, \chi_{\beta^0}) = \frac{1}{|G|} \sum_{g \in G} \chi_{\beta^0}(g) \chi_\alpha(g^{-1})$$

The induced character χ_{β^0} is given and allows us to write

$$m_2 = \frac{1}{|G|} \sum_{g \in G} \frac{1}{|H|} \sum_{s \in G} \chi_\beta'(s^{-1} g s) \chi_\alpha(g^{-1})$$

where, as before,

$$\chi_\beta'(g) = \begin{cases} \chi_\beta(g), & g \in H \\ 0, & g \notin H \end{cases}$$

and since the character is a class function,

$$m_2 = \frac{1}{|H||G|} \sum_{g \in G} \sum_{s \in G} \chi_\beta'(s^{-1}gs)\chi_\alpha(s^{-1}gs)$$

$$= \frac{1}{|G|} \sum_{s \in G} \left[\frac{1}{|H|} \sum_{g \in H} \chi_\beta'(g)\chi_\alpha(g) \right]$$

$$= \frac{1}{|G|} \sum_{s \in G} \left[\frac{1}{|H|} \sum_{g \in H} \chi_\beta(g)\chi_\alpha(g) \right]$$

$$= \frac{1}{|G|} |G| m_1 = m_1$$

as required. □

There are many interesting results available on induced characters and representations, particularly in the case when the subgroup H is normal in G, which we denote by $H \triangleleft G$. For example, let ρ be an irreducible representation of G and let ρ_0 be the corresponding subduced representation of $H \triangleleft G$. We define two representations ρ_1 and ρ_2 of H to be conjugate relative to G if there exists an element $a \in G$ such that $\rho_1(h) = \rho_2(a^{-1}ha)$ for every $h \in H$. Then it is possible to show that ρ_0 is either an irreducible representation of H or else is a direct sum of irreducible conjugate representations of H relative to G, i.e., if γ is one irreducible component of ρ_0, then every other irreducible component of ρ_0 is conjugate to γ and every conjugate of γ appears in ρ_0 and to the same multiplicity.

There also exist criteria to establish the equivalence and irreducibility of induced representations. These results tend to require lengthy notations and computations and we omit them [e.g., see Curtis and Reiner (1966)].

We consider now some elementary facts on monomial representations. A representation ρ of a group G is a monomial representation if $\rho(g)$ has exactly one nonzero entry in each row and each column for every $g \in G$. Clearly if γ is a one-dimensional representation of the proper subgroup H of G, then γ^0, the induced representation, is a monomial representation of G. In particular, let G be a transitive permutation group of degree n and ρ_N the natural representation on n letters. Let H be the subgroup of G that fixes the letter 1. Then ρ_N is equivalent to γ^0, where γ is some one-dimensional representation of H. Monomial representations are particularly important in the study of metacyclic groups, where a metacyclic group G is one that contains a cyclic normal subgroup H such that G/H is also cyclic. Necessary and sufficient conditions are known on such groups for every irreducible representation to be either one dimensional or equivalent to an induced monomial representation [see Curtis and Reiner (1966, Section 47)].

5.6 Direct and Semidirect Products

The concepts of direct and semidirect products are often useful in any investigation involving group structure. Let G_1 and G_2 be two groups with elements $\{g_i^{(1)}, g_i^{(1)} \in G_1\}$ and $\{g_i^{(2)}, g_i^{(2)} \in G_2\}$. We define the direct product $G_1 \times G_2$ as the set of ordered pairs $\{(g_i^{(1)}, g_j^{(2)}), g_i^{(1)} \in G_1, g_j^{(2)} \in G_2\}$ with the multiplication

$$(g_i^{(1)}, g_j^{(2)})(g_k^{(1)}, g_l^{(2)}) = (g_i^{(1)}g_k^{(1)}, g_j^{(2)}g_l^{(2)})$$

which clearly makes $G_1 \times G_2$ a group of order $|G_1||G_2|$. The subgroups G_1' and G_2' of $G_1 \times G_2$ defined by

$$G_1' = \{(g_i^{(1)}, 1)\}, \qquad G_2' = \{(1, g_j^{(2)})\}$$

are normal subgroups of $G_1 \times G_2$. If G_1 has s_1 conjugacy classes $C_i^{(1)}$, $i = 1, \ldots, s_1$, and G_2 has s_2 conjugacy classes $C_i^{(2)}$, $i = 1, \ldots, s_2$, then $G_1 \times G_2$ has $s_1 s_2$ conjugacy classes and they are all of the form

$$C_i^{(1)} \times C_j^{(2)} = \{(g_k^{(1)}, g_l^{(2)}), g_k^{(1)} \in C_i^{(1)}, g_l^{(2)} \in C_j^{(2)}\}$$

The subgroups G_1' and G_2' just defined have the properties that the elements of G_1' and G_2' commute (pairwise), $G_1' \cap G_2' = \{(1, 1)\}$, and

$$G = G_1' G_2' = \{(g_i^{(1)}, 1)(1, g_j^{(2)})\}$$

Conversely, it can be shown that if an arbitrary group G contains two subgroups G_1 and G_2 such that $G_1 \cap G_2 = \{1\}$, $G = G_1 G_2$, and the elements of G_1 and G_2 commute pairwise, then G is isomorphic to the direct product of G_1 and G_2.

Another construction of a group from two given groups is the semidirect product, which is a little more involved than the direct product. Let H and K be groups and let aut H be the group of automorphisms of H. Let ψ be a homomorphism of K into aut H. Then we define the semidirect product of H and K with respect to ψ as the set of pairs $\{(h, k), h \in H, k \in K\}$ with the multiplication

$$(h_1, k_1)(h_2, k_2) = (h_1 \cdot \psi(k_1)(h_2), k_1 k_2)$$

It is not difficult to verify that this is a group and that the sets

$$H' = \{(h, 1), h \in H\}, \qquad K' = \{(1, k), k \in K\}$$

are the subgroups with the properties (i) $G = H'K'$, (ii) $H' \cap K' = \{1\}$, and (iii) $(1, k)(h, 1) = (\psi(k)(h), 1)(1, k)$, and, indeed, H' is a normal subgroup of G. Conversely, if we are given a group G with subgroups H and K and a mapping

ψ of K into the set of all mappings of H into itself which have the properties that (i) $G = HK$, (ii) $H \cap K = \{1\}$, and (iii) $kh = \psi(k)(h)k$, then ψ must be a homomorphism of K into aut H and G is the semidirect product of H and K with respect to ψ.

Rather than pursuing the interesting subject of group structure further, we return to the simple direct product $G_1 \times G_2 = G$ and determine all the irreducible representations of G in terms of those of G_1 and G_2. If ρ_1 and ρ_2 are representations of G_1 and G_2, respectively, then if $g = (g_1, g_2) \in G$, the set of matrices

$$\rho(g) = \rho_1(g_1) \otimes \rho_2(g_2), \qquad g \in G$$

forms a representation of G since, if $g = (g_1, g_2)$, $h = (h_1, h_2) \in G$, then

$$\rho(g)\rho(h) = (\rho_1(g_1) \otimes \rho_2(g_2))(\rho_1(h_1) \otimes \rho_2(h_2))$$
$$= (\rho_1(g_1)\rho_1(h_1)) \otimes (\rho_2(g_2)\rho_2(h_2)) = \rho(gh)$$

which follows from the elementary properties of the tensor products of matrices discussed previously. The character of $\rho(g)$ is $\chi^1(g)\chi^2(g)$, where χ^1 is the character of ρ_1 and χ^2 is that of ρ_2. Furthermore, for any two representations ρ_1 and ρ_2 of G_1 and G_2, respectively, we can calculate

$$(\chi, \chi) = \frac{1}{|G|} \sum_{(g_1, g_2) \in G} \chi^1(g_1)\chi^2(g_2)\chi^1(g_1^{-1})\chi^2(g_2^{-1})$$
$$= \left[\frac{1}{|G_1|}\sum_{g_1 \in G_1} \chi^1(g_1)\chi^1(g_1^{-1})\right]\left[\frac{1}{|G_2|}\sum_{g_2 \in G_2} \chi^2(g_2)\chi^2(g_2^{-1})\right]$$
$$= (\chi^1, \chi^1)_{G_1}(\chi^2, \chi^2)_{G_2}$$

which is equal to unity if and only if ρ_1 and ρ_2 are irreducible representations of G_1 and G_2, respectively. The subscripts on the inner product indicate the group over which the calculations are made. Similarly, let ρ_i^1 and ρ_j^1 be two irreducible representations of G_1 and let ρ_k^2 and ρ_l^2 be two irreducible representations of G_2 and let the characters be χ_i^1, χ_j^1, χ_k^2, and χ_l^2, respectively. Let $\rho' = \rho_i^1 \otimes \rho_k^2$ and $\rho'' = \rho_j^1 \otimes \rho_l^2$ with characters χ' and χ''. As before, we can calculate

$$(\chi', \chi'') = (\chi_i^1, \chi_j^1)(\chi_k^2, \chi_l^2)$$

and thus ρ' and ρ'', which are irreducible representations of G, are equivalent iff ρ_i^1 is equivalent to ρ_j^1 and ρ_k^2 is equivalent to ρ_l^2. If G_i has s_i conjugacy classes, $i = 1, 2$, then it also has s_i irreducible representations, $i = 1, 2$. Also, $G = G_1 \times G_2$ has $s_1 s_2$ conjugacy classes and hence $s_1 s_2$ irreducible representations. Since we have just observed how to construct $s_1 s_2$ such irreducible

5.6 Direct and Semidirect Products

representations, we conclude that every irreducible representation of $G = G_1 \times G_2$ is equivalent to the Kronecker product of an irreducible representation of G_1 and an irreducible representation of G_2.

Example Suppose ρ^0 is a representation of a group G which is obtained from inducing the representation ρ of H, where $G = H \times K$. We show that ρ^0 is isomorphic to $\rho \otimes \rho_{R_K}$ where ρ_{R_K} is the regular representation of K. It is both necessary and sufficient to show that the characters of the two representations be the same.

Let

$$g = (g_1, g_2) \in G = H \times K, \qquad g_1 \in H, \quad g_2 \in K$$

We identify H and K with the subgroups $\{(h, 1),\ h \in H\}$ and $\{(1, k),\ k \in K\}$, respectively. Then, by definition

$$\rho \otimes \rho_{R_K}(g_1, g_2) = \rho(g_1) \otimes \rho_{R_K}(g_2)$$

and

$$\chi((g_1, g_2)) = \operatorname{tr}(\rho \otimes \rho_{R_K}(g_1, g_2)) = \chi_\rho(g_1)\chi_{\rho_{R_K}}(g_2)$$
$$= \begin{cases} 0 & \text{if } g_2 \neq 1 \\ |K|\chi_\rho((g_1, 1)) & \text{if } g_2 = 1 \end{cases}$$

On the other hand, we can calculate the character of the induced representation ρ^0 with this formula as

$$\chi_{\rho^0}((g_1, g_2)) = \frac{1}{|H|} \sum_{(h_i, k_j) \in G} \chi_\rho'((h_i^{-1}, k_j^{-1})(g_1, g_2)(h_i, k_j))$$

where

$$\chi_\rho'((a, b)) = \begin{cases} 0 & \text{if } b \neq 1 \\ \chi_\rho((a, 1)) & \text{if } b = 1 \end{cases}$$

Now the element $(h_i^{-1}, k_j^{-1})(g_1, g_2)(h_i, k_j) \in H$ iff $g_2 = 1$ and thus we have

$$\chi_\rho'((g_1, g_2)) = \frac{1}{|H|} \sum_{(h_i, k_j) \in G} \chi_\rho((h_i^{-1} g_1 h_i, 1))$$
$$= \frac{|K|}{|H|} \sum_{(h_i, 1) \in G} \chi_\rho((h_i^{-1} g_1 h_i, 1))$$
$$= |K|\chi_\rho((g_1, 1))$$

and we conclude that $\chi = \chi_{\rho^0}$ and hence that ρ^0 and $\rho \otimes \rho_{R_K}$ are isomorphic representations.

5.7 Real Representations

All the previous work in this chapter has been concerned with representations over the complex numbers. For our purposes in the next chapter, however, we shall require representations over the real numbers and the purpose of this section is to indicate how to obtain these representations from complex representations. By a real or complex representation we shall of course mean a matrix representation which has either real or complex entries, respectively. We shall denote the conjugate of a matrix $\mathbf{A}$ by $\overline{\mathbf{A}}$ and note that if ρ is a representation, then so also is $\overline{\rho}$.

Let ρ be an irreducible representation of G with character χ. We say that:

(i) ρ is of the first kind if it is equivalent to a real representation.

(ii) ρ is of the second kind if it is equivalent to $\overline{\rho}$ but is not equivalent to a real representation.

(iii) ρ is of the third kind if it is not equivalent to $\overline{\rho}$.

Notice that ρ is of the third kind iff it has a complex character. Recall that we assume all representations are in terms of unitary (or, in the real case, orthogonal) matrices and that if two representations are equivalent, then they are unitarily equivalent. We prove a slightly stronger result now.

Theorem 7.1 Let ρ be an irreducible representation of G that is (unitarily) equivalent to $\overline{\rho}$, i.e., $\overline{\rho}(g) = \mathbf{U}^{-1}\rho(g)\mathbf{U}$ for some unitary matrix $\mathbf{U}$. Then $\mathbf{U}^{\mathrm{T}} = +\mathbf{U}$ iff ρ is a representation of the first kind and $\mathbf{U}^{\mathrm{T}} = -\mathbf{U}$ iff ρ is a representation of the second kind.

Proof From the theorem statement we have that

$$\overline{\rho}(g) = \mathbf{U}^{-1}\rho(g)\mathbf{U} \quad \text{or} \quad \rho(g) = \overline{\mathbf{U}}^{-1}\overline{\rho}(g)\overline{\mathbf{U}}$$

Since $\mathbf{U}$ is unitary, we have $\mathbf{U}^{-1} = \overline{\mathbf{U}}^{\mathrm{T}}$ and so

$$\rho(g) = \mathbf{U}^{\mathrm{T}}\overline{\rho}(g)(\mathbf{U}^{\mathrm{T}})^{-1}$$

Substituting this into the relationship above gives

$$\overline{\rho}(g) = \mathbf{U}^{-1}\mathbf{U}^{\mathrm{T}}\overline{\rho}(g)(\mathbf{U}^{\mathrm{T}})^{-1}\mathbf{U}$$

and from Schur's lemma we conclude that $\mathbf{U}^{-1}\mathbf{U}^{\mathrm{T}} = a\mathbf{I}$, $a \in \mathbb{C}$, i.e., $\mathbf{U}^{\mathrm{T}} = a\mathbf{U}$ or $\mathbf{U} = a\mathbf{U}^{\mathrm{T}} = a^2\mathbf{U}$ and we conclude that $a = \pm 1$. Since we have assumed ρ is equivalent to $\overline{\rho}$, it must be either of the first or second kind. If it is of the first

5.7 Real Representations

kind, then there exists a unitary matrix $\mathbf{V}$ such that $\mathbf{V}^{-1}\rho(g)\mathbf{V}$ is a real matrix for every $g \in G$. Thus, using the above observations, we have

$$\mathbf{V}^{-1}\rho(g)\mathbf{V} = (\overline{\mathbf{V}^{-1}})\bar{\rho}(g)\overline{\mathbf{V}} = (\overline{\mathbf{V}^{-1}})\mathbf{U}^{-1}\rho(g)\mathbf{U}\overline{\mathbf{V}}$$

and, again using Schur's lemma, we conclude that

$$\mathbf{U}\overline{\mathbf{V}}\mathbf{V}^{-1} = b\mathbf{I} \quad \text{or} \quad \mathbf{U} = b\mathbf{V}(\overline{\mathbf{V}^{-1}}) = b\mathbf{V}\mathbf{V}^T$$

which implies that $\mathbf{U} = \mathbf{U}^T$.

Hence if ρ is of the first kind, then we must have $\mathbf{U}^T = +\mathbf{U}$.

Conversely, suppose $\mathbf{U}^T = +\mathbf{U}$. Since $\mathbf{U}$ is unitary, it is unitarily equivalent to a diagonal matrix $\mathbf{D}$ and suppose

$$\mathbf{A}^{-1}\mathbf{U}\mathbf{A} = \mathbf{D}$$

where $\mathbf{A}$ is unitary. It follows that

$$\mathbf{U} = \mathbf{A}\mathbf{D}\mathbf{A}^{-1} = \mathbf{U}^T = (\mathbf{A}^T)^{-1}\mathbf{D}\mathbf{A}^T$$

or $\mathbf{A}^T\mathbf{A}$ commutes with $\mathbf{D}$. Now let $\mathbf{E}$ be a unitary diagonal matrix, with the property that $\mathbf{E}^2 = \mathbf{D}$, and that $\mathbf{E}$ commutes with $\mathbf{A}^T\mathbf{A}$. We define $\mathbf{W} = \mathbf{A}\mathbf{E}\mathbf{A}^{-1}$. By direct calculation we observe that $\mathbf{W}$ is unitary and that $\mathbf{W}^2 = \mathbf{A}\mathbf{E}^2\mathbf{A}^{-1} = \mathbf{U}$. Also, we have

$$\mathbf{W}^T = (\mathbf{A}^T)^{-1}\mathbf{E}\mathbf{A}^T = (\mathbf{A}^T)^{-1}\mathbf{E}\mathbf{A}^T\mathbf{A}\mathbf{A}^{-1}$$

But recall that $\mathbf{E}$ commutes with $\mathbf{A}^T\mathbf{A}$ and we can write

$$\mathbf{W}^T = (\mathbf{A}^T)^{-1}\mathbf{A}^T\mathbf{A}\mathbf{E}\mathbf{A}^{-1} = \mathbf{A}\mathbf{E}\mathbf{A}^{-1} = \mathbf{W}$$

Since $\mathbf{W}$ is unitary and symmetric, $\mathbf{W}^{-1} = \overline{\mathbf{W}}$ and thus

$$\mathbf{W}^{-1}\rho(g)\mathbf{W} = \mathbf{W}\mathbf{U}^{-1}\rho(g)\mathbf{U}\mathbf{W}^{-1} = \mathbf{W}\bar{\rho}(g)\mathbf{W}^{-1} = (\overline{\mathbf{W}^{-1}})(\overline{\rho(g)})(\overline{\mathbf{W}})$$

(since $\mathbf{W}^2 = \mathbf{U}$ or $\mathbf{W} = \mathbf{W}^{-1}\mathbf{U}$) and we conclude that $\mathbf{U}^T = +\mathbf{U}$ iff ρ is equivalent to a representation of the first kind. Equivalently, $\mathbf{U}^T = -\mathbf{U}$ iff ρ is equivalent to a representation of the second kind. □

The following theorem gives a simple criterion for determining which of three kinds a given representation is.

Theorem 7.2 Let χ be an irreducible character of a group G and define

$$h = \frac{1}{|G|}\sum_{g \in G}\chi(g^2)$$

Then $h = +1, -1, 0$ according to whether χ is an irreducible representation of the first, second, or third kind, respectively.

Proof Since we assume all representations are unitary, we can write

$$\chi(g^2) = \mathrm{tr}(\boldsymbol{\rho}(g)\boldsymbol{\rho}(g)) = \sum_{i=1}^{n}\left[\sum_{j=1}^{n} \rho(g)_{ij}\rho(g)_{ji}\right], \qquad \dim \boldsymbol{\rho} = n$$

Since $\boldsymbol{\rho}(g) = \overline{(\boldsymbol{\rho}(g)^{-1})^{\mathrm{T}}}$, it follows that $\rho(g)_{ji} = \overline{(\rho(g^{-1}))_{ij}}$ and hence

$$h = \frac{1}{|G|} \sum_{g\in G} \sum_{i,j=1}^{n} \rho(g)_{ij}\overline{(\rho(g^{-1})_{ij})}$$

If ρ is of the third kind, then ρ is not equivalent to $\bar{\rho}$ and from the orthogonality of characters we have $h = 0$. If ρ is of the first kind, then $\overline{\rho(g^{-1})_{ij}} = \rho(g^{-1})_{ij}$ and

$$h = \frac{1}{|G|}\sum_{g\in G}\sum_{i,j=1}^{n}\rho(g)_{ij}\rho(g^{-1})_{ij} = \sum_{i,j=1}^{n}\frac{1}{n}\delta_{ij} = 1$$

Finally, if ρ is of the second kind, then from Theorem 7.1 there exists a unitary matrix $\mathbf{U} = -\mathbf{U}^{\mathrm{T}}$ such that

$$\mathbf{U}^{-1}\boldsymbol{\rho}(g)\mathbf{U} = \overline{\boldsymbol{\rho}(g)} = -\overline{\mathbf{U}}\boldsymbol{\rho}(g)\mathbf{U}$$

It follows that

$$\overline{\rho(g^{-1})_{ij}} = -\sum_{s,t=1}^{n}\overline{U}_{is}\rho(g^{-1})_{st}U_{tj} \qquad \text{where} \quad \mathbf{U} = (U_{ij})$$

and from the formula for h,

$$h = \frac{1}{|G|}\sum_{g\in G}\sum_{i,j=1}^{n}\rho(g)_{ij}\left[-\sum_{s,t=1}^{n}\overline{U}_{is}\rho(g^{-1})_{st}U_{tj}\right]$$

$$= -\sum_{s,t=1}^{n}\sum_{i,j=1}^{n}\overline{U}_{is}U_{tj}\left[\frac{1}{|G|}\sum_{g\in G}\rho(g)_{ij}\rho(g^{-1})_{st}\right]$$

which, from the orthogonality property, we can write as

$$h = -\sum_{s,t=1}^{n}\sum_{i,j=1}^{n}\frac{1}{n}\overline{U}_{is}U_{tj}\delta_{it}\delta_{js} = -\frac{1}{n}\sum_{i,j}\overline{U}_{ij}U_{ij} = -1$$

which completes the proof of the theorem. □

Suppose that ρ is an irreducible representation of the second kind of the group G. Since ρ is equivalent to $\bar{\rho}$, there exists a unitary matrix U such that $U^{-1}\boldsymbol{\rho}(g)U = \bar{\boldsymbol{\rho}}(g)$ for each $g \in G$. If $\dim \rho = n$, then we have $U^{\mathrm{T}} = -U$ and $\det U = (-1)^n \det U$, implying that n is even. Thus every irreducible representation of G of the second kind is of even order. If G is a group of odd order, then it can have no irreducible representations of the second kind.

5.7 Real Representations

If C_i is a conjugacy class of G and $g \in C_i$, then we define the class C_{i-1} as the conjugacy class containing g^{-1}. The definition is not ambiguous, since it is independent of the element $g \in C_i$. Furthermore, it is a simple matter to check that $c_i = |C_i| = |C_{i-1}| = c_{i-1}$. If $C_i = C_{i-1}$, we say the conjugacy class is ambivalent.

We now show that the number of ambivalent conjugacy classes of G is equal to the number of columns in the character table that are real, which is also equal to the number of rows of the character table that are real. Let $C_i = C_{i-1}$ be an ambivalent conjugacy class and χ_j an irreducible character of G. In general, we have $\chi_j(g) = \bar{\chi}_j(g^{-1})$ and since, if $g, g^{-1} \in C_i$, they have identical characters, we conclude that $\chi_j(g) = \chi_j(g^{-1}) = \bar{\chi}_j(g)$ and the character of any ambivalent conjugacy class is real. On the other hand, suppose that C_i is not an ambivalent conjugacy class, so that $C_i \neq C_{i-1}$, and suppose that χ_j is real on C_i for $j = 1, \ldots, s$. Once again we have that $\chi_j(g^{-1}) = \bar{\chi}_j(g) = \chi_j(g)$ from the assumption of reality on the characters of C_i. But this implies that we have two distinct columns of the character table identical. This is impossible from the orthogonality relations on the columns of the character table. Thus the number of real columns is equal to the number of ambivalent conjugacy classes.

To show that the number of real columns of the character table is equal to the number of real rows, we proceed as follows. Define the matrix $\mathbf{U} = (U_{ij})$, $U_{ij} = (c_j/|G|)^{1/2}\chi_i(C_j)$, and consider the mapping $\chi_i \to \bar{\chi}_i$ that permutes the rows of $\mathbf{U}$. We account for this permutation by premultiplying by the appropriate permutation matrix $\mathbf{A}$ and denote $\mathbf{U}' = \mathbf{AU}$. Similarly, consider the mapping $C_i \to C_{i-1}$ that permutes the columns of $\mathbf{U}$, and we account for this by postmultiplying by the appropriate permutation matrix $\mathbf{B}$ and letting $\mathbf{U}'' = \mathbf{UB}$. Notice that $c_i = |C_i| = |C_{i-1}| = c_{i-1}$. Now the (i,j)th element of $\mathbf{U}'$ is $(c_j/|G|)^{1/2}\bar{\chi}_i(C_j)$ and that of $\mathbf{U}''$ is $(c_{j-1}/|G|)^{1/2}\chi_i(C_{j-1})$. But $c_j = c_{j-1}$ and $\bar{\chi}_i(C_j) = \chi_i(C_{j-1})$ and we conclude that $\mathbf{U}' = \mathbf{U}''$ or

$$\mathbf{AU} = \mathbf{UB} \quad \text{or} \quad \mathbf{B} = \mathbf{U}^{-1}\mathbf{AU}$$

This implies that $\text{tr}(\mathbf{A}) = \text{tr}(\mathbf{B})$. But the trace of $\mathbf{A}$ is simply the number of rows of $\mathbf{U}$ that the permutation $\chi_i \to \bar{\chi}_i$ leaves fixed. The rows that this permutation leaves fixed are precisely the real rows. Similarly the trace of $\mathbf{B}$ is the number of columns left fixed by the permutation $C_i \to C_{i-1}$, which we see is the number of real columns of $\mathbf{U}$. We conclude that the number of ambivalent conjugate classes of G is the same as the number of real columns of $\mathbf{U}$, which is the same as the number of real rows of $\mathbf{U}$, and the proof is complete.

If the order of the group G is odd, however, then it can have only one ambivalent conjugacy class, that consisting of the identity element. To see this, suppose $|G|$ is odd and that $hgh^{-1} = g^{-1}$ or $g = hg^{-1}h^{-1}$. Then $h^2gh^{-2} = hg^{-1}h^{-1} = g$ and h^2 commutes with g, which implies that any even power of

h commutes with g. However, h must have odd order since $H = \langle h \rangle$ is such that $|H| \,|\, |G|$, and so h^2 also generates H. Thus for some k, $h^{2k} = h$ and h also commutes with g, which implies that $g = g^{-1}$ or $g^2 = 1$, which implies that $g = 1$ since G contains no elements of even order. Thus a group of odd order has precisely one representation of the first kind and all others are of the third kind. For a general group the number of representations of the first and second kinds is equal to the number of ambivalent conjugacy classes and, as shown, to the number of real rows of the character table. For other interesting properties of the character table see Lomont (1959, pp. 63–65).

Suppose we are given an irreducible representation of the second or third kind of dimension n. We wish to construct a real representation which is irreducible over the reals, i.e., a real irreducible representation. Write the given matrices as $\mathbf{\rho}(g) = \mathbf{u}(g) + i\mathbf{v}(g)$, where $\mathbf{u}(g)$ and $\mathbf{v}(g)$ are real matrices, for every $g \in G$. If we let

$$\mathbf{U} = \frac{1}{\sqrt{2}} \begin{bmatrix} I_n & iI_n \\ iI_n & I_n \end{bmatrix}$$

then the following relationship is easily verified:

$$\mathbf{U}^{-1}(\mathbf{\rho}(g) \oplus \bar{\mathbf{\rho}}(g))\mathbf{U}$$
$$= \frac{1}{\sqrt{2}} \begin{bmatrix} I_n & -iI_n \\ -iI_n & I_n \end{bmatrix} \begin{bmatrix} \mathbf{u}(g) + i\mathbf{v}(g) & 0 \\ 0 & \mathbf{u}(g) - i\mathbf{v}(g) \end{bmatrix} \frac{1}{\sqrt{2}} \begin{bmatrix} I_n & iI_n \\ iI_n & I_n \end{bmatrix}$$
$$= \begin{bmatrix} \mathbf{u}(g) & -\mathbf{v}(g) \\ \mathbf{v}(g) & \mathbf{u}(g) \end{bmatrix} = \mathbf{\rho}'(g).$$

That $\mathbf{\rho}'(g)$ is indeed a real irreducible representation is easily seen by calculating the multiplicity of any irreducible representation in it. Notice that the term "real irreducible representation" now has a somewhat ambiguous meaning since it may be an irreducible representation over the complex numbers that happens to be real, or a real representation that is irreducible over the reals but reducible over the complex numbers. It will be clear from the context which is meant.

Consider the decomposition of the regular representation of G

$$\bigoplus_i n_i \mathbf{\rho}_i(g)$$

where $\mathbf{\rho}_i$ is an irreducible representation of G of dimension n_i. To decompose the regular representation into real representations, irreducible over the reals, we consider the three types of representations possible. If $\mathbf{\rho}_i$ is a representation of the first kind, then it is already equivalent to a real representation. If $\mathbf{\rho}_i$ is a representation of the second kind, then it has an even dimension and

5.7 Real Representations

appears in the above decomposition an even number of times. Since ρ_i is equivalent to its conjugate in this case, we may pair each ρ_i with a $\bar{\rho}_i$ to obtain $n_i/2$ real irreducible representations of dimension $2n_i$. Similarly, if ρ_i is a representation of the third kind, then $\bar{\rho}_i$ is also a representation and both of these representations appear in the above decomposition. Again pairing ρ_i with $\bar{\rho}_i$ we form a real irreducible representation of dimension $2n_i$ and obtain n_i such representations. Thus, from the complex decomposition of the regular representation we can obtain the real decomposition. In the above discussion we assumed equivalent versions of representations were available when needed.

In the next chapter we will require the information contained in the following lemma.

Lemma Let V be a vector space of dimension n affording the real irreducible representation ρ of the finite group G. Let D be the division ring of $n \times n$ matrices that commute with all the matrices of ρ. Then we have the following three situations:

(i) If ρ is absolutely irreducible (i.e., real but also irreducible over the complex numbers), then $D \cong \mathbb{R}$.
(ii) If $\rho \cong \rho' \oplus \bar{\rho}'$, where ρ' is a complex irreducible representation of the second kind, then $D \cong \mathbb{H}$, where $\mathbb{H}$ is the set of real quaternions.
(iii) If $\rho \cong \rho' \oplus \bar{\rho}'$, where ρ' is a complex irreducible representation of the third kind, then $D \cong \mathbb{C}$.

Proof From a previous lemma we know that the set of all matrices that commute with all the matrices of a real irreducible representation is a division ring D. Clearly real scalar multiples of the identity are in D for each of the three cases and $\mathbb{R} \subset D$. From Herstein (1964) there are precisely three division rings containing $\mathbb{R}$, namely $\mathbb{R}$, $\mathbb{C}$, and $\mathbb{H}$. By Schur's lemma D is precisely $\mathbb{R}$ for case (i). Suppose now that $D \cong \mathbb{H}$. The vector space V can be viewed as a vector space of dimension $n/4$ over $\mathbb{H}$, where scalar multiplication is

$$q \cdot x = q(x), \quad q \in \mathbb{H}, \quad x \in V$$

Let $u_1, \ldots, u_{n/4}$ be a fixed $\mathbb{H}$ basis for V and let $\{\rho_{\mathbb{H}}(g), g \in G\}$ be a matrix representation for G over $\mathbb{H}$ of dimension $n/4$ obtained by the action of g on the basis.

Now consider V as a vector space over $\mathbb{C}$ and choose as a basis the elements $u_1, \ldots, u_{n/4}, ju_1, \ldots, ju_{n/4}$. If $\rho_{\mathbb{H}}(g) = (q_{ij})$ as an $\mathbb{H}$ representation, then

$$g: u_s \to \sum_{l=1}^{n/4} q_{ls} u_l = \sum_{l=1}^{n/4} [(\alpha_{ls} + i\beta_{ls}) + (\gamma_{ls} + i\varepsilon_{ls})j]u_l$$

and
$$g: ju_s \to j\sum_{l=1}^{n/4} q_{ls}u_l = \sum_{l=1}^{n/4}[(\alpha_{ls} - i\beta_{ls})ju_l - (\gamma_{ls} - i\varepsilon_{ls})u_l]$$

where $\{1, j\}$ is a basis of $\mathbb{H}$ over $\mathbb{C}$.

Thus if $\rho_\mathbb{H}(g) = (q_{ij}) = \mathbf{A} + \mathbf{B}j$, where $\mathbf{A}, \mathbf{B} \in M_{n/4}(\mathbb{C})$, the set of $n/4 \times n/4$ matrices over $\mathbb{C}$, then

$$\rho_\mathbb{C}(g) = \begin{pmatrix} \mathbf{A} & \mathbf{B} \\ -\bar{\mathbf{B}} & \bar{\mathbf{A}} \end{pmatrix}$$

is an $n/2 \times n/2$ complex representation afforded by V as a vector space over $\mathbb{C}$. From the assumptions on V, $\rho_\mathbb{C}$ is a complex irreducible representation and from its construction its trace is real. Thus V affords the real representation $\rho = \rho' \oplus \bar{\rho}'$, where ρ' is a complex irreducible representation of the second kind when $D = \mathbb{H}$. When ρ is of the first kind, then, as mentioned, from Schur's lemma it follows that $D \cong \mathbb{R}$. If $D \cong \mathbb{H}$, then we have shown that the trace of ρ must be real, implying ρ is either of the first or second kind. Since it cannot be of the first kind, ρ is of the second kind. Finally, since every real representation obtained from a complex irreducible representation of the third kind can be assumed of the form

$$\begin{pmatrix} u(g) & -v(g) \\ v(g) & u(g) \end{pmatrix}$$

it is clear that scalar multiples of the matrices

$$\begin{pmatrix} I & 0 \\ 0 & I \end{pmatrix} \quad \text{and} \quad \begin{pmatrix} 0 & -I \\ I & 0 \end{pmatrix}$$

commute with the representation. Thus D contains a set of matrices isomorphic to $\mathbb{C}$, and since D cannot be $\mathbb{H}$ we conclude that $D \cong \mathbb{C}$. □

Notice that the result of a lemma in Section 5.3, that the only real symmetric matrices that commute with all the matrices of a real irreducible representation are real scalar multiples of the identity matrix, follows readily from the much stronger result contained in this last lemma.

The result concerning commutating real symmetric matrices can be used to obtain orthogonality relationships for real irreducible representations. The proof parallels the proof of Section 5.4 for complex irreducible representations and we omit it. If ρ_1 and ρ_2 are two nonequivalent real irreducible representations, then we can show that

$$\sum_{g \in G} \rho_1(g^{-1})_{ij} \rho_2(g)_{kl} = 0 \quad \text{any} \quad i, j = 1, \ldots, n_1, \quad l, k = 1, \ldots, n_2 \quad (7.1a)$$

$$\sum_{g \in G} [\rho_1(g^{-1})_{ki} \rho_1(g)_{jl} + \rho_1(g^{-1})_{kj} \rho_1(g)_{il}] = \frac{2}{n} \delta_{ij} \delta_{kl} \quad (7.1b)$$

$$i, j, k, l = 1, \ldots, n_1$$

5.7 Real Representations

The corresponding relations for characters are

$$\sum_{g \in G} \chi_1(g)\chi_2(g) = 0 \tag{7.2a}$$

and

$$\sum_{g \in G} [\chi_1(g)\chi_1(g) + \chi_1(g^2)] = 2|G| \tag{7.2b}$$

We will require some modifications of these orthogonality relationships for the next chapter. We adopt the $4n \times 4n$ matrix representation of the quaternions over the reals as

$$e_0 \to \begin{bmatrix} I_n & & & \\ & I_n & & 0 \\ & 0 & I_n & \\ & & & I_n \end{bmatrix}, \quad e_1 \to \begin{bmatrix} 0 & -I_n & & 0 \\ I_n & 0 & & \\ \hline & & 0 & -I_n \\ 0 & & I_n & 0 \end{bmatrix},$$

$$e_2 \to \begin{bmatrix} 0 & & -I_n & 0 \\ & & 0 & I_n \\ \hline I_n & 0 & & \\ 0 & -I_n & & 0 \end{bmatrix}, \quad e_3 \to \begin{bmatrix} & 0 & & 0 & -I_n \\ & & & -I_n & 0 \\ \hline 0 & -I_n & & & \\ I_n & 0 & & 0 & \end{bmatrix} \tag{7.3}$$

If ρ is a real irreducible representation of the second kind, then without loss of generality we can assume the representation commutes with these matrices, implying that the matrices of the representation must be in the form

$$\rho(g) = \begin{bmatrix} U_1(g) & U_2(g) & V_1(g) & V_2(g) \\ -U_2(g) & U_1(g) & -V_2(g) & V_1(g) \\ -V_1(g) & V_2(g) & U_1(g) & -U_2(g) \\ -V_2(g) & -V_1(g) & U_2(g) & U_1(g) \end{bmatrix}$$

Similarly, if ρ is a real irreducible representation of the third kind of dimension $2n$, from the real matrix representation of the quaternions we choose the real matrix representation of the complex numbers

$$e_0 \to \begin{bmatrix} I & 0 \\ 0 & I \end{bmatrix}, \quad e_3 \to \begin{bmatrix} 0 & -I \\ I & 0 \end{bmatrix} \tag{7.4}$$

implying that the matrices of the group representation must be of the form

$$\rho(g) = \begin{bmatrix} U(g) & -V(g) \\ V(g) & U(g) \end{bmatrix}$$

For the remainder of the section we assume that real irreducible representations are of the second or third kind, and hence matrix bases for their commuting matrices are as just given.

Let ρ be a real irreducible representation and define the matrix

$$\mathbf{B} = \sum_{g \in G} \rho^{-1}(g) \mathbf{A} \rho(g)$$

where $\mathbf{A}$ is an arbitrary real $n \times n$ matrix. Clearly $\mathbf{B}$ commutes with all the matrices of the representation. If $\mathbf{A} = (a_{rs})$, then the (i, l) element of $\mathbf{B}$ can be expressed as

$$b_{il} = \sum_{g \in G} \sum_{r, s} \rho^{-1}(g)_{ir} a_{rs} \rho(g)_{sl}$$

Let $\mathbf{A}$ be the matrix with unity in the (j, k) position and zeros elsewhere and denote the corresponding matrix $\mathbf{B}$ by $\mathbf{B}^{jk} = (b_{uv}^{jk})$. Then we have that

$$b_{il}^{jk} = \sum_{g \in G} \rho^{-1}(g)_{ij} \rho(g)_{kl} = \sum_{g \in G} \rho(g)_{ji} \rho(g)_{kl}$$

since ρ is assumed orthogonal.

If ρ is a real irreducible representation of the first kind, then $\mathbf{B}^{jk}$ is a real scalar multiple of the identity matrix and

$$b_{il}^{jk} = \sum_{g \in G} \rho(g)_{ji} \rho(g)_{kl} = \lambda^{jk} \delta_{il} = \frac{|G|}{n} \delta_{jk} \delta_{il} \qquad (7.5)$$

If ρ is a real irreducible representation of the second kind, then b_{il}^{jk} is the (i, l) element of a real linear combination of the matrices e_0, e_1, e_2, and e_3, i.e.,

$$b_{il}^{jk} = \sum_{g \in G} \rho(g)_{ji} \rho(g)_{kl} = (\lambda_0^{jk} e_0 + \lambda_1^{jk} e_1 + \lambda_2^{jk} e_2 + \lambda_3^{jk} e_3)_{il} \qquad (7.6)$$

$$\lambda_r^{jk} \in \mathbb{R}, \quad r = 0, 1, 2, 3$$

If ρ is a real irreducible representation of the third kind, then

$$b_{il}^{jk} = \sum_{g \in G} \rho(g)_{ji} \rho(g)_{kl} = (\lambda_0^{jk} e_0 + \lambda_3^{jk} e_3)_{il}, \qquad \lambda_r^{jk} \in \mathbb{R}, \quad r = 0, 3 \qquad (7.7)$$

Equations (7.5)–(7.7) are actually modified orthogonality relationships, but they are in a form that will be needed in the next chapter.

5.8 Modules, Group Algebras, and Representations

In Chapter 4 the structure of the group algebra was studied in some detail and it was indicated that such a structure was useful for investigating cyclic codes and Abelian codes and for group representations. We discuss this latter connection in this section.

5.8 Modules, Group Algebras, and Representations

Let ρ be a representation of the group G with representation space V, i.e.,

$$\rho: \quad G \to GL(V)$$

We extend this by linearity to an algebra homomorphism ρ^* from KG to $\text{end}_K(V)$, where KG is the group algebra of G over K and $\text{end}_K(V)$ the set of K endomorphisms (isomorphic to K_n, the set of $n \times n$ matrices over K) by defining

$$\rho^*: \quad KG \to \text{end}_K(V), \qquad \sum_{g \in G} \alpha_g g \to \sum_{g \in G} \alpha_g \rho(g)$$

That ρ^* is an algebra homomorphism (i.e., a K linear map such that $\rho^*(aa') = \rho^*(a)\rho^*(a')$, $a,a' \in KG$) is easily checked by showing that the defining relations of such a map are satisfied, i.e.,

$$\rho^*(a+b) = \rho^*(a) + \rho^*(b), \qquad \rho^*(ab) = \rho^*(a)\rho^*(b)$$
$$\rho^*(\alpha a) = \alpha \rho^*(a), \qquad \rho^*(1) = i = \text{the identity map},$$
$$a, b \in KG, \quad \alpha \in K$$

Definition Let R be a ring with identity. A left R module M is an additive Abelian group together with a mapping $R \times M \to M$ that satisfies:

(i) $r(m_1 + m_2) = rm_1 + rm_2$.
(ii) $(r_1 + r_2)m = r_1 m + r_2 m$.
(iii) $1 \cdot m = m$.
(iv) $r_1(r_2 m) = (r_1 r_2)m$, $m, m_1, m_2 \in M$, $r, r_1, r_2 \in R$.

If we define the mapping $KG \times V \to V$ by

$$\left(\sum_{g \in G} \alpha_g g \right) \cdot v = \sum \alpha_g \rho(g) v$$

then it is clear that V is a KG module, where we interpret KG as an associative ring with identity. Conversely, suppose we are given a KG module V. We can then define $\rho^* \in \text{end}_K(V)$ by

$$\rho^*(g): \quad V \to V$$
$$x \mapsto gx$$

and it is readily checked that this yields a representation of the group G, the representation afforded by the KG module V. Thus there is a one-to-one correspondence between the representations of a group G and KG modules V. The structure of a general R module M is then a natural one to study for group representations.

In this chapter we have restricted our attention to representations over the field $\mathbb{C}$ (and thence over $\mathbb{R}$). Many of the theorems hold in any field K with characteristic zero and some hold for a field K with characteristic p where $(p, |G|) = 1$. Others require algebraic closure of the field. In these cases, however, it was shown in Chapter 4 that the group algebra is semisimple. Thus the study of modules over semisimple rings is of particular importance in group representation theory.

By a submodule N of the R module M is meant an additive subgroup of M such that if $m \in N$, then $rm \in N$ for all $r \in R$. If V is a KG module, then the subspaces of V invariant under a representation are simply the submodules of V. An irreducible module is one that contains no nontrivial submodules. If R is semisimple, then it can be shown that the R module M can be written as a direct sum of irreducible submodules, and if M is the KG module V, then these irreducible submodules correspond to irreducible representations of G over K. In all of this the structure of the group algebra KG, as given in Chapter 4, plays a role.

In particular let us consider the algebra KG as a KG module. We then define the representation ρ by

$$\rho(g): KG \to KG, \qquad \rho(g) \in \text{end}_K(KG)$$
$$\sum_{h \in G} \alpha_h h \mapsto \sum_{h \in G} \alpha_h gh$$

If we choose the group elements as a basis of KG, then the resulting matrices will be the permutation matrices of the regular representation.

It is easily checked that an irreducible submodule of the above KG module is a minimal subring of the group algebra and vice versa. In general an irreducible submodule of any R module is isomorphic to some minimal left ideal of R. However, in Chapter 4 we gave an extensive discussion of the decomposition of the group algebra into a direct sum of minimal left ideals.

We recall some facts from Chapter 4. Any semisimple ring with identity that satisfies the minimum condition can be written as a direct sum of simple two-sided ideals. The decomposition is unique up to ordering and each two-sided ideal is generated by a central idempotent. Now each such minimal two-sided ideal, or simple component, can be decomposed into a direct sum of minimal left ideals, each such minimal left ideal having an idempotent generator, and, for any given simple component of KG, these minimal left ideals are isomorphic, although the decomposition is not unique. Minimal left ideals appearing in the decomposition of distinct simple components are non-isomorphic.

If the field K is the complex numbers, then KG is semisimple. The regular representation defined above in terms of KG as a KG module can be written as

$$\rho_R = \bigoplus_{i=1}^{s} n_i \rho_i$$

5.8 Modules, Group Algebras, and Representations

where ρ_i is an irreducible representation of dimension n_i and G has s conjugacy classes and s nonequivalent irreducible representations. If the group algebra has the decomposition

$$KG = \bigoplus_{i=1}^{s} S_i$$

where S_i, $i = 1, \ldots, s$, are the simple components of KG, then a representation equivalent to the representation $n_i \rho_i$ of dimension n_i^2 is obtained from the S_i by

$$n_i \rho_i(g): \quad S_i \to S_i$$
$$a \mapsto ga$$

Similarly, if L_i is a minimal left ideal of S_i, then a representation equivalent to the irreducible representation ρ_i can be obtained by defining

$$\rho_i(g): \quad L_i \to L_i$$
$$a \mapsto ga$$

If we decompose S_i into $\sum_{j=1}^{n_i} L_{i_j}$, L_{i_j} a set of isomorphic minimal left ideals, then by choosing the correct basis in each L_{i_j} we can obtain the representation $n_i \rho_i$, i.e., a direct sum of identical representations. We shall call such a representation a full homogeneous component and a representation of the form $k\rho_i$, $k < n_i$, a homogeneous component.

We shall demonstrate the statement concerning the connection between full homogeneous and simple components. Recall Theorem 6.2 of Chapter 4, which stated that the elements

$$\eta_i = \sum_{g \in C_i} g, \quad i = 1, \ldots, s$$

span the center of the group algebra KG as a vector space over K.

For this paragraph let $K = \mathbb{C}$. Consider the decomposition

$$KG = \bigoplus_{i=1}^{s} S_i$$

of KG into a direct sum of minimal two-sided ideals and let

$$1 = e_1 + \cdots + e_s$$

be the corresponding (unique) decomposition of the group algebra unit as a sum of central idempotents. Let the irreducible characters of G be denoted by χ_i, $i = 1, \ldots, s$, as before. For convenience we assume that χ_i is associated with the KG module KGe_i, $i = 1, \ldots, s$. From Chapter 4, we have that $e_i e_j = \delta_{ij} e_i$. Since e_i acts as a unit on KGe_i and annihilates all elements with null components in KGe_i, in any matrix representation $\boldsymbol{\rho}_i$ of KG afforded by KGe_i we must have that

$$\boldsymbol{\rho}_i(e_j) = \delta_{ij} \mathbf{I}_{n_i}$$

However, e_i is in the center of the group algebra and thus can be written as a $\mathbb{C}$ linear function of the η_i. Since η_j commutes with all elements, its matrix with respect to any base of KGe_i must be a scalar multiple of the identity matrix

$$\rho_i(\eta_j) = \lambda \mathbf{I}_{n_i} = \left(\sum_{g \in C_j} \rho_i(g)\right)$$

Taking traces on both sides yields

$$\lambda n_i = c_j \chi_i(C_j)$$

where, as before, $|C_j| = c_j$ and χ_i is the character (invariant under a change of basis) afforded by KGe_i. Thus we have

$$\rho_i(\eta_j) = \frac{c_j \chi_i(C_j)}{n_i} \mathbf{I}_{n_i}$$

and it follows that we can write this as

$$\rho_i(\eta_j) = \sum_{k=1}^{s} \frac{c_j \chi_k(C_j)}{n_k} \rho_i(e_k)$$

and hence

$$\eta_j = \sum_{k=1}^{s} \frac{c_j \chi_k(C_j)}{n_k} e_k$$

where χ_k is the irreducible character obtained from KGe_k. Now consider the sum

$$\frac{n_i}{|G|} \sum_{j=1}^{s} \bar{\chi}_i(C_j)\eta_j = \frac{n_i}{|G|} \sum_{j=1}^{s} \bar{\chi}_i(C_j) \sum_{k=1}^{s} \frac{c_j \chi_k(C_j)}{n_k} e_k$$

$$= \frac{n_i}{|G|} \sum_{k=1}^{s} \frac{e_k}{n_k} \sum_{j=1}^{s} c_j \bar{\chi}_i(C_j)\chi_k(C_j)$$

$$= \frac{n_i}{|G|} \sum_{k=1}^{s} \frac{e_k}{n_k} \delta_{ik} |G| = e_i$$

and thus we have the formula

$$e_i = \frac{n_i}{|G|} \sum_{j=1}^{s} \bar{\chi}_i(C_j)\eta_j \qquad (8.1)$$

for the central idempotent that generates the minimal two-sided ideal KGe_i, any minimal left ideal of which gives rise to an irreducible representation isomorphic to ρ_i with character χ_i.

5.8 Modules, Group Algebras, and Representations

Serre (1971, Theorem 8, p. 84) gives an interesting theorem which is similar to the above discussion in some respects. We outline this result here. Let ρ be a representation of G with representation space V and suppose ρ has the decomposition

$$\rho \cong \bigoplus_{i=1}^{s} m_i \rho_i$$

into irreducible representations ρ_i (m_i may be zero). Now let the corresponding decomposition of V into invariant subspaces be given by

$$V = U_1 \oplus \cdots \oplus U_m$$

Group the U_i that are isomorphic together and call their direct sum V_i, which gives rise to the representation $m_i \rho_i$. Thus we have

$$V = V_1 \oplus \cdots \oplus V_s$$

It can then be shown that this decomposition is unique up to ordering and that the operator

$$p_i = \frac{n_i}{|G|} \sum_{g \in G} \bar{\chi}_i(g) \rho(g)$$

is a projection operator of V onto V_i. If we let V be the group algebra, then V_i becomes a simple component S_i. The central idempotent generator e_i of S_i then gives rise to the projection map

$$p_i': \quad KG \to S_i$$
$$a \mapsto a e_i$$

and in this case $p_i' = p_i$.

The decomposition of the V_i or S_i into irreducible subspaces or modules is discussed further by Serre (1971, Section 2.7).

Finally, we mention that there are two sets of structure constants associated with each group. If we consider the elements

$$\eta_i = \sum_{g \in C_i} g \in C(KG)$$

then, since $\eta_i \eta_j$ is also in $C(KG)$ and since η_i, $i = 1, \ldots, s$, span $C(KG)$, then we can write

$$\eta_i \eta_j = \sum_{k=1}^{s} c_{ijk} \eta_k, \quad 1 \leq i, j \leq s$$

where $c_{ijk} \in K$. Similarly, if ρ_i and ρ_j are irreducible representations of G, then $\rho_i \otimes \rho_j$ is also a representation with character $\chi_i \chi_j$ and, since $\rho_i \otimes \rho_j$ is equivalent to a direct sum of irreducible representations, i.e.,

$$\rho_i \otimes \rho_j \cong \sum_{k=1}^{s} g_{ijk} \rho_k$$

then we can write for characters

$$\chi_i \chi_j = \sum_{k=1}^{s} g_{ijk} \chi_k, \qquad 1 \leqslant i, j \leqslant s$$

The sets of constants $\{c_{ijk}\}$ and $\{g_{ijk}\}$ can be used to give a great deal of information about G and its representations.

The above discussion gives a brief indication of the role of modules and group algebras in group representation theory. The treatise of Curtis and Reiner (1966) on the subject gives a complete account of this approach.

5.9 Comments

There are many books on group representation theory written by mathematicians, chemists, and physicists. The monograph by Serre (1971) is particularly lucid and direct and the early part of this chapter tends to follow his approach. The books by Boerner (1963), Lomont (1959), and Jansen and Boon (1967) take an applied view of the subject, being concerned mainly with applications of the theory to quantum mechanics and molecular structure. The books by Curtis and Reiner (1966), Burrow (1965), and van der Waerden (1970) take a mathematical approach. The presentation here is a survey of the theory including many, but by no means all, of the basic results.

Exercises

1 Let G be a group with subgroups K and H, $H \subset K \subset G$. Show that if ρ is a representation of H, then ρ^G is isomorphic to $(\rho^K)^G$, where ρ^G is the representation ρ induced to G and $(\rho^K)^G$ the representation ρ induced first to K, then to G.

2 Show that a group G is simple if and only if for every $g \in G$ and for every irreducible character χ_i (except the identity character) it is true that $\chi_i(g) \neq \chi_i(1)$, $g \neq 1$.

3 If χ is a character of an irreducible representation ρ, $\deg \rho > 1$, show that $\chi(g) = 0$ for some element $g \in G$.

4 Let G be a group containing an Abelian normal subgroup H of index two in G. Show that the multiplicity of any irreducible representation of G in the tensor product of any two irreducible representations of G cannot be greater than one.

Exercises

5 Determine which of the irreducible characters for A_5 are induced from irreducible characters of A_4.

6 One set of structure constants for a group G is defined in the group algebra KG by

$$\eta_i \eta_j = \sum_{k=1}^{s} c_{ijk} \eta_k, \qquad \eta_i = \sum_{g \in C_i} g$$

Show that if χ is an irreducible character of G, then we have

$$c_i \chi(C_i) c_j \chi(C_j) = n \sum_{k=1}^{s} c_{ijk} c_k \chi(C_k)$$

where n is the dimension of the associated representation.

7 Let ρ be an irreducible representation of G that is faithful (the kernel of the representation $= \{1\}$). Show that every irreducible representation appears as a component of the representation $\rho^k = \rho \otimes \rho \otimes \cdots \otimes \rho$ (k times) for some integer k.

8 Let H be a subgroup of G and let ρ be the representation of G obtained by inducing the identity representation of H.

(i) Show that the character χ of ρ is given by

$$\chi(C_j) = \frac{|G| \, |C_j \cap H|}{|H| c_j}$$

(ii) Show that ρ contains the identity representation of G exactly once.

(iii) Show that ρ is a component of the regular representation of G.

(iv) Show that ρ is faithful if and only if neither H nor any of its subgroups is normal in G.

9 Let χ_i and χ_j be two irreducible characters of a finite group G of dimension n_i and n_j, respectively. From Theorem 3.4 we have that

$$(\chi_i, \chi_j) = \frac{1}{|G|} \sum \chi_i(g) \chi_j(g^{-1}) = \delta_{ij} = \frac{1}{|G|} \sum_{k=1}^{s} c_k \chi_i(C_k) \bar{\chi}_j(C_k)$$

From the corollary to Theorem 4.1 we have

$$\sum_{i=1}^{s} \chi_i(C_j) \bar{\chi}_i(C_k) = \frac{|G|}{c_k} \delta_{jk}$$

Show that

$$\sum_{g \in G} \chi_i(hg) \bar{\chi}_j(g) = \chi_i(h) \frac{|G|}{n_i} \delta_{ij}$$

6 Group Codes for the Gaussian Channel

6.1 Introduction

Packing problems of various types have long been of interest in mathematics. Perhaps the most investigated is that of packing spheres into n-dimensional Euclidean space E_n. Recently algebraic codes over finite fields have been used to construct some particularly dense lattice packings in E_n (Leech and Sloane, 1970). The problem considered in this chapter is that of packing spherical caps on the unit sphere in E_n. The connection between this problem and the equal-energy signaling problem of communications is established in the next section.

As the dimensionality of the problem increases and the number of caps packed onto the sphere increases, it becomes important to have a theory that we can use to predict the essential properties of the packing without actually constructing it. This is the role that group representation theory will play. The situation is somewhat similar to that of algebraic coding theory, where we were able to provide techniques to describe codes with a lower bound on the minimum distance without actually constructing the code. Unfortunately, the development of the application of group representation theory to constructing group codes is still in its infancy. The structures seem to be promising but much work remains to be done.

6.2 Codes for the Gaussian Channel

We first consider the relevance of the problem of packing spherical caps on a sphere in Euclidean space to communication theory. In a typical M-ary communication system the receiver selects one of M possible messages to transmit every T seconds. We assume that each such message has the same energy, i.e., if the signal set is denoted by $\{m_i(t), i = 1, \ldots, M\}$, then

$$\int_0^T m_i^2(t)\, dt = E_m$$

The received message at the receiver is then assumed to be of the form

$$y(t) = s(t) + n(t), \qquad 0 \leq t \leq T$$

where $s(t) = m_i(t)$ for some i and $n(t)$ is a sample function from white Gaussian noise with a power spectral density

$$S(f) = N_0/2, \qquad -\infty < f < \infty$$

The receiver must decide on the value of i, preferably in some optimum manner, to determine which message was transmitted during the time interval $(0, T)$.

Now it is always possible to express each member of the signal set $\{m_i(t)\}$ as

$$m_i(t) = \sum_{j=1}^n m_{ij} g_j(t), \qquad 0 \leq t \leq T$$

where $\{g_j(t), j = 1, \ldots, n\}$ is a set of orthonormal functions over the period $(0, T)$, i.e.,

$$\int_0^T g_i(t) g_j(t)\, dt = \delta_{ij}$$

That this can always be done follows from the Gram–Schmidt orthogonalization process for such functions. Thus for $g_1(t)$ we choose $(1/\sqrt{E_m}) m_1(t)$. For $g_2(t)$ we take

$$g_2(t) = K_2 \left[m_2(t) - \frac{a}{\sqrt{E_m}} m_1(t) \right]$$

where

$$a = \frac{1}{\sqrt{E_m}} \int_0^T m_1(t) m_2(t)\, dt$$

and K_2 normalizes the energy of $g_2(t)$ to unity. The process is continued until we can describe every signal in the set by such a linear combination of orthonormal functions. The functions $\{g_j(t), j = 1, \ldots, n\}$ are thought of as an orthonormal basis of a signal space of dimension n. Each signal $m_i(t)$ can be written as

$$m_i(t) = \sum_{j=1}^{n} m_{ij} g_j(t)$$

and thus to each signal we can make the association

$$m_i(t) \leftrightarrow (m_{i1}, m_{i2}, \ldots, m_{in}) = \mathbf{m}_i$$

where

$$m_{ij} = \int_0^T m_i(t) g_j(t) \, dt$$

At the receiver we form the vector

$$\mathbf{y} = \mathbf{m}_i + \mathbf{n}$$

where

$$y_j = \int_0^T y(t) g_j(t) \, dt$$

and the function of the receiver is to determine $\mathbf{m}_i$ or i optimally, where we have yet to define the criterion for optimality. In the absence of noise the received vector would be just the transmitted signal. Since

$$\int_0^T m_i^2(t) \, dt = \int_0^T \left[\sum_{j=1}^{n} m_{ij} g_j(t) \right]^2 dt = \sum_{j,k=1}^{n} m_{ij} m_{ik} \int_0^T g_j(t) g_k(t) \, dt$$

$$= \sum_{j=1}^{n} m_{ij}^2 = E_m$$

we interpret the signal vector $\mathbf{m}_i$ as a point on the sphere of radius $\sqrt{E_m}$ in n dimensions. The received point $\mathbf{y} = \mathbf{m}_i + \mathbf{n}$ is the signal vector perturbed by the noise vector $\mathbf{n}$. Since the noise was assumed white Gaussian, we note that n_i has mean zero and variance

$$E[n_i^2] = \int_0^T \int_0^T g_i(t) g_i(s) E[n(t) n(s)] \, dt \, ds$$

$$= \int_0^T g_i^2(t) \tfrac{1}{2} N_0 \, dt = \tfrac{1}{2} N_0$$

Furthermore, n_i and n_j are independent random variables. The received vector $\mathbf{y}$ is then a Gaussian vector with mean vector $\mathbf{m}_i$ if the ith message was transmitted and each component of $\mathbf{y}$ has variance $N_0/2$ and is independent of other components.

6.2 Codes for the Gaussian Channel

Our decoding criterion is that of maximum likelihood, where we decide that message i was transmitted if

$$P(i|\mathbf{y}) \geq P(j|\mathbf{y}) \quad \text{for all } j \neq i$$

Using Bayes's theorem on conditional probabilities and assuming that all messages are equally likely to be transmitted, it is readily found that this maximum likelihood criterion reduces to defining decision regions in Euclidean n-space E_n by

$$R_i = \{\mathbf{x} \in E_n | \; |\mathbf{x} - \mathbf{m}_i| \leq |\mathbf{x} - \mathbf{m}_j|, \; j \neq i\}$$

and then assuming that message i was transmitted if $\mathbf{y} \in R_i$. The decision regions are convex regions bounded by hyperplanes passing through the origin. The intersection of any two regions is at most a portion of a common bounding hyperplane and the union of all the R_i is E_n.

Now denote the probability of deciding in error when $m_i(t)$ is transmitted by P_{ei}. The average probability of error is then defined as

$$P_e = \frac{1}{M} \sum_{i=1}^{M} P_{ei}$$

That the above decoding rule minimizes this average probability of error follows from the decision regions defined and the fact that

$$P(\mathbf{y}|i) = \exp\left[-\frac{1}{N_0}(\mathbf{y} - \mathbf{m}_i)(\mathbf{y} - \mathbf{m}_i)^T\right] \bigg/ (\pi N_0)^{n/2}$$

The equivalence between the continuous signaling and vector approach has now been established. To make the equivalence complete, however, requires the fact that the vector $\mathbf{y}$ is a sufficient statistic for the information $y(t), 0 \leq t \leq T$, on which to base a decision as to which signal was transmitted. In other words, no other information contained in the sample path $\{y(t), 0 \leq t \leq T\}$ will decrease the average probability of error. For verification of this see Wozencraft and Jacobs (1965). As a final note on this equivalence we observe that for the case where all signals are of equal energy it is sufficient to assume an energy of unity since all signaling schemes with different energies are radial projections of a signaling scheme with unit energy.

The expression for the probability of error P_{ei} is given by

$$P_{ei} = \int_{\bar{R}_i} P(\mathbf{y}|i) \, d\mathbf{y}$$

where $\bar{R}_i$ is the complement of the region R_i and the equation for $P(\mathbf{y}|i)$ is given above. Define the minimum distance d of the code dictionary $\{\mathbf{m}_i, i = 1, \ldots, M\}$ as

$$d = \min_{\substack{i,j \\ i \neq j}} \sqrt{d_{ij}}$$

where

$$d_{ij} = (\mathbf{m}_i - \mathbf{m}_j)(\mathbf{m}_i - \mathbf{m}_j)^{\mathrm{T}} = |\mathbf{m}_i - \mathbf{m}_j|^2$$

The precise relationship between P_e and d is complex. The probability of error is, of course, a function of M, n, and N_0 and for very small N_0 it will generally be lower for codes with larger d, all other parameters being equal. Thus while the problem of most interest is to design codes for a given M, n, and N_0 that minimize the average probability of error, our concern in this chapter is exclusively with the design of codes that maximize the minimum distance d. To describe such codes in spaces of large dimension can be a formidable task. We consider a few simple, but important, examples of such codes before placing the problem in the more general framework of group representation theory.

If we choose any orthonormal basis in E_n, we can describe the basis vectors by the n-tuples

$$\mathbf{m}_i = (0, 0, \ldots, 0, 1, 0, \ldots, 0), \qquad i = 1, \ldots, n$$

where the ith n-tuple has exactly one one in the ith position. These n-tuples form an orthogonal code containing n code vectors in E_n with a minimum distance of $\sqrt{2}$. In general, a code with M codewords in E_n with minimum distance d will be designated an (M, n, d) code. An expression for the probability of error of this code is easily derived if equally likely signals are assumed.

If in addition to the above codewords we include their negatives, we obtain the biorthogonal code which contains $2n$ codewords with minimum distance $\sqrt{2}$. Again, an expression for the probability of error is readily found.

It is clear that the probability of error is invariant under a rotation of the signal set under the assumptions made. It is also invariant under a translation, although this will alter the signal energies involved. If we consider the orthogonal signal set in E_n, we observe that the average of the signal vectors is given by

$$\mathbf{m} = \frac{1}{n} \sum_{i=1}^{n} \mathbf{m}_i = \frac{1}{n}(1, 1, \ldots, 1)$$

By translating the orthogonal set by the vector $-\mathbf{m}$, we obtain a signal set whose ith vector $\mathbf{m}_i'$ is

$$\mathbf{m}_i' = \left(-\frac{1}{n}, -\frac{1}{n}, \ldots, -\frac{1}{n}, 1 - \frac{1}{n}, -\frac{1}{n}, \ldots, -\frac{1}{n}\right), \qquad i = 1, \ldots, n$$

6.3 Group Codes for the Gaussian Channel

and these vectors all lie on the sphere of radius $[(n-1)/n]^{1/2}$. Projecting these vectors onto the unit sphere yields the signal set $\{\mathbf{m}_i'', i = 1, \ldots, n\}$ where

$$\mathbf{m}_i'' = \left(-\frac{1}{[n(n-1)]^{1/2}}, \ldots, -\frac{1}{[n(n-1)]^{1/2}}, \left(\frac{n-1}{n}\right)^{1/2},\right.$$
$$\left. -\frac{1}{[n(n-1)]^{1/2}}, \ldots, -\frac{1}{[n(n-1)]^{1/2}}\right)$$

The signal set $\{\mathbf{m}_i'\}$ has the same error probability as the orthogonal code yet has lower signal energy since

$$\mathbf{m}_i' \cdot \mathbf{m}_j' = \begin{cases} (n-1)/n & \text{if } i = j \\ -1/n & \text{if } i \neq j \end{cases}$$

Thus the signal set $\{\mathbf{m}_i''\}$ has a lower probability of error than the orthogonal signal set with the same energy and

$$\mathbf{m}_i'' \cdot \mathbf{m}_j'' = \begin{cases} 1 & \text{if } i = j \\ -1/(n-1) & \text{if } i \neq j \end{cases}$$

These last two signal sets $\{\mathbf{m}_i'\}$ and $\{\mathbf{m}_i''\}$ are referred to as simplex signal sets. Since the sum of all signal vectors is zero, the vectors all lie on a hyperplane through the origin and hence actually span $n - 1$ dimensions. Thus the code set $\{\mathbf{m}_i'\}$ is, for example, an $(n, n - 1, \sqrt{2})$ code. While it is not difficult to show that the simplex maximizes the minimum distance between n points on the unit sphere in E_{n-1} (signal set $\{\mathbf{m}_i''\}$), it is considerably more difficult to prove that it minimizes the probability of error for n points on the unit sphere in E_{n-1}. The attempt to prove this by Landau and Slepian (1966) was unfortunately invalid, as shown by Farber (1968).

Fundamental bounds on the probability of error for optimal codes for the Gaussian channel have been found by Shannon (1959). Slepian (1963) investigates the relationship between M, n, and N_0 and presents the information graphically. A thorough discussion of the signal design problem is given by Weber (1968).

6.3 Group Codes for the Gaussian Channel

For the remainder of this chapter we shall explore the fundamental results of Slepian on group codes for the Gaussian channel. We have observed that rotation of a signal set of vectors leaves the probability of error unchanged. We call two codes that are related by a rotation, equivalent codes.

Let $X = \{x_i, i = 1, \ldots, M\}$ be a collection of unit vectors in E_n that spans E_n and let $O = \{O_i, i \in I\}$, I some index set, be the group of orthogonal matrices that leave the collection of vectors X invariant, i.e., the action of any particular O_i is to permute the elements of X. The group O is clearly finite since the action of an element of O on X is completely defined by its action on a subset of X that spans E_n and $|O| \leq M!$.

Definition An $[M, n]$ group code $X = \{x_i, i = 1, \ldots, M\}$ is a collection of M unit vectors in E_n that span E_n such that there exists in the group O a set of matrices $O_{i_1}, O_{i_2}, \ldots, O_{i_m}$ such that

$$x_j = O_{i_j} x_i, \quad j = 1, \ldots, M$$

for each $i = 1, \ldots, M$,

Since orthogonal matrices preserve distances, it follows from this definition that the set of distances of all codewords of X from a particular codeword x_i is independent of i. Thus if $x_{i_1}, \ldots, x_{i_k}$ are the nearest neighbors of a codeword x_i and O_j, $j \in I$, takes x_i into x_l, then the nearest neighbors of x_l are $O_j x_{i_1}, \ldots, O_j x_{i_k}$. However, by the same reasoning, if O_j, $j \in I$, transforms x_i to x_k, then the maximum likelihood region R_i is transformed into R_k by O_j and the two regions are congruent. It follows that the probability of error for a given codeword x_i is independent of i. We gather these facts in the following:

Lemma 3.1 The set of distances from x_i to all other codewords of the group code X and the probability of decoding in error when x_i is transmitted are independent of the particular word x_i.

Now we may view the group of matrices O as a permutation group P of degree M and order less than or equal to $M!$. It is clear from the definition of a group code that this permutation group is transitive. From the previous chapter it is also seen that the group of matrices O forms a faithful representation of P. Conversely, suppose we are given a real representation ρ of dimension n of a finite group G and for a given $x \in E_n$ consider the set of vectors

$$X = \{\rho(g)x \mid g \in G\}$$

This set of vectors may not span E_n and $|X|$ may, of course, be less than $|G|$. Indeed if $|X| < |G|$, then for some two group elements g_i and g_j we must have that

$$\rho(g_i)x = \rho(g_j)x \quad \text{or} \quad \rho(g_j^{-1} g_i)x = x$$

6.3 Group Codes for the Gaussian Channel

and x is an eigenvector of $\rho(g_j^{-1}g_i)$ with eigenvalue unity. Thus, by avoiding the choice of x as an eigenvector of any of the group matrices, we can always be assured of $|G|$ distinct code vectors. If $|X| < |G|$, then let H be the subgroup of G such that

$$H = \{g \in G \mid \rho(g)x = x\}$$

Letting $g_1, \ldots, g_k$ be a complete set of left coset representatives, we have the decomposition

$$G = \bigcup_{i=1}^{k} g_i H$$

and $\rho(g)x = \rho(g')x$ if and only if g and g' are in the same coset. Thus $|X| = k$ and, by Lagrange's theorem for cosets, $k \mid |G|$. It should be noted, however, that a collection of vectors $\{x_i, i = 1, \ldots, n\}$ with the property that the set of distances

$$L(i) = \{|x_i - x_j|, j = 1, \ldots, n\}$$

is independent of i need not be a group code. This was shown using a counter-example by Slepian (1971).

The question of whether the elements of X span E_n is rather more difficult. There is one case we can dispose of quite readily. Let the representation ρ be irreducible over the real numbers as discussed in the previous chapter. If the set of vectors X does not span E_n, then they lie in some invariant subspace of E_n, say V. Then, from Maschke's theorem there exists an invariant subspace W of E_n such that $E_n = V \oplus W$, which implies that ρ is reducible, contrary to assumption. We conclude that a real irreducible representation of a group G generates a group code with M vectors and if $M < |G|$, then $M \mid |G|$.

When the representation is reducible over the reals there will in general exist vectors $x \in E_n$ such that X does not span E_n. However, we are able to prove the following existence result.

Theorem 3.1 There exists a vector $x \in E_n$ such that the set of vectors $\{\rho(g)x, g \in G\}$ spans E_n if and only if every complex irreducible representation contained in ρ appears with multiplicity less than or equal to its dimension.

Proof Let ρ_R be the left regular representation of the group G, whose matrices are permutation matrices. The action of these $|G| \times |G|$ matrices on the vector $x = (1, 0, 0, \ldots, 0)$ is to permute the location of the unit element in this vector. The set of vectors $\{\rho_R(g)x, g \in G\}$ thus spans $E_{|G|}$. Let **O** be a real orthogonal matrix that transforms each $\boldsymbol{\rho}_R(g)$, $g \in G$, to a block diagonal form of real irreducible representations, i.e.,

$$\mathbf{O}\boldsymbol{\rho}_R(g)\mathbf{O}^{-1} = \bigoplus_{i \in I} \boldsymbol{\rho}_i(g) = \boldsymbol{\rho}_R'(g)$$

where ρ_i is a real irreducible representation for each $i \in I$, I some index set. Then the set of vectors $\{\boldsymbol{\rho}_R'(g)\mathbf{O}x, g \in G\}$ spans $E_{|G|}$. If we let $(\mathbf{O}x)_i$ be the projection of $\mathbf{O}x$ onto the representation space V_i of ρ_i, where

$$E_{|G|} = \bigoplus_{i \in I} V_i$$

then if the given representation ρ can be expressed as

$$\boldsymbol{\rho}(g) = \bigoplus_{i \in J} \boldsymbol{\rho}_i(g)$$

where $J \subset I$, the action of ρ on the vector

$$x_n' = \bigoplus_{i \in J} (\mathbf{O}x)_i$$

will produce a set of vectors $\{\boldsymbol{\rho}(g)x_n' | g \in G\}$ which span E_n, dim $\rho = n$, as required.

For the converse suppose that ρ is a representation such that $\{\boldsymbol{\rho}(g)x | g \in G\}$ spans the representation space V. We assume, as before, that

$$\boldsymbol{\rho}(g) = \bigoplus_{i \in J} \boldsymbol{\rho}_i(g)$$

and dim $V =$ dim $\rho = n$. Viewing V as an $\mathbb{R}G$ module, the above statement implies that V has a single generator x, i.e., V is cyclic. Define the left ideal A of $\mathbb{R}G$ by

$$A = \{\beta \in \mathbb{R}G \,|\, \beta x = 0\}$$

and consider the mapping

$$\varphi: \mathbb{R}G \to V$$
$$\alpha \mapsto \alpha x$$

Regarding $\mathbb{R}G$ as a left module over itself, i.e., $_{\mathbb{R}G}\mathbb{R}G$, it is readily verified that φ is an $\mathbb{R}G$-linear mapping of the $\mathbb{R}G$ module $_{\mathbb{R}G}\mathbb{R}G$ onto V. Since, by the definition of A, ker $\varphi = A$, it follows that $\mathbb{R}G/A \cong V$ as $\mathbb{R}G$ modules. But this implies that V is isomorphic to a submodule of the left regular module. We conclude that ρ cannot contain any real irreducible representation to a multiplicity higher than it has in the real decomposition of the left regular representation, and the theorem is proved. □

The theorem will also be demonstrated later in connection with configuration matrices of group codes.

6.3 Group Codes for the Gaussian Channel

Before continuing with some general properties of group codes for the Gaussian channel we consider what little there is known about the existence of such codes. In particular, we consider the work of Biglieri and Elia (1972) and Slepian (1971).

Suppose that $\{\boldsymbol{\rho}(g), g \in G\}$ is a representation of the group G of dimension n. A group code $\{\boldsymbol{\rho}(g)x, g \in G\}$ is generated by its action on the initial vector $x \in E_n$. Let $H \subset G$ be the subgroup of G that leaves x invariant. Suppose that H is normal in G and consider a set $\{\boldsymbol{\rho}(g_i)x, g_i \in G, i = 1, \ldots, n\}$ that spans E_n, as there must be by the definition of a group code. We can express an arbitrary vector $z \in E_n$ by

$$z = \sum_{i=1}^{n} a_i \boldsymbol{\rho}(g_i)x$$

and consider the action of $\boldsymbol{\rho}(h)$ on z where h is an arbitrary element of H:

$$\boldsymbol{\rho}(h)z = \sum_{i=1}^{n} a_i \boldsymbol{\rho}(h)\boldsymbol{\rho}(g_i)x$$

$$= \sum_{i=1}^{n} a_i \boldsymbol{\rho}(g_i)\boldsymbol{\rho}(h')x, \quad h' \in H$$

$$= \sum_{i=1}^{n} a_i \boldsymbol{\rho}(g_i)x = z$$

and thus $\boldsymbol{\rho}(h) = \mathbf{I}_n$, $h \in H$.

Now suppose that the number of distinct vectors in the set $\{\boldsymbol{\rho}(g)x, g \in G\}$ is a prime number p and that the p distinct vectors span E_n, i.e., form a $[p, n]$ group code. We assume that $\boldsymbol{\rho}(g) \neq \mathbf{I}_n$ for $g \neq 1$ since if this is not true, then the set of elements $\{g \in G \mid \boldsymbol{\rho}(g) = \mathbf{I}_n\}$ is a normal subgroup of G which we may factor out. Since $p \mid |G|$, there exists a cyclic subgroup of G, say C, of order p such that $C = \langle g \rangle$. If we label the vectors of the group code x_i, $i = 1, \ldots, p$, then there must exist an x_i for which $\boldsymbol{\rho}(g)x_i \neq x_i$, for if this is not true, then $\boldsymbol{\rho}(g)$ leaves E_n invariant and $\boldsymbol{\rho}(g) = \mathbf{I}_n$, contrary to assumption.

Suppose in particular that $\boldsymbol{\rho}(g)x_1 \neq x_1$ and consider the set of vectors $\{\boldsymbol{\rho}(g^i)x_1 = \boldsymbol{\rho}^i(g)x_1, i = 1, \ldots, p\}$. We claim that these p vectors are distinct and hence form the group code. For suppose that

$$\boldsymbol{\rho}^k(g)x_1 = x_1, \quad k < p$$

then, since $(k, p) = 1$, there exists an l, $1 \leq l < p$, such that $\boldsymbol{\rho}^{kl}(g) = \boldsymbol{\rho}(g)$ and hence

$$\boldsymbol{\rho}(g)x_1 = (\boldsymbol{\rho}^k(g))^l x_1 = x_1$$

contrary to assumption. Thus the group code is generated by a cyclic subgroup of G.

Definition An $[M, n]$ group code whose vectors terminate in an $(n - 1)$-dimensional hyperplane of E_n is called a planar group code.

These planar group codes are clearly inefficient, for if we translate the hyperplane so that it passes through the origin, then we can obtain a code in $n - 1$ dimensions. Clearly any representation containing the identity representation generates a code that is equivalent to a planar group code, where two group codes $\{\rho(g)x\}$ and $\{\rho'(g)x\}$ are equivalent if $\{\rho(g)x\} = \{O\rho'(g)x\}$, where **O** is an orthogonal matrix.

Now from the previous chapter every group of odd order has precisely one representation of the first kind, namely the principal or identity representation. Every other real representation is obtained from a representation of the third kind and thus has even dimension. Thus an $[M, n]$ group code which is nonplanar must be generated by a group of even order if n is odd. It follows that if M is a prime and n is odd, then an $[M, n]$ nonplanar group code cannot exist since it is generated by a cyclic group of order M.

Finally, if G is a cyclic group of order $|G|$ and ρ a representation of G of dimension n, then we can use this representation to generate a nonplanar $[|G|, n]$ group code if (i) n is even and $|G| > 2$ and (ii) n odd and $|G|$ even, and in both cases $n < |G|$ [see Biglieri and Elia (1972)].

Suppose we are given an $[M, n]$ group code. We ask the question as to whether it is always possible to generate this code with a group of $n \times n$ matrices of order M. That it is not was shown by Slepian (1971) with the following counterexample.

Consider the five-dimensional representation of the symmetric group on five letters S_5 generated by the following matrices:

$$\rho(12) = \begin{pmatrix} 1 & & & & \\ & 1 & & & 0 \\ & & -1 & & \\ & & & 1 & \\ 0 & & & & -1 \end{pmatrix} \quad \rho(23) = \begin{pmatrix} 1 & 0 & 0 & & 0 \\ 0 & -\frac{1}{2} & \sqrt{\frac{3}{2}} & & \\ 0 & \sqrt{\frac{3}{2}} & \frac{1}{2} & & \\ & & & -\frac{1}{2} & \sqrt{\frac{3}{2}} \\ & & 0 & \sqrt{\frac{3}{2}} & \frac{1}{2} \end{pmatrix}$$

$$\rho(34) = \begin{pmatrix} -\frac{1}{3} & 2\sqrt{\frac{2}{3}} & & & \\ 2\sqrt{\frac{2}{3}} & \frac{1}{3} & & 0 & \\ & & 1 & & \\ & 0 & & 1 & \\ & & & & 1 \end{pmatrix} \quad \rho(45) = \begin{pmatrix} 1 & 0 & 0 & 0 & 0 \\ 0 & -\frac{1}{2} & 0 & \sqrt{\frac{3}{2}} & 0 \\ 0 & 0 & -\frac{1}{2} & 0 & \sqrt{\frac{3}{2}} \\ 0 & \sqrt{\frac{3}{2}} & 0 & \frac{1}{2} & 0 \\ 0 & 0 & \sqrt{\frac{3}{2}} & 0 & \frac{1}{2} \end{pmatrix}$$

6.3 Group Codes for the Gaussian Channel

The action of these matrices on the transpose of the row vector $(1, 0, 0, 0, 0)$ produces ten distinct vectors which we label x_i, $i = 1, \ldots, 10$, given by

$$x_1 = (1, 0, 0, 0, 0),$$
$$x_2 = (-\tfrac{1}{3}, -\tfrac{1}{3}\sqrt{2}, -\tfrac{1}{3}\sqrt{6}, 0, 0),$$
$$x_3 = (-\tfrac{1}{3}, \tfrac{1}{6}\sqrt{2}, \tfrac{1}{6}\sqrt{6}, -\tfrac{1}{6}\sqrt{6}, -\tfrac{1}{2}\sqrt{2}),$$
$$x_4 = (-\tfrac{1}{3}, -\tfrac{1}{3}\sqrt{2}, \tfrac{1}{3}\sqrt{6}, 0, 0),$$
$$x_5 = (-\tfrac{1}{3}, \tfrac{1}{6}\sqrt{2}, -\tfrac{1}{6}\sqrt{6}, -\tfrac{1}{6}\sqrt{6}, \tfrac{1}{2}\sqrt{2}),$$
$$x_6 = (-\tfrac{1}{3}, \tfrac{2}{3}\sqrt{2}, 0, 0, 0)$$
$$x_7 = (-\tfrac{1}{3}, -\tfrac{1}{3}\sqrt{2}, 0, \tfrac{1}{3}\sqrt{6}, 0)$$
$$x_8 = (\tfrac{1}{3}, -\tfrac{1}{6}\sqrt{2}, -\tfrac{1}{6}\sqrt{6}, -\tfrac{1}{6}\sqrt{6}, -\tfrac{1}{2}\sqrt{2})$$
$$x_9 = (\tfrac{1}{3}, -\tfrac{1}{6}\sqrt{2}, \tfrac{1}{6}\sqrt{6}, -\tfrac{1}{6}\sqrt{6}, \tfrac{1}{2}\sqrt{2})$$
$$x_{10} = (\tfrac{1}{3}, \tfrac{1}{3}\sqrt{2}, 0, \tfrac{1}{3}\sqrt{6}, 0)$$

Since these vectors span E_5, they form a [10, 5] code. To show that there does not exist a group of ten 5×5 matrices that generates the code we assume first that such a group G of matrices exists. By Sylow's theorem such a group of matrices would have to contain an element of order two, but we will show that this is not possible, which will imply that an $[M, n]$ group code need not not possess a transitive group of symmetries of order M.

Each matrix of the group G, except the identity, will permute all the vectors, leaving none invariant. This implies that the proposed element of G of order two, viewed now as a permutation on ten letters, must be a product of five transpositions, since otherwise it could not be of order two. Denote this element by σ, and the corresponding orthogonal matrix by σ. Since this matrix preserves angles, we have that

$$(x_i, x_j) = (\sigma x_i, \sigma x_j)$$

TABLE X

$(x_i, x_j) = \tfrac{1}{3}$

i	j	i	j
1	8, 9, 10	6	3, 5, 10
2	5, 7, 8	7	2, 4, 10
3	4, 6, 8	8	1, 2, 3
4	3, 7, 9	9	1, 4, 5
5	2, 6, 9	10	1, 6, 7

To facilitate computation, we have a table of inner products (Table X). Now if σ transposes x_1 and x_2 [$\sigma(1) = 2$, $\sigma(2) = 1$], then, since the only vectors having an inner product of one-third with x_1 are x_8, x_9, x_{10} and with x_2 are x_5, x_7, and x_8, then σ must send elements of $\{8, 9, 10\}$ into $\{5, 7, 8\}$. Thus $\sigma(8)$ must be either 5 or 7, implying that either 9 or 10 is sent into 8. Thus 8 cannot be involved in any transposition and thus the transposition (12) cannot be one of the five transpositions expressing σ. Precisely the same reasoning can be used to show that $(1, i)$, $i = 1, \ldots, 7$, is not a possible transposition of σ. From Table X there are precisely six remaining possibilities for σ given as follows:

$$\sigma_1 = (1, 8)(9,2)(10, 3)(\ )(\ )$$
$$\sigma_2 = \ \vdots \quad \vdots \quad \vdots$$
$$\sigma_6 = (1, 10)(8, 7)(9, 6)(\ )(\ )$$

But each of these may be ruled out in turn. For example, σ_1 transposes 9 and 2, implying that 5 must go to 1 or 4 and either 7 or 8 must go into 5, whence 5 cannot be in a transposition. Thus the [10, 5] code does not possess a transitive group of symmetries of order ten.

We have established in this section some of the basic properties of group codes for the Gaussian channel. In the next three sections we will consider more specific properties of these codes concerning their structure and distance properties.

6.4 The Configuration Matrix

We consider first the general situation where we are given a collection of unit vectors $x_1, \ldots, x_M$ which span E_n. We define the configuration matrix $\mathbf{C}$ of this set by $\mathbf{C} = (C_{ij})$, $C_{ij} = (x_i, x_j)$, and observe that it is symmetric, real, nonnegative definite, and, since it is the Gram matrix of this set of vectors, it must be of rank n. It is closely related to the distance matrix $\mathbf{D} = (d_{ij})$ of this set of vectors where $d_{ij} = |x_i - x_j|^2$. The diagonal elements of $\mathbf{C}$ are all unity and the off-diagonal elements have magnitude less than or equal to unity. The converse of these statements is contained in a lemma due to Slepian:

Lemma 4.1 Every real, symmetric, $M \times M$, nonnegative-definite matrix $\mathbf{C}$ of rank n with diagonal elements unity and off-diagonal elements of magnitude less than unity is the configuration matrix of a set of n vectors which span E_n.

6.4 The Configuration Matrix

Proof Since **C** is real and symmetric there exists an orthogonal nonsingular matrix **T** which diagonalizes it so that

$$\mathbf{TCT}^{-1} = \Omega$$

where Ω has precisely n positive diagonal elements and all other elements are zero. Suppose that $\Omega = \mathrm{diag}\{\lambda_1, \lambda_2, \ldots, \lambda_n, 0, \ldots, 0\}$, $\lambda_i > 0$, $i = 1, \ldots, n$, and consider the set of vectors x_i, $i = 1, \ldots, M$, in E_n defined by

$$x_i = (\sqrt{\lambda_1} t_{1i}, \sqrt{\lambda_2} t_{2i}, \ldots, \sqrt{\lambda_n} t_{ni})$$

where $T = (t_{ij})$. Then we have that

$$(x_i, x_j) = \sum_{k=1}^{n} \lambda_k t_{ki} t_{kj} = C_{ij}$$

and thus this set of vectors has **C** as a configuration matrix. Since **C** has rank n by assumption, and is the Gram determinant of the set of vectors, there must exist n linearly independent vectors in the set and hence they span E_n. □

We consider now that case when the set of vectors is a group code and denote $x_i = \rho(g_i)x$, $g_i \in G$, $i = 1, \ldots, M = |G|$, where ρ is an orthogonal representation of dimension n and $g_1 = $ identity and $x_1 = x$. If we define the function $\theta(g_i)$ by

$$\theta(g_i) = (x_i, x_1)$$

then we have that $\theta(g_i) = \theta(g_i^{-1})$ and

$$(x_i, x_j) = \theta(g_i^{-1} g_j)$$

All rows are permutations of the first row and if there are h elements in the first row with a value of unity, then $h | M$ and the code has M/h distinct vectors. Surprisingly, a converse to this situation can be formulated.

Lemma 4.2 Let θ be a real-valued function defined on the elements g_i, $i = 1, \ldots, M$, of a group G, $|G| = M$. Let $\theta(g_1) = 1$, where g_1 is the identity element of the group, and let $\theta(g_i) = \theta(g_i^{-1})$ $i = 1, \ldots, M$. If the $M \times M$ matrix $\mathbf{C} = (C_{ij})$, $C_{ij} = \theta(g_i^{-1} g_j)$, is nonnegative definite and of rank n, then there exists an $[M', n]$ group code generated by an orthogonal representation of G of dimension n which has **C** as its configuration matrix. Here $M' = M/h$, where h is the number of different values of j for which $\theta(g_j) = 1$.

Proof From Lemma 4.1 we can find M unit vectors x_i that are not necessarily distinct but that span E_n such that $(x_i, x_j) = C_{ij}$. We assume that $x_1, x_2, \ldots, x_n$ span E_n. If $g_i \cdot g_j = g_{l(j, i)}$, then we define the matrix $\rho(g_i)$ by its action on $x_1, \ldots, x_n$, i.e.,

$$\rho(g_i) x_j = x_{l(j, i)}, \qquad j = 1, \ldots, n$$

and this completely determines $\rho(g_i)$. Since

$$(x_i, x_j) = C_{ij} = \theta(g_i^{-1} g_j) = \theta(g_i^{-1} g_k^{-1} g_k g_j) = \theta(g_{l(i,k)}^{-1} g_{l(j,k)})$$
$$= (x_{l(i,k)}, x_{l(j,k)}) = (\rho(g_k) x_i, \rho(g_k) x_j), \quad 1 \le i, j \le n$$

it follows that $\rho(g_k)$ preserves angles between any two vectors and is thus an orthogonal matrix. We show now that it is also a representation. It is necessary to show that

$$\rho(g_k) x_i = x_{l(i,k)},$$

for all i, not just $i \le n$ as in the foregoing. We first verify that $\rho(g)^{-1} = \rho(g^{-1})$. Consider the expression

$$(\rho(g_k)^{-1} x_i, x_j) = (x_i, \rho(g_k) x_j) = C_{il(j,k)} = \theta(g_i^{-1} g_{l(j,k)}) = \theta(g_i^{-1} g_k g_j)$$
$$1 \le i, j \le n$$

and if $g_k^{-1} = g_{k'}$, then

$$(\rho(g_k^{-1}) x_i, x_j) = (\rho(g_{k'}) x_i, x_j) = (x_{l(i,k')}, x_j) = C_{l(i,k')j}$$
$$= \theta(g_{l(i,k')}^{-1} g_j) = \theta((g_k^{-1} g_i)^{-1} g_j)$$
$$= \theta(g_i^{-1} g_k g_j)$$

and we conclude that $\rho(g_k)^{-1} = \rho(g_k^{-1})$. Using the same notation as before, we have that for a fixed j

$$(\rho(g_k) x_j, x_i) = (x_j, \rho(g_k^{-1}) x_i) = (x_j, \rho(g_{k'}) x_i)$$
$$= (x_j, x_{l(i,k')})$$
$$= \theta(g_j^{-1} g_{l(i,k')}) = \theta(g_j^{-1} g_{k'} g_i)$$
$$= \theta(g_j^{-1} g_k^{-1} g_i) = \theta((g_k g_j)^{-1} g_i)$$
$$= \theta(g_{l(k j,)}^{-1} g_i) = C_{l(j,k)i}$$
$$= (x_{l(j,k)}, x_i), \quad j > n, \quad 1 \le i \le n$$

Since this holds for all i, $1 \le i \le n$, we conclude that

$$\rho(g_k) x_i = x_{l(i,k)}$$

for all i, $1 \le i \le M$. It follows that the matrices $\rho(g_k)$, $1 \le k \le M$, so defined form a representation of G. □

The configuration matrix of a group code can thus be used to actually construct a group representation if the group multiplication table is known.

Now we prove that the configuration matrix of a group code can be written as a linear sum of orthogonal idempotent matrices each of which is

6.4 The Configuration Matrix

the configuration matrix of some group code. In other words, we obtain a type of spectral resolution theorem for configuration matrices. While the implications of this theorem for the construction of good group codes are not yet clear, some important facts are brought out in its proof.

Assume the group G has s nonequivalent real irreducible representations, which we denote by $\{\rho^i(g), g \in G\}$, $i = 1, \ldots, s$. An arbitrary real representation is then orthogonally equivalent to

$$\underbrace{\rho^1 \oplus \cdots \oplus \rho^1}_{d_1} \oplus \underbrace{\rho^2 \oplus \cdots \oplus \rho^2}_{d_2} \oplus \cdots \oplus \underbrace{\rho^s \oplus \cdots \oplus \rho^s}_{d_s}$$

which we write as

$$\bigoplus_{\alpha=1}^{s} d_\alpha \rho^\alpha$$

Without loss of generality we assume that the representation ρ is in this form and refer to $d_i \rho^i$ as the homogeneous component of ρ^i in ρ. If x is an n-tuple in $V_n(\mathbb{R})$, $\dim \rho = n$, we have the corresponding decomposition

$$x = \bigoplus_{\alpha=1}^{s} \bigoplus_{\beta=1}^{d_\alpha} x^{\alpha\beta}$$

where $x^{\alpha\beta}$ is of dimension n_α, $\beta = 1, \ldots, d_\alpha$, $\alpha = 1, \ldots, s$.

If $\mathbf{C}$ is the configuration matrix of the code $\{\rho(g)x, g \in G\}$, $\mathbf{C}^\alpha$ that of the code $\{d_\alpha \rho^\alpha(g)(\oplus_{\beta=1}^{d_\alpha} x^{\alpha\beta}), g \in G\}$, and $\mathbf{C}^{\alpha\beta}$ that of the code $\{\rho^\alpha(g)x^{\alpha\beta}, g \in G\}$, then we have the identities

$$\mathbf{C} = \sum_{\alpha=1}^{s} \mathbf{C}^\alpha = \sum_{s=1}^{\alpha} \sum_{\beta=1}^{d_\alpha} \mathbf{C}^{\alpha\beta}$$

For the remainder of this section we use the notation of the previous chapter and denote by e_0, e_1, e_2, e_3 and e_0, e_3 real matrix representations of $\mathbb{H}$ and $\mathbb{C}$. The following lemma is an intermediate step to our main result:

Lemma 4.3 Let ρ^i and ρ^j be real irreducible representations of the finite group G of dimensions n_i and n_j, respectively, and $\mathbf{C}^i$ and $\mathbf{C}^j$ the configuration matrices of the group codes $\{\rho^i(g)x^i, g \in G\}$ and $\{\rho^j(g)x^j, g \in G\}$, respectively. Then:

(i) If ρ^i and ρ^j are not equivalent, then $\mathbf{C}^i \cdot \mathbf{C}^j = 0$ for any x^i, x^j.
(ii) If $\rho^i = \rho^j$ and (a) ρ^i is of the first kind and $(x^i, x^j) = 0$, then $\mathbf{C}^i \mathbf{C}^j = 0$; (b) ρ^i is of the second kind and $(x^i, e_k x^j) = 0$, $k = 0, 1, 2, 3$, then $\mathbf{C}^i \mathbf{C}^j = 0$; (c) ρ^i is of the third kind and $(x^i, e_k x^j) = 0$, $k = 0, 3$, then $\mathbf{C}^i \mathbf{C}^j = 0$.
(iii) If $\rho^i = \rho^j$ and $x^i = x^j$, then $(\mathbf{C}^i)^2 = (|G|/n_i)|x^i|^2 \mathbf{C}^i$.

Proof Consider the (g, h) element of $\mathbf{C}^i\mathbf{C}^j$:

$$(\mathbf{C}^i\mathbf{C}^j)_{gh} = \sum_{f \in G} (\boldsymbol{\rho}^i(g)x^i, \boldsymbol{\rho}^i(f)x^i)(\boldsymbol{\rho}^j(f)x^j, \boldsymbol{\rho}^j(h)x^j)$$

$$= \sum_{f \in G} \left\{ \sum_{a} \left[\sum_{b} \rho^i(g)_{ab} x_b^i \right] \left[\sum_{c} \rho^i(f)_{ac} x_c^i \right] \right\}$$

$$\times \left\{ \sum_{r} \left[\sum_{s} \rho^j(f)_{rs} x_s^j \right] \left[\sum_{t} \rho^j(h)_{rt} x_t^j \right] \right\}$$

$$= \sum_{a,b,r,t} \rho^i(g)_{ab} \rho^j(h)_{rt} \left\{ \sum_{c,s} \left[\sum_{f \in G} \rho^i(f)_{ac} \rho^j(f)_{rs} \right] x_c^i x_s^j \right\} x_b^i x_t^j \quad (4.1)$$

where x_l^i is the lth component of x^i. If ρ^i and ρ^j are not equivalent, then the inner summation over the group elements is, by the orthogonality relationships (7.1a) and (7.1b) of the previous chapter, zero. This proves part (i). If $\rho^i = \rho^j$ and $x^i = x^j$, then $\mathbf{C}^i = \mathbf{C}^j$. The coefficient of $x_u^i x_v^i$ in the (c, s) summation of Eq. (4.1) is then

$$\sum_{f \in G} [\rho^i(f)_{au} \rho^i(f)_{rv} + \rho^i(f)_{av} \rho^i(f)_{ru}]$$

By the orthogonality relationships (7.1a) and (7.1b) of Chapter 5 this last equation reduces to

$$\frac{2|G|}{n_i} \delta_{ar} \delta_{uv}$$

Thus

$$(\mathbf{C}^i)^2_{gh} = \sum_{a,b,t} \rho^i(g)_{ab} \rho^i(h)_{at} \frac{|G|}{n_i} |x^i|^2 x_b^i x_t^i$$

$$= \frac{|G|}{n_i} |x^i|^2 (\boldsymbol{\rho}^i(g)x^i, \boldsymbol{\rho}^i(h)x^i) = \frac{|G|}{n_i} |x^i|^2 C^i_{gh}$$

and we conclude that

$$(\mathbf{C}^i)^2 = \frac{|G|}{n_i} |x^i|^2 \mathbf{C}^i$$

which proves part (iii).

For part (ii) suppose that $\rho^i = \rho^j$ and $x^i \neq x^j$. We consider the three cases according to whether ρ^i is a real irreducible representation of the first, second, or third kind.

6.4 The Configuration Matrix

Suppose first that ρ^i is a real irreducible representation of the first kind and that $(x^i, x^j) = 0$. From Eq. (4.1) we have

$$\sum_{c,s} \left[\sum_{f \in G} \rho^i(f)_{ac}\, \rho^i(f)_{rs} \right] x_c^i x_s^j = \sum_{c,s} \frac{|G|}{n_i} \delta_{ar}\, \delta_{cs}\, x_c^i x_s^j$$

$$= \delta_{ar}\, \frac{|G|}{n_i}\, (x^i, x^j) = 0$$

and so $\mathbf{C}^i \mathbf{C}^j = 0$.

Suppose now that ρ^i is a real irreducible representation of the second kind and that $(x^i, e_k x^j) = 0$, $k = 0, 1, 2, 3$. From Eq. (7.6) of Chapter 5 we have that

$$\sum_{f \in G} \rho^i(f)_{ac}\, \rho^i(f)_{rs} = (\lambda_0^{ar} e_0 + \lambda_1^{ar} e_1 + \lambda_2^{ar} e_2 + \lambda_3^{ar} e_3)_{cs}, \quad \lambda_l^{ar} \in \mathbb{R}, \quad l = 0, 1, 2, 3$$

It follows that

$$\sum_{c,s} \left[\sum_{f \in G} \rho^i(f)_{ac}\, \rho^i(f)_{rs} \right] x_c^i x_s^j$$

$$= \sum_c x_c^i (\lambda_0^{ar} e_0 x^j + \lambda_1^{ar} e_1 x^j + \lambda_2^{ar} e_2 x^j + \lambda_3^{ar} e_3 x^j)_c$$

$$= \sum_{l=0}^{3} \lambda_l^{ar}(x^i, e_l x^j) = 0$$

The procedure is similar for the third kind of representation, which completes the lemma. □

We now state and prove the main theorem on the decomposition of configuration matrices (Blake, 1974).

Theorem 4.1 Let $\mathbf{C}$ be the configuration matrix of the $[M, n]$ group code $\{\rho(g)x, g \in G\}$ where

$$\rho(g) = \bigoplus_{\alpha=1}^{s} d_\alpha \rho^\alpha(g), \quad \dim \rho = n$$

is the decomposition of ρ into real irreducible and orthogonal representations and

$$x = \bigoplus_{\alpha=1}^{s} \bigoplus_{\beta=1}^{d_\alpha} x^{\alpha\beta}, \quad x^{\alpha\beta} \in V_{n_\alpha}(\mathbb{R})$$

Then we can find a vector $\hat{x} \in V_n(\mathbb{R})$

$$\hat{x} = \bigoplus_{\alpha=1}^{s} \bigoplus_{\beta=1}^{d_\alpha} \hat{x}^{\alpha\beta}, \quad \hat{x}^{\alpha\beta} \in V_{n_\alpha}(\mathbb{R})$$

such that if $\hat{\mathbf{C}}^{\alpha\beta}$ is the configuration matrix of the group code $\{\rho^{\alpha}(g)\hat{x}^{\alpha\beta}, g \in G\}$, then

$$\mathbf{C} = \sum_{\alpha=1}^{s} \sum_{\beta=1}^{d_\alpha} \hat{\mathbf{C}}^{\alpha\beta}$$

and

$$\hat{\mathbf{C}}^{\alpha\beta}\hat{\mathbf{C}}^{\gamma\delta} = \frac{|G|}{n_\alpha} \lambda_\beta^\alpha \, \delta_{\alpha\gamma} \, \delta_{\beta\delta} \, \hat{\mathbf{C}}^{\alpha\beta}$$

Proof First notice that in the decomposition

$$\mathbf{C} = \sum_{\alpha=1}^{s} \mathbf{C}^{\alpha}$$

the matrices $\mathbf{C}^\alpha$ are diagonalizable and have the property that $\mathbf{C}^\alpha \cdot \mathbf{C}^{\alpha'} = 0$ if $\alpha \neq \alpha'$. It follows that $\mathbf{C}$ is of rank n if and only if $\mathbf{C}^\alpha$ is of rank $d_\alpha n_\alpha$, $\alpha = 1, \ldots, s$, i.e., the group code $\{(\rho(g)x, g \in G\}$ spans $V_n(\mathbb{R})$ if and only if the group code

$$\left\{ d_\alpha \rho^\alpha(g) \left(\bigoplus_{\beta=1}^{d_\alpha} x^{\alpha\beta} \right), \; g \in G \right\}$$

spans $V_{d_\alpha n_\alpha}(\mathbb{R})$ for each $\alpha = 1, \ldots, s$. In this sense the homogeneous components act quite independently of one another. Of course if $\mathbf{C}$ is of rank n, then each component $\mathbf{C}^\alpha$ is of rank $d_\alpha n_\alpha$, but the converse is not so clear.

The (g, h) element of the matrix $\mathbf{C}^\alpha$ is given by

$$C_{gh}^\alpha = \sum_{\beta=1}^{d_\alpha} C_{gh}^{\alpha\beta} = \sum_{\beta=1}^{d_\alpha} (\rho^\alpha(g)x^{\alpha\beta}, \rho^\alpha(h)x^{\alpha\beta}) \quad (4.2)$$

The argument now depends on the kind of irreducible representation that ρ is. Assume it to be of the first kind. Equation (4.2) is rewritten as,

$$C_{gh}^\alpha = \sum_{\beta=1}^{d_\alpha} \sum_{i,j=1}^{n_\alpha} \rho^\alpha(g^{-1}h)_{ij} x_i^{\alpha\beta} x_j^{\alpha\beta} = \sum_{i,j=1}^{n_\alpha} \rho^\alpha(g^{-1}h)_{ij} \left(\sum_{\beta=1}^{d_\alpha} x_i^{\alpha\beta} x_j^{\alpha\beta} \right)$$

$$= \sum_{i,j=1}^{n_\alpha} \rho^\alpha(g^{-1}h)_{ij} A_{ij}^\alpha \quad (4.3)$$

where

$$A_{ij}^\alpha = \sum_{\beta=1}^{d_\alpha} x_i^{\alpha\beta} x_j^{\alpha\beta}$$

is the (i, j) element of the $n_\alpha \times n_\alpha$, real, symmetric, nonnegative-definite matrix $\mathbf{A}^\alpha$. Let $\mathbf{X}$ be the $n_\alpha \times d_\alpha$ matrix whose βth column is $x^{\alpha\beta}$ and suppose that $\mathbf{X}$ has rank $e_\alpha \leq d_\alpha$. Then $\mathbf{A}^\alpha = \mathbf{X}\mathbf{X}^T$ is the Gram matrix of the n_α vectors

6.4 The Configuration Matrix

that form the rows of $\mathbf{X}$ and hence $\mathbf{A}^\alpha$ is of rank e_α also. Denote the nonzero eigenvalues of $\mathbf{A}^\alpha$ by $\lambda_1^\alpha, \lambda_2^\alpha, \ldots, \lambda_{e_\alpha}^\alpha$ where $\lambda_i^\alpha > 0$, $i = 1, \ldots, e_\alpha$, and let $S^{\alpha\beta}$ be an eigenvector of unit length with eigenvalue λ_β^α. Regardless of the multiplicity of a particular eigenvalue, it is always possible to determine a set of e_α orthogonal eigenvectors $\{S^{\alpha\beta}, \beta = 1, \ldots, e_\alpha\}$ and it follows easily that we can write

$$A_{ij}^\alpha = \sum_{\beta=1}^{e_\alpha} \lambda_\beta^\alpha S_i^{\alpha\beta} S_j^{\alpha\beta}$$

where $S_i^{\alpha\beta}$ is the ith component of $S^{\alpha\beta}$. Substituting into Eq. (4.3),

$$C_{gh}^\alpha = \sum_{i,j=1}^{n_\alpha} \rho^\alpha(g^{-1}h)_{ij} \left(\sum_{\beta=1}^{e_\alpha} \lambda_\beta^\alpha S_i^{\alpha\beta} S_j^{\alpha\beta} \right)$$

$$= \sum_{\beta=1}^{e_\alpha} (\rho^\alpha(g)(\lambda_\beta^\alpha)^{1/2} S^{\alpha\beta}, \rho^\alpha(h)(\lambda_\beta^\alpha)^{1/2} S^{\alpha\beta})$$

Thus we can write

$$\mathbf{C}^\alpha = \sum_{\beta=1}^{e_\alpha} \hat{\mathbf{C}}^{\alpha\beta}$$

where $\hat{\mathbf{C}}^{\alpha\beta}$ is the configuration matrix of the group code $\{\rho^\alpha(g)(\lambda_\beta^\alpha)^{1/2} S^{\alpha\beta}, g \in G\}$. Since $\mathbf{C}^\alpha$ is generated by $d_\alpha \rho^\alpha$ acting on the vector

$$\bigoplus_{\beta=1}^{e_\alpha} (\lambda_\beta^\alpha)^{1/2} S^{\alpha\beta}$$

and this vector satisfies the conditions of part (ii,a) of Lemma 4.3, we have

$$\hat{\mathbf{C}}^{\alpha\beta} \hat{\mathbf{C}}^{\alpha\beta'} = \frac{|G|}{n_\alpha} \lambda_\beta^\alpha \delta_{\beta\beta'} \hat{\mathbf{C}}^{\alpha\beta}$$

If we choose $\hat{x}^{\alpha\beta} = (\lambda_\beta^\alpha)^{1/2} S^{\alpha\beta}$, the result of the theorem follows. If $\mathbf{C}^\alpha$ is of rank $d_\alpha n_\alpha$, then it follows that $e_\alpha = d_\alpha$. We conclude that $\mathbf{C}^\alpha$ is of rank $d_\alpha n_\alpha$ if and only if the set of vectors $\{x^{\alpha\beta}, \beta = 1, \ldots, d_\alpha\}$ is linearly independent.

Suppose now that ρ^α is a real irreducible representation of the second kind and write the (g, h) element of $\mathbf{C}^\alpha$ as

$$C_{gh}^\alpha = \sum_{\beta=1}^{d_\alpha} (\rho^\alpha(g) x^{\alpha\beta}, \rho^\alpha(h) x^{\alpha\beta})$$

$$= \frac{1}{4} \sum_{\beta=1}^{d_\alpha} [(\rho^\alpha(g) e_0 x^{\alpha\beta}, \rho^\alpha(h) e_0 x^{\alpha\beta}) + (\rho^\alpha(g) e_1 x^{\alpha\beta}, \rho^\alpha(h) e_1 x^{\alpha\beta})$$

$$+ (\rho^\alpha(g) e_2 x^{\alpha\beta}, \rho^\alpha(h) e_2 x^{\alpha\beta}) + (\rho^\alpha(g) e_3 x^{\alpha\beta}, \rho^\alpha(h) e_3 x^{\alpha\beta})] \quad (4.4)$$

a result which follows from the commutativity results of Chapter 5, where the e_i are the matrices of Eq. (7.3) of that chapter. As with representations of the first kind, we write Eq. (4.4) in the form

$$C^\alpha_{gh} = \frac{1}{4} \sum_{i,j=1}^{n_\alpha} \rho^\alpha(g^{-1}h)_{ij} A^\alpha_{ij}$$

where, if $\mathbf{X}$ is the $n_\alpha \times d_\alpha$ matrix whose βth column is $x^{\alpha\beta}$, then the matrix $\mathbf{A}^\alpha$ is equal to $\mathbf{YY}^T$, where $\mathbf{Y}$ is the $n_\alpha \times 4d_\alpha$ matrix

$$\mathbf{Y} = [\mathbf{X} \vdots e_1\mathbf{X} \vdots e_2\mathbf{X} \vdots e_3\mathbf{X}]$$

Notice that from a result mentioned previously $4d_\alpha$ must be less than $n_\alpha = \dim \rho^\alpha$ if the code is to span the space. Again $\mathbf{A}^\alpha$ is a real, symmetric, and nonnegative-definite matrix. Suppose λ_β^α is a nonzero eigenvalue and $S^{\alpha\beta}$ a corresponding eigenvector. It is readily verified that the matrices e_i commute with $\mathbf{A}^\alpha$, $i = 1, 2, 3$. For example, we have

$$\mathbf{A}^\alpha = \sum_{i=0}^{3} e_i \mathbf{XX}^T e_i^T$$

so that

$$\mathbf{A}^\alpha e_1 = \mathbf{XX}^T e_1 + e_1 \mathbf{XX}^T + e_2 \mathbf{XX}^T e_3 - e_3 \mathbf{XX}^T e_2$$

and

$$e_1 \mathbf{A}^\alpha = e_1 \mathbf{XX}^T + \mathbf{XX}^T e_1 - e_3 \mathbf{XX}^T e_2 + e_2 \mathbf{XX}^T e_3$$

Thus $e_i S^{\alpha\beta}$, $i = 1, 2, 3$, are also eigenvectors of $\mathbf{A}^\alpha$ corresponding to the eigenvalue λ_β^α. Since these four eigenvectors are linearly independent and, from the form of the matrices e_i, orthogonal, we conclude that the rank of $\mathbf{A}^\alpha$ is divisible by four and suppose it to be $4e_\alpha$. Let $\{e_i S^{\alpha\beta}, i = 0, 1, 2, 3, \beta = 1, 2, \ldots, e_\alpha\}$ be a set of $4e_\alpha$ eigenvectors of unit length. We can always construct such a set with the property that $(e_i S^{\alpha\beta}, e_j S^{\alpha\beta'}) = 0$ if $\beta \neq \beta'$, $i, j = 0, 1, 2, 3$ or $i \neq j$. We can thus write

$$C^\alpha_{gh} = \frac{1}{4} \sum_{\beta=1}^{e_\alpha} \sum_{i=0}^{3} (\boldsymbol{\rho}^\alpha(g)(\lambda_\beta^\alpha)^{1/2} e_i S^{\alpha\beta}, \boldsymbol{\rho}^\alpha(h)(\lambda_\beta^\alpha)^{1/2} e_i S^{\alpha\beta})$$

$$= \sum_{\beta=1}^{e_\alpha} (\boldsymbol{\rho}^\alpha(g)(\lambda_\beta^\alpha)^{1/2} S^{\alpha\beta}, \boldsymbol{\rho}^\alpha(h)(\lambda_\beta^\alpha)^{1/2} S^{\alpha\beta})$$

and hence we have that

$$\mathbf{C}^\alpha = \sum_{\beta=1}^{e_\alpha} \hat{\mathbf{C}}^{\alpha\beta}$$

6.4 The Configuration Matrix

where $\hat{\mathbf{C}}^{\alpha\beta}$ is the configuration matrix of the group code $\{\boldsymbol{\rho}^\alpha(g)(\lambda_\beta^\alpha)^{1/2} S^{\alpha\beta}, g \in G\}$. It follows from part (ii,b) of Lemma 4.3 that

$$\hat{\mathbf{C}}^{\alpha\beta} \hat{\mathbf{C}}^{\alpha\beta'} = \frac{|G|}{n_\alpha} \lambda_\beta^\alpha \delta_{\beta\beta'} \hat{\mathbf{C}}^{\alpha\beta}$$

since the vectors $S^{\alpha\beta}$, $\beta = 1, \ldots, e_\alpha$, satisfy the condition in the proposition. Again, by comparing ranks, we have that $e_\alpha = d_\alpha$ if $\mathbf{C}^\alpha$ is of rank $d_\alpha n_\alpha$. This occurs if and only if the matrix A^α is of rank $4d_\alpha$, which is true if and only if the set of vectors $\{e_i x^{\alpha\beta}, i = 0, 1, 2, 3, \beta = 1, \ldots, d_\alpha\}$ is linearly independent. If ρ^α is a real irreducible representation of the third kind, the argument proceeds along similar lines and is omitted. □

It is not yet clear if or how the decomposition of the configuration matrix contained in this theorem can be used to assist in the solution to the initial vector problem. However, the following two corollaries, whose proofs are contained in the proof of the theorem, are of some interest in relation to the initial vector problem.

Corollary 1 Let $\mathbf{C}$ and x be as in Theorem 4.1. Then $\mathbf{C}$ is of rank n iff the following conditions are satisfied for each $\alpha = 1, \ldots, s$:

(i) If ρ^α is of the first kind, then $\{x^{\alpha\beta}, \beta = 1, \ldots, d_\alpha\}$ is a linearly independent set.

(ii) If ρ^α is of the second kind, then $\{e_i x^{\alpha\beta}, \beta = 1, \ldots, d_\alpha, i = 0, 1, 2, 3\}$ is a linearly independent set.

(iii) If ρ^α is of the third kind, then $\{e_i x^{\alpha\beta}, \beta = 1, \ldots, d_\alpha, i = 0, 3\}$ is a linearly independent set.

Thus we can determine whether a code spans the space or not, simply by inspection of the initial vector, assuming the representation is in block diagonal form.

Corollary 2 If $\mathbf{C}^\alpha$ is the configuration matrix of the group code

$$\left\{ d_\alpha \boldsymbol{\rho}^\alpha(g) \left(\bigoplus_{\beta=1}^{d_\alpha} x^{\alpha\beta} \right), \quad g \in G \right\}$$

we can without loss of generality assume that $(x^{\alpha\beta}, x^{\alpha\beta'}) = 0$ if $\beta = \beta'$.

This corollary is both practically and conceptually of interest in choosing an initial vector since it restricts the search considerably and often simplifies the manipulations.

The nonzero eigenvalues of $\mathbf{C}$ are simply $(|G|/n_\alpha)\lambda_\beta^\alpha$, $\alpha = 1, \ldots, s$, $\beta = 1, \ldots, d_\alpha$, and the multiplicity of this eigenvalue is n_α. The theorem is actually a refinement of the usual spectral resolution theorem for diagonalizable operators, which requires distinct eigenvalues [e.g., Hoffman and Kunze (1961)]. Notice that implicit in the proof of Theorem 4.1 is the fact that if $\rho = d_\alpha \rho^\alpha$, ρ^α is real irreducible representation of dimension n_α, and $n_\alpha < d_\alpha$, then any group code generated using this representation will not span the space.

6.5 Distance Properties of Group Codes

In this section we present two theorems of Slepian (1968) on distance properties of group codes which are independent of the initial vector. The results are, unfortunately, weak in the sense that they do not assist in the design of good codes. They are nevertheless interesting and can be used to determine the quality of a given code.

Let ρ be a representation of dimension n of the group G with elements $g_1, g_2, \ldots, g_M$. We assume for the moment that ρ is a real irreducible representation and denote the square of the Euclidean distance between two vectors x_i and x_j by $d^2(x_i, x_j)$.

Theorem 5.1 Let a group code be generated by the action of the real irreducible representation of the finite group G, $|G| = M$, ρ on the initial vector x. Then we have for every integer m that

$$\sum_{i=1}^{M} d^2(\rho(g_i^m)x, x) = 2M(1 - \mu_m)$$

where

$$\mu_m = \frac{1}{nM} \sum_{i=1}^{M} \chi(g_i^m)$$

is a constant, independent of the initial vector x, and χ is the character of ρ.

Proof Let **T** be the matrix

$$\mathbf{T} = \sum_{g \in G} \rho(g^m)$$

That **T** is symmetric follows from the fact that

$$\rho(g^{-1}) = \rho(g)^\mathrm{T}$$

and hence

$$\mathbf{T}^\mathrm{T} = \sum_{g \in G} \rho(g^m)^\mathrm{T} = \sum_{g \in G} \rho(g^{-m}) = \mathbf{T}$$

6.5 Distance Properties of Group Codes

It is also clear that **T** commutes with the matrices $\rho(g)$ since

$$\mathbf{T}\rho(g_i) = \sum_{g \in G} \rho(g^m)\rho(g_i) = \sum_{g \in G} \rho(g^m g_i)$$

and if $gg_i = g_i h$, then $g^m g_i = g_i h^m$ and

$$\mathbf{T}\rho(g_i) = \sum_{h \in G} \rho(g_i h^m) = \rho(g_i)\mathbf{T}$$

From a result of the previous chapter, the only real symmetric matrices that commute with real irreducible representations (constructed in the manner discussed in the previous section) are real scalar multiples of the identity matrix. Thus $\mathbf{T} = \alpha \mathbf{I}_n$, where

$$\text{tr}(\mathbf{T}) = \alpha n = \sum_{g \in G} \chi(g^m)$$

It follows that

$$\sum_{g \in G} d^2(\rho(g^m)x, x) = 2M - 2 \sum_{i,j=1}^{n} x_i x_j \sum_{g \in G} \rho(g^m)_{ij}$$

$$= 2\left[M - \frac{1}{n}\sum_{g \in G} \chi(g^m)\right] = 2M(1 - \mu_m)$$

as required, where $x = (x_1, \ldots, x_n)$. □

Notice that if we set $m = 1$ in the above theorem, we obtain

$$\sum_{g \in G} d^2(\rho(g)x, x) = 2M$$

if ρ is not the identity representation.

Let G be a finite group with conjugacy classes C_i, $i = 1, \ldots, s$, $|C_i| = c_i$, and let ρ be a real irreducible representation of G of dimension n, as before, with character χ.

Theorem 5.2 Let the real irreducible representation ρ of the finite group G generate a group code by its action on the initial vector x. Then we have

$$\sum_{g \in C_i} d^2(\rho(g)x, x) = 2c_i\left[1 - \frac{1}{n}\chi(C_i)\right]$$

a formula which is independent of the initial vector x.

Proof The proof considers two cases according to whether or not ρ is also complex irreducible. For both cases we need the fact that if ρ' is any complex irreducible representation of dimension n', then

$$\sum_{g \in C_i} \rho'(g) = \frac{c_i}{n'}\chi(C_i)\mathbf{I}_{n'}$$

a fact which can be demonstrated by using Schur's Lemma.

Case (a) Suppose ρ is real and complex irreducible (i.e., a representation of the first kind). Then

$$\sum_{g \in C_i} d^2(\rho(g)x, x) = 2c_i - 2 \sum_{j,k=1}^{n} x_j x_k \left[\sum_{g \in C_i} \rho(g)_{jk} \right]$$

$$= 2c_i \left[1 - \frac{\chi(C_i)}{n} \right]$$

where $x = (x_1, \ldots, x_n)$ and $|x| = 1$.

Case (b) Suppose ρ is obtained from a complex irreducible representation of the second or third kind and thus we assume it to be in the form

$$\rho(g) = \begin{bmatrix} \mathbf{u}(g) & \mathbf{v}(g) \\ -\mathbf{v}(g) & \mathbf{u}(g) \end{bmatrix}$$

where $\boldsymbol{\rho}'(g) = \mathbf{u}(g) + i\mathbf{v}(g)$ is the complex irreducible representation of dimension n', character χ', and $2n' = n = \dim \rho$. Define the matrices $\mathbf{T}$ and $\mathbf{U}$ by

$$\mathbf{T} = \sum_{g \in C_i} \boldsymbol{\rho}(g), \qquad \mathbf{U} = \frac{1}{\sqrt{2}} \begin{bmatrix} \mathbf{I}_{n'} & i\mathbf{I}_{n'} \\ i\mathbf{I}_{n'} & \mathbf{I}_{n'} \end{bmatrix}$$

Since we have

$$\mathbf{U}^{-1} \boldsymbol{\rho}(g) \mathbf{U} = \begin{bmatrix} \boldsymbol{\rho}'(g) & 0 \\ 0 & \bar{\boldsymbol{\rho}}'(g) \end{bmatrix}$$

we conclude that

$$\mathbf{U}^{-1} \mathbf{T} \mathbf{U} = \frac{c_i}{n'} \begin{bmatrix} \chi'(C_i) \mathbf{I}_{n'} & 0 \\ 0 & \bar{\chi}'(C_i) \mathbf{I}_{n'} \end{bmatrix}$$

and

$$\mathbf{T} = \frac{c_i}{n'} \begin{bmatrix} [\operatorname{Re} \chi'(C_i)] \mathbf{I}_{n'} & [\operatorname{Im} \chi'(C_i)] \mathbf{I}_{n'} \\ -[\operatorname{Im} \chi'(C_i)] \mathbf{I}_{n'} & [\operatorname{Re} \chi'(C_i)] \mathbf{I}_{n'} \end{bmatrix}$$

Recalling that

$$\sum_{g \in C_i} d^2(\rho(g)x, x) = 2c_i - 2 \sum_{j,k=1}^{n} x_j x_k \left[\sum_{g \in C_i} \rho(g)_{jk} \right]$$

$$= 2c_i - 2 \sum_{j,k=1}^{n} x_j x_k t_{jk}, \qquad (t_{jk}) = \mathbf{T}$$

we calculate easily that

$$\sum_{g \in C_i} d^2(\rho(g)x, x) = 2c_i \left[1 - \frac{\operatorname{Re} \chi'(C_i)}{n'} \right] = 2c_i \left[1 - \frac{\chi(C_i)}{n} \right]$$

as for case (a), which completes the proof. $\square$

6.5 Distance Properties of Group Codes

This formula can be of some use, for notice that if $d^2(\rho(g)x, x)$ is the same for all group elements in a conjugacy class, for each conjugacy class in the group, then the code must have the maximum minimum distance possible. This is discussed further in the next section.

The above situation generalizes readily. Let ρ be an arbitary real representation of the finite group G with representation space V. Let

$$V = V_1 \oplus \cdots \oplus V_h$$

be a decomposition of V such that the representation ρ_i, afforded by V_i is equivalent to a direct sum of some number of copies of the same irreducible representation and the component representations of V_i and V_j are nonequivalent if $i \neq j$.

Definition The submodules V_i of V are called the homogeneous components of V and the representations ρ_i are called homogeneous representations. If $\dim V_i = n_i^2$, where n_i is the dimension of an irreducible component of the representation ρ_i afforded by V_i, then we say that ρ_i is a full homogeneous component or representation. In other words, a representation is a full homogeneous component if it is the direct sum of s copies of a real irreducible representation, where s is the multiplicity of this representation in a real decomposition of the left regular representation.

Corollary Let ρ be the real representation afforded by the $\mathbb{R}G$ module V, where

$$V = V_1 \oplus \cdots \oplus V_h$$

where the V_i are the homogeneous components of V. Let

$$\rho = \bigoplus_{i=1}^{h} k_i \rho_i \quad \text{and} \quad x = x_1 + \cdots + x_h, \quad x_i \in V_i$$

Then we have the equations

$$\frac{1}{c_i} \sum_{g \in C_i} (\rho(g)x, x) = \sum_{j=1}^{h} \frac{|x_i|^2}{n_j} \chi_j(C_i)$$

and

$$\frac{1}{c_i} \sum_{g \in c_i} d^2(\rho(g)x, x) = 2|x|^2 - 2 \sum_{j=1}^{h} \frac{|x_j|^2}{n_j} \operatorname{Re} \chi_j(C_i)$$

where n_i is the dimension of the real irreducible representation ρ_i ($\dim V_i = k_i n_i$) and χ_i is the character of ρ_i.

The proof of this corollary follows readily from the previous theorem.

6.6 The Initial Vector Problem

The previous sections indicate that group codes possess some very interesting properties. A fundamental problem remains in the construction of such codes, however; namely, if we are given a representation ρ of dimension n of a finite group G, what is the vector $x \in E_n$ that maximizes the minimum distance of the code? This is the initial vector problem of group codes and, except for a few cases, is unsolved. We describe one such case here following the work of Djoković and Blake (1972) and Blake (1972). Again, the treatment will be slightly more matrix-oriented than module-oriented. Ingemarsson (1968) has obtained results for certain Abelian groups. Also, it should be noted that if the group G is of order M, then we want to choose $\mathbf{x} \in E_n$ to yield M distinct vectors. This is a little restrictive in that good codes can arise even when the M vectors are not distinct. However, for the present we shall adopt this restriction.

We first require some definitions. Let ρ be a real representation of dimension n of the finite group G and let V be its representation space, an $\mathbb{R}G$ module. Let

$$V = V_1 \oplus V_2 \oplus \cdots \oplus V_h$$

be the decomposition of V into homogeneous components and let ρ_i be the complex irreducible representation afforded by some irreducible submodule of V_i. Let χ_i be the character of ρ_i and $\dim \rho_i = n_i = \chi_i(1)$.

Definition We say that the vector $x \in V$ is balanced (strongly balanced) if $\operatorname{Re}(\rho(g)x, x)[(\rho(g)x, x)]$ depends only on the conjugacy class of g.

From the distance properties determined in the previous section it readily follows that $x \in V$ is balanced iff

$$\operatorname{Re}(\rho(g)x, x) = \sum_{i=1}^{h} \frac{1}{n_i} |x_i|^2 \operatorname{Re} \chi_i(g)$$

and $x \in V$ is strongly balanced iff

$$(\rho(g)x, x) = \sum_{i=1}^{h} \frac{1}{n_i} |x_i|^2 \chi_i(g)$$

where $x = x_1 + x_2 + \cdots + x_h$, $x_i \in V_i$.

If we restrict our attention to real modules and their decomposition over the reals, then the concepts of balanced and strongly balanced coincide. Furthermore, it is clear that these balanced vectors are optimal since no other initial vectors yield a greater minimum distance.

6.6 The Initial Vector Problem

For the remainder of this section we restrict attention to the case where V is a full homogeneous component. For this case we are able to determine completely all balanced vectors. As a preliminary, we prove a theorem of Wedderburn (Lang, 1971).

Let W be an irreducible $\mathbb{R}G$ module. By Schur's lemma, $\text{end}_{\mathbb{R}G}(W) = D$ is a division ring and is either $\mathbb{R}$, $\mathbb{H}$, or $\mathbb{C}$, depending on whether the real irreducible representation afforded by W contains an absolutely irreducible representation of the first, second, or third kind. Let $\phi \in D$; then

$$\phi((\textstyle\sum \alpha_g g)x) = (\textstyle\sum \alpha_g g)\phi(x), \quad x \in W$$

by definition and thus all the elements of D commute with all the transformations of the algebra $\mathbb{R}G$, which is another statement of Schur's lemma. We can also view W as a D module with the operation

$$\phi \cdot x = \phi(x), \quad \phi \in G \quad \text{and} \quad x \in W$$

If the dimension of D over $\mathbb{R}$ is s and the dimension of W over D is k, then the dimension of W over $\mathbb{R}$ is ks. Considering W as a left D module, we can also consider the ring $\text{end}_D(W)$. Notice that if $\phi \in \text{end}_{\mathbb{R}G}(W) = D$ and we define

$$f_\alpha: W \to W$$
$$x \mapsto \alpha x, \quad \alpha \in \mathbb{R}G, \quad x \in W$$

then $f_\alpha \in \text{end}_D(W)$ since

$$f_\alpha(\phi(x)) = \alpha\phi(x) = \phi(\alpha x) = \phi(f_\alpha(x))$$

The map

$$\mathbb{R}G \to \text{end}_D(W), \quad \alpha \mapsto f_\alpha$$

is easily shown to be a ring homomorphism. We want to show that it is actually a ring isomorphism and conclude that every D endomorphism of W is realized by the left multiplication on W by a fixed element of $\mathbb{R}G$.

Now let W be an arbitrary semisimple module over $\mathbb{R}G$, and let $K = \text{end}_{\mathbb{R}G}(W)$, which is a division ring if and only if W is irreducible. If $f \in \text{end}_K(W)$, where we view W as a K module in the same manner as in the foregoing, and if $x \in W$, then we claim there must exist $\alpha \in \mathbb{R}G$ such that $f(x) = \alpha x$. Since W is semisimple, we can write

$$W = \mathbb{R}Gx \oplus Y$$

for some submodule Y. Let π be the projection of W on $\mathbb{R}Gx$. Then $\pi \in \text{end}_{\mathbb{R}G}(W)$ and so

$$f(x) = f(\pi x) = \pi f(x) = \alpha x$$

for some $\alpha \in \mathbb{R}G$ since $\pi f(x) \in \mathbb{R}Gx$.

We now extend this situation to n elements and assume that W is irreducible. Let $W^{(n)} = W \times \cdots \times W$ be the product of n copies of W and define the map

$$f^{(n)}: W^{(n)} \to W^{(n)}$$
$$(y_1, \ldots, y_n) \mapsto (f(y_1), \ldots, f(y_n)), \quad y_i \in W$$

where $f \in \text{end}_D(W)$ and $D = \text{end}_{\mathbb{R}G}(W)$. If we let $K = \text{end}_{\mathbb{R}G}(W^{(n)})$, then K is simply the ring of $n \times n$ matrices over D, i.e., any $\mathbb{R}G$ endomorphism of $W^{(n)}$ is of the form

$$\phi = \begin{bmatrix} \phi_{11} & \cdots & \phi_{1n} \\ \vdots & & \vdots \\ \phi_{n1} & \cdots & \phi_{nn} \end{bmatrix}$$

where $\phi_{ij} \in \text{end}_{\mathbb{R}G}(W)$. However, if $f \in \text{end}_D(W)$, then its action on W commutes with elements of D by definition. By direct calculaton we then have for $\phi \in K$

$$f^{(n)}(\phi(y_1, \ldots, y_n)) = f^{(n)}\left(\sum_i \phi_{1i}(y_i), \ldots, \sum_i \phi_{ni}(y_i)\right)$$
$$= \left(\sum_i \phi_{1i}f(y_i), \ldots, \sum_i \phi_{ni}f(y_i)\right)$$
$$= \phi(f(y_1), \ldots, f(y_n))$$
$$= \phi \circ f^{(n)}(y_1, \ldots, y_n)$$

and we conclude that $f^{(n)} \in \text{end}_K(W^{(n)})$. But we are now back to the single-variable case and since $W^{(n)}$ is an $\mathbb{R}G$ module, we have that for any $(x_1, \ldots, x_n) \in W^{(n)}$ there exists an element $\alpha \in \mathbb{R}G$ such that if $f \in \text{end}_D(W)$, then

$$(f(x_1), \ldots, f(x_n)) = (\alpha x_1, \ldots, \alpha x_n)$$

Now let W be an irreducible module over $\mathbb{R}G$ again and $x_1, \ldots, x_n$ a D-basis for W, where we view W as a left D-vector space. Let $f \in \text{end}_D(W)$; then, by the foregoing discussion we have that there exists $\alpha \in \mathbb{R}G$ such that

$$\alpha x_i = f(x_i), \quad i = 1, \ldots, n, \quad \alpha \in \mathbb{R}G$$

and since f commutes with elements of D, the map extends by linearity to W. Thus the map

$$\eta: \mathbb{R}G \to \text{end}_D(W)$$
$$\alpha \mapsto f_\alpha$$

6.6 The Initial Vector Problem

is surjective. Since its kernel is zero, it is an isomorphism and we conclude that every D endomorphism of W is obtained by left multiplication by elements of $\mathbb{R}G$, i.e., we have proved the following theorem, which is a weaker version of a theorem due to Wedderburn:

Theorem 6.1 Let W be an irreducible module over the group algebra $\mathbb{R}G$. If D is the division ring $\text{end}_{\mathbb{R}G}(W)$ and W is finite dimensional over D, then $\mathbb{R}G = \text{end}_D(W)$.

We shall require this theorem in finding all balanced vectors.

Let V be a full homogeneous component and $\mathbb{C}G$ module. We shall say that V is of the first, second, or third kind depending on whether the complex irreducible representation which an irreducible submodule of it affords is of the first, second, or third kind. We will now define all balanced vectors for the real representation afforded by V. The definition is quite straightforward but depends on the kind of module which V is. We first define the vectors and then show that they are indeed all the balanced vectors that exist for full homogeneous components.

Suppose V is an $\mathbb{R}G$ module and a full homogeneous component of the first kind. If an irreducible submodule of V yields the representation ρ, then we choose a basis in V so that V decomposes as

$$V = V_1 \oplus \cdots \oplus V_n$$

$\dim V_i = n$, $\dim V = n^2$, and the real matrix representation γ afforded by V is expressed as the block diagonal sum of n copies of ρ. We shall define a principal vector for V to be one of the form

$$x = x_1 + \cdots + x_n, \qquad x_i \in V_i$$

where, if we view x_i as an n-tuple over $\mathbb{R}$ rather than as an element of V_i (the isomorphism is clear), we have

$$(x_i, x_j) = \frac{1}{n}\delta_{ij}$$

Suppose V is an $\mathbb{R}G$ module and a full homogeneous component of the second kind. Let W be an irreducible submodule of V over R and ρ the real irreducible representation afforded by W. Since V is of the second kind, $\text{end}_{\mathbb{R}G}(W) \cong \mathbb{H}$. We can always choose a basis in V with the properties that

$$V = V_1 \oplus \cdots \oplus V_{n/2}$$

$\dim V_i = 2n$, $\dim V = n^2$, and the representation afforded by V is a block diagonal sum of $n/2$ copies of ρ. The basis can also be chosen so that the matrices of $\text{end}_{\mathbb{R}G}(W)$ corrresponding to the quaternion basis e_0, e_1, e_2, e_3 used in the previous chapter are given by

$$e_0 \to \begin{bmatrix} I_{n/2} & 0 & & 0 & \\ 0 & I_{n/2} & & & \\ \hline & & I_{n/2} & 0 & \\ & 0 & 0 & I_{n/2} \end{bmatrix}, \quad e_1 \to \begin{bmatrix} 0 & -I_{n/2} & & & \\ I_{n/2} & 0 & & 0 & \\ \hline & & & 0 & -I_{n/2} \\ & 0 & & I_{n/2} & 0 \end{bmatrix}$$

$$e_2 \to \begin{bmatrix} & 0 & -I_{n/2} & 0 \\ & & 0 & I_{n/2} \\ \hline I_{n/2} & 0 & & \\ 0 & -I_{n/2} & & 0 \end{bmatrix}, \quad e_3 \to \begin{bmatrix} & & 0 & -I_{n/2} \\ & 0 & -I_{n/2} & 0 \\ \hline & 0 & I_{n/2} & & \\ I_{n/2} & 0 & & 0 \end{bmatrix}$$

Applying these matrices to the real irreducible representation ρ implies that these matrices must be of the form

$$\rho \to \begin{bmatrix} \mathbf{u}_1 & \mathbf{u}_2 & \mathbf{v}_1 & \mathbf{v}_2 \\ -\mathbf{u}_2 & \mathbf{u}_1 & -\mathbf{v}_2 & \mathbf{v}_1 \\ -\mathbf{v}_1 & \mathbf{v}_2 & \mathbf{u}_1 & -\mathbf{u}_2 \\ -\mathbf{v}_2 & -\mathbf{v}_1 & \mathbf{u}_2 & \mathbf{u}_1 \end{bmatrix}$$

Again we write $x \in V$ as

$$x = x_1 + \cdots + x_{n/2}$$

and interpret x_i as a $2n$-tuple over $\mathbb{R}$. For convenience we shall indicate the application of elements of $\text{end}_{\mathbb{R}G}(W)$ on x_i by their corresponding elements of $\mathbb{H}$. The vector x is then a principal vector if and only if the vectors

$$x_1, \ldots, x_{n/2}, \quad e_1 x_1, \ldots, e_1 x_{n/2}, \quad e_2 x_1, \ldots, e_2 x_{n/2}, \quad e_3 x_1, \ldots, e_3 x_{n/2}$$

form an orthogonal basis of the set of $2n$-tuples over $\mathbb{R}$ and $(x_j, x_i) = (2/n)\delta_{ij}$.

Finally suppose that V is an $\mathbb{R}G$ module and a full homogeneous component of the third kind. Let W be an irreducible submodule of V over $\mathbb{R}$ and ρ the real irreducible representation afforded by W. Since V is of third kind, we have that $\text{end}_{\mathbb{R}G}(W) \cong \mathbb{C}$. We can always choose a basis in V with the properties that

$$V = V_1 \oplus \cdots \oplus V_n$$

where $\dim V_i = 2n$, $\dim V = 2n^2$, and the representation afforded by V is a block diagonal sum of n copies of ρ. The basis can also be chosen so that the matrices of $\text{end}_{\mathbb{R}G}(W)$, corresponding to the basis of 1 and i of $\mathbb{C}$ over $\mathbb{R}$, are given by

$$1 \to \begin{bmatrix} I_n & 0 \\ 0 & I_n \end{bmatrix}, \quad e_3 \to \begin{bmatrix} 0 & -I_n \\ I_n & 0 \end{bmatrix}$$

6.6 The Initial Vector Problem

Let a vector $x \in V$ be written as

$$x = x_1 + \cdots + x_n, \qquad x_i \in V_i$$

and again interpret x_i as a $2n$-tuple over $\mathbb{R}$. Then, using the same notation as before, we say that x is a prinicpal vector if and only if the vectors,

$$x_1, \ldots, x_n, e_3 x_1, \ldots, e_3 x_n$$

form an orthogonal basis of the set of $2n$-tuples over $\mathbb{R}$ and $(x_i, x_j) = (1/n)\delta_{ij}$. The importance of these principal vectors stems from the following theorem.

Theorem 6.2 Let V be a full homogeneous component. Then $x \in V$ is balanced if and only if it is principal.

Proof Much of the proof depends on which kind of full homogeneous component V is. For convenience we prove the theorem for V of the second kind, the proof for the other two kinds being parallel. Thus we will assume that V is a full homogeneous component of the second kind and an $\mathbb{R}G$ module and has a decomposition.

$$V = V_1 \oplus \cdots \oplus V_{n/2}$$

$\dim_\mathbb{R}(V_i) = 2n$, $\dim V = n^2$, V_i an irreducible submodule, and $\mathrm{end}_{\mathbb{R}G}(V_i) \cong \mathbb{H}$. We assume that the representation afforded by V is a block diagonal sum of $n/2$ copies of ρ, the real irreducible representation afforded by V_1. If

$$x = x_1 + \cdots + x_{n/2}, \qquad x_i \in V_i$$

then we have to show that x is strongly balanced if and only if x is principal. In other words

$$\sum_{i=1}^{n/2} (x_i, \rho(g)x_i) = \frac{\chi(g)}{2n}$$

if and only if x is principal, where $\mathrm{tr}(\rho(g)) = \chi(g)$.

Suppose first that x is principal. Form the matrix $\mathbf{X}$ with rows

$$x_1, \ldots, x_{n/2}, \; e_1 x_1, \ldots, e_1 x_{n/2}, \; e_2 x_1, \ldots, e_2 x_{n/2}, \; e_3 x_1, \ldots, e_3 x_{n/2}$$

It follows that

$$\sum_{l=1}^{n/2} [(x_l, \rho(g)x_l) + (e_1 x_l, \rho(g)e_1 x_l) + (e_2 x_l, \rho(g)e_2 x_l) + (e_3 x_l, \rho(g)e_3 x_l)]$$

$$= \mathrm{tr}(\mathbf{X}\rho(g)\mathbf{X}^T) = \frac{2}{n} \mathrm{tr}\, \rho(g) = \frac{2\chi(g)}{n}$$

However, by definition, $\rho(g)$ commutes with e_1, e_2, and e_3 for all $g \in G$ and thus

$$(x_l, \rho(g)x_l) = (e_1 x_l, \rho(g) e_1 x_l)$$
$$= (e_2 x_l, \rho(g) e_2 x_l)$$
$$= (e_3 x_l, \rho(g) e_3 x_l)$$

and thus

$$\sum_{l=1}^{n/2} (x_l, \rho(g)x_l) = \frac{\chi(g)}{2n}$$

as required.

Now suppose x is strongly balanced, which implies that

$$\sum_{l=1}^{n/2} (x_l, \rho(g)x_l) = \frac{\chi(g)}{2n}$$

We want to show that it must also be principal. From the form of Wedderburn's theorem given previously we have that every D (in this case $\mathbb{H}$) endomorphism of an irreducible $\mathbb{R}G$ submodule of V, say V_i, is realized by left multiplication by an element of $\mathbb{R}G$. Let $x_1, \ldots, x_{n/2}$ be elements of the set of n-tuples over $\mathbb{R}$, which we view as an $\mathbb{H}$-vector space using the $2n \times 2n$ matrix representation of $\mathbb{H}$ given previously. We denote this $\mathbb{H}$-vector space by W and let $W = X \oplus Y$, where X is the subspace spanned by $x_1, \ldots, x_{n/2}$. Let P be the projection of W onto Y with kernel X, i.e., $P(x) = 0$, $x \in X$, $P(y) = y$, $y \in Y$, and $P(z) = y$ if $z = x + y$, $x \in X$, $y \in Y$. However, P can be written as a linear combination of the matrices of the representation afforded by W and since, by assumption, we have that

$$\sum_{l=1}^{n/2} (x_l, \rho(g)x_l) = \frac{\chi(g)}{2n}$$

we must also have that

$$\sum_{l=1}^{n/2} (x_l, P(x_l)) = \frac{\operatorname{tr} P}{2n}$$

Since $P(x_l) = 0$, $l = 1, \ldots, n/2$, we have that $\operatorname{tr} P = 0$, implying that $P = 0$ and $W = X$. It follows that $x_1, \ldots, x_{n/2}$ is an $\mathbb{H}$ basis of W.

Now the vectors

$$x_1, \ldots, x_{n/2}, \quad e_1 x_1, \ldots, e_1 x_{n/2}, \quad e_2 x_1, \ldots, e_2 x_{n/2}, \quad e_3 x_1, \ldots, e_3 x_{n/2}$$

form an $\mathbb{R}$ basis for W. Let A be a linear transformation of W as an $\mathbb{H}$-vector space and we write

$$A x_l = \sum_{m=1}^{n/2} \sigma_{ml} x_m$$

6.6 The Initial Vector Problem

where $\sigma_{ml} \in \mathbb{H}$ and we let

$$\sigma_{ml} = \alpha_{ml}^0 + \alpha_{ml}^1 i + \alpha_{ml}^2 j + \alpha_{ml}^3 k$$

where $\{1, i, j, k\}$ is a basis of $\mathbb{H}$ over $\mathbb{R}$. From the assumption that A is an $\mathbb{H}$ linear transformation of W, we have that

$$A(e_1 x) = e_1 A x_l = e_1 \sum_{r=1}^{n/2} (\sigma_{lr}) x_r$$

However, by assumption, we also have that

$$\sum_{l=1}^{n/2} (x_l, A x_l) = \frac{1}{2n} \operatorname{tr} A$$

which implies that, when we view A as a real transformation with basis

$$x_1, \ldots, x_{n/2}, \quad e_1 x_1, \ldots, e_1 x_{n/2}, \quad e_2 x_1, \ldots, e_2 x_{n/2}, \quad e_3 x_1, \ldots, e_3 x_{n/2}$$

we have

$$\sum_{r,l=1}^{n/2} [\alpha_{rl}^0(x_r, x_l) + \alpha_{rl}^1(e_1 x_r, x_l) + \alpha_{rl}^2(e_2 x_r, x_l) + \alpha_{rl}^3(e_3 x_r, x_l)] = \frac{1}{2n} \sum_{l=1}^{n/2} \alpha_{ll}^0$$

Since we may choose the elements α_{rl}^i arbitrarily, this implies that $(x_l, x_m) = (2/n)\delta_{lm}$ and

$$(e_1 x_l, x_m) = (e_2 x_l, x_m) = (e_3 x_l, x_m) = 0$$

for any l, m. We conclude that x is a principal vector. As mentioned, the proofs for the other cases follow in a similar manner. $\square$

Corollary Let V be as in the above theorem. If x is strongly balanced, then V is generated by x as a G module. If V is homogeneous, but not a full homogeneous component, then no strongly balanced vectors for V exist.

Proof The first part of the corollary follows immediately from the theorem and the definition of principal vectors. The second part follows from the fact that if x is a strongly balanced vector for V, where V is homogeneous, but not a full homogeneous component, then the configuration matrix of the resulting group code is the same as that for a strongly balanced vector in a full homogeneous component. This implies that the dimensions of the representation spaces are the same in each case, which is contrary to assumption. $\square$

6.7 Comments

This chapter contains most of the significant results known on group codes for the Gaussian channel, although there is some very interesting additional material given by Slepian (1968), who considers the problem of finding the optimal vector for the direct sum of two representations. This appears to be an extremely difficult problem, even when optimal vectors for the components of the direct sum are known. The fundamental region of the group G also appears to have significance in determining optimal vectors. The precise relationship is, again, difficult to establish. There is, however, an interesting section in the work of Slepian (1968) on the problem. It is clear that there are many problems yet to be solved in connection with these group codes. Unfortunately, there is no indication that these problems will have closed-form solutions.

Exercises

1 Find an optimal vector for the irreducible representation contained in the natural representation of S_n, the symmetric group on n letters. Show that the maximum minimum distance possible for this case is

$$[24/[(n-1)n(n+1)]]^{1/2}$$

[See Slepian (1951) and Blake (1972)].

2 An existence theorem of coding theory states that it is possible to find M unit vectors in E_n such that in the limit as $n \to \infty$, $d^2(\mathbf{x}_i, \mathbf{x}_j) \geq \delta > 0$, $i \neq j$, and $R(\delta) = (\ln M)/n$ is independent of n [e.g., see Wyner (1965)]. Show that the irreducible representations of S_n cannot be used to construct such a set of vectors. (Blake, 1972).

3 Show that group codes for the binary symmetric channel may be interpreted as group codes for the Gaussian channel (Slepian, 1968).

4 Permutation Codes (Slepian, 1965) Let $x_1, \ldots, x_m$ be n-tuples over $\mathbb{R}$ that form a code. We consider two types of codes. For the variant I code we choose x_1 as an arbitrary sequence of n real numbers, not necessarily distinct. For the remaining codewords we take all the distinct n-tuples that can be formed by permuting the elements of x_1. For a variant II code we take for x_1 an arbitrary sequence of n positive real numbers. For the remaining codewords we take all distinct n-tuples that can be formed by permuting the order and/or changing the sign of the elements of x_1. A suggested decoder for variant I codes operates as follows: if $Z = (z_1, \ldots, z_n)$ is a received word, then we replace the ith largest number in Z with the ith largest number in x_1, $i = 1, \ldots, n$. For variant II codes we replace the ith largest number of Z in absolute value with the ith largest number in x_1. The algebraic signs in Z are then restored. Show that this decoding scheme is, in each case, maximum likelihood. Find the size of each code. These permutation codes are actually group codes generated by finite reflection groups of various types acting on appropriate initial vectors.

APPENDIX A
The Möbius Inversion Formula

The Möbius inversion function $\mu(n)$ defined on the positive integers is given by

$$\mu(1) = 1$$
$$\mu(p_1 p_2 \cdots p_k) = (-1)^k \quad \text{if} \quad p_i \neq p_j, \quad i \neq j, \quad p_i \text{ prime}$$
$$\mu(n) = 0 \quad \text{if} \quad p^2 | n \text{ for some prime } p$$

If $n = p_1^{e_1} p_2^{e_2} \cdots p_k^{e_k} > 1$, then the only nonzero terms in the summation

$$\sum_{d|n} \mu(d)$$

are those for which $d = p_{i_1} p_{i_2} \cdots p_{i_j}$, $p_{i_a} \neq p_{i_b}$, $a \neq b$. It follows that

$$\sum_{d|n} \mu(d) = \mu(1) + \sum_{i=1}^{k} \mu(p_i) + \sum_{i \neq j} \mu(p_i p_j) + \cdots$$
$$= 1 + (-1)k + (-1)^2 \binom{k}{2} + \cdots + (-1)^k \binom{k}{k}$$
$$= (1-1)^k = 0$$

When $n = 1$ then $\mu(n) = 1$. The Möbius inversion formula states that if

$$h(n) = \sum_{d|n} g(d)$$

then

$$g(n) = \sum_{d|n} \mu\left(\frac{n}{d}\right) h(d) = \sum_{d|n} \mu(d) h\left(\frac{n}{d}\right)$$

To show this, consider the sum

$$\sum_{d|n} \mu(d) h\left(\frac{n}{d}\right) = \sum_{d|n} \mu(d) \left[\sum_{b|n/d} h(b)\right] = \sum_{bd|n} \mu(d) h(b) = \sum_{b|n} h(b) \left[\sum_{d|n/b} \mu(d)\right]$$

and, by a previous expression, this last bracket is zero unless $n/b = 1$, in which case the last equality reduces to $h(n)$, as required.

APPENDIX B
Lucas's Theorem

Let m and n be positive integers, $m > n$, and p a prime. If

$$m = \sum_i m_i p^i, \qquad n = \sum_i n_i p^i$$

are the p-ary expansions of m and n, respectively, $0 \leqslant m_i < p$, $0 \leqslant n_i < p$, then

$$\binom{m}{n} \equiv \prod_i \binom{m_i}{n_i} \bmod p$$

The proof of this statement is by induction and we suppose $m = m'p + m_0$, $n = n'p + n_0$, and $0 \leqslant m_0, n_0 < p$. We have then that

$$(1+x)^m = \sum_{k=0}^{m} \binom{m}{k} x^k = (1+x)^{m'p+m_0} = (1+x)^{m'p}(1+x)^{m_0}$$

$$\equiv (1+x^p)^{m'}(1+x)^{m_0} \bmod p$$

$$\equiv \left[\sum_{j=0}^{m'} \binom{m'}{j} x^{jp}\right]\left[\sum_{l=0}^{m_0} \binom{m_0}{l} x^l\right] \bmod p$$

Comparing the coefficient of x^n in the first and last equations yields

$$\binom{m}{n} = \binom{m'}{n'}\binom{m_0}{n_0} \bmod p$$

Repeating the argument on m' and n' yields, inductively, the result. The proof used here is that of Berlekamp (1968).

APPENDIX C
The Mathieu Groups

The Mathieu groups have been mentioned in several places in the text, mainly in connection with combinatorial designs and perfect codes. In this appendix we gather some information of a general nature on these remarkable groups.

We recall first some definitions. Let G be a permutation group of degree n acting on $\Delta = \{1, \ldots, n\}$. We say that G is k-transitive if for any two ordered k subsets $\delta_1 = \{i_1, \ldots, i_k\}$ and $\delta_2 = \{j_i, \ldots, j_k\}$ of Δ there exists an element $\sigma \in G$ such that

$$\sigma(i_s) = j_s, \quad s = 1, \ldots, k$$

We say that G is sharply k-transitive if there exists only one such element. Equivalently, G is sharply k-transitive if it is k-transitive and only the identity fixes k points.

If G is k-transitive and of degree n then it can be shown that the order of G, $|G|$, is divisible by $n(n-1) \cdots (n-k+1)$, while if G is sharply k-transitive of degree n, then $|G| = n(n-1) \cdots (n-k+1)$. Denote by G_i the subgroup of G that leaves i invariant. Then if G is sharply k-transitive of degree n, then G_i is sharply $(k-1)$-transitive on $\Delta \setminus \{i\}$. Clearly the symmetric group on n letters, S_n, is sharply n-transitive of degree n. It is also sharply $(n-1)$-transitive. Also, if $n \geq 3$, then A_n, the alternating group on n letters, is sharply $(n-2)$-transitive and of degree n. Apart from these well-known examples, highly transitive groups are very sparse. Indeed, no k-transitive groups other than A_n and S_n are known when $k \geq 6$ and we shall discuss all 4- and 5-transitive groups known here. A theorem of Jordan (Passman, 1968, p. 283) states that if G is a nontrivial, sharply k-transitive group of degree n, then if $k \geq 4$, there are only two possible cases, namely $k = 4$ and $n = 11$ or $k = 5$ and $n = 12$. The Mathieu groups M_{11} and M_{12} are such groups. We

331

give the presentation of these groups due to Passman (1968, p. 290) and consider the elements

$$x_1 = (4\ 5\ 6)(7\ 8\ 9)(10\ 11\ 12) \qquad x_5 = (5\ 11\ 6\ 9)(7\ 12\ 10\ 8)$$
$$x_2 = (4\ 7\ 10)(5\ 8\ 11)(6\ 9\ 12) \qquad x_6 = (14)(7\ 8)(9\ 11)(10\ 12)$$
$$x_3 = (5\ 6\ 7\ 10)(8\ 9\ 12\ 11) \qquad x_7 = (12)(7\ 10)(8\ 11)(9\ 12)$$
$$x_4 = (5\ 8\ 6\ 12)(7\ 11\ 10\ 9) \qquad x_8 = (23)(7\ 12)(8\ 10)(9\ 11)$$

Then $M_{11} = \langle x_1, x_2, x_3, x_4, x_5, x_6, x_7 \rangle$ is a sharply 4-transitive group of degree 11 acting on $\{1, 2, 4, 5, 6, \ldots, 12\}$ and $|M_{11}| = 11 \cdot 10 \cdot 9 \cdot 8 = 7920$. The group $\langle M_{11}, x_8 \rangle = M_{12}$ is sharply 5-transitive of degree 12 and $|M_{12}| = 12 \cdot |M_{11}| = 95{,}040$.

The three other Mathieu groups, M_{22}, M_{23}, and M_{24}, which are not sharply transitive, are, respectively, 3-transitive, 4-transitive, and 5-transitive. Their orders are given by $|M_{22}| = 443{,}520$, $|M_{23}| = 10{,}200{,}960$, and $|M_{24}| = 244{,}823{,}040$. M_{23} is the stabilizer of an element on which M_{24} acts and M_{11} is the stabilizer of an element on which M_{12} acts. All five of the Mathieu groups are, furthermore, simple. While there are no other 4- or 5 transitive groups known and no 6-transitive groups known (apart from S_n and A_n), there are other 3-transitive groups known, as shown by the linear fractional groups constructed in Chapter 1. Until recently these Mathieu groups were the only finite simple groups known which did not fit into a known infinite family. However, this is no longer true, since several other finite groups are now known with this property.

The character tables of M_{11} through M_{24} are given in Tables C-1 to C-5 taken from the following sources: M_{11} and M_{22} from Burgoyne and Fong (1966), M_{12} from Whitelaw (1966), and M_{23} and M_{24} from Todd (1966). In each case we have standardized notation to coincide with that of Todd (1966). In this notation the first line of the table gives the type of element which is contained in the conjugacy class corresponding to that column. Thus 5^2 indicates the conjugacy class consisting of products of two five cycles, all other elements being fixed. In the second line of the table the order of the group divided by the order of the conjugacy class is given as a more convenient number than the order of the conjugacy class. The first column of the table gives, of course, the dimension of the irreducible representation whose characters are given in that row.

Some papers that consider further properties of the Mathieu groups and their relationship to coding and combinatorics are Assmus and Mattson (1966a, b, 1967b), Berlekamp (1971), Paige (1956), Garbe and Mennicke (1964), Coxeter (1958), Hall (1962), Stanton (1951), and Todd (1959, 1966) as well as the fundamental work of Witt (1938a, b).

TABLE C-1

$M_{11}: z = (-1 \pm i\sqrt{11})/2$

1^{11}	2^4	4^2	3^3	5^2	8·2	8·2	6.3.2	11	11
g	48	8	18	5	8	8	6	11	11
1	1	1	1	1	1	1	1	1	1
10	2	2	1	0	0	0	−1	−1	−1
11	3	−1	2	1	−1	−1	0	0	0
55	−1	−1	1	0	1	1	−1	0	0
45	−3	1	0	0	−1	−1	0	1	1
44	4	0	−1	−1	0	0	1	0	0
16	0	0	−2	1	0	0	0	z	$\bar{z}$
10	−2	0	1	0	$i\sqrt{2}$	$-i\sqrt{2}$	1	−1	−1
16	0	0	−2	1	0	0	0	$\bar{z}$	z
10	−2	0	1	0	$-i\sqrt{2}$	$i\sqrt{2}$	1	−1	−1

TABLE C-2

$$M_{12}: z = (-1 \pm i\sqrt{11})/2$$

1^{12}	2^4	4^2	3^3	5^2	$2\cdot 8$	$2\cdot 3\cdot 6$	11	11	2^6	$2\cdot 10$	$2^3 4^2$	3^4	6^2	$4\cdot 8$
g	192	32	54	10	8	6	11	11	240	10	32	36	12	8
1	1	1	1	1	1	1	1	1	1	1	1	1	1	1
11	3	3	2	1	1	1	0	0	−1	−1	−1	−1	−1	−1
11	3	−1	2	1	−1	0	0	0	−1	−1	3	−1	−1	−1
55	−1	3	1	0	−1	−1	0	0	−5	0	−1	1	1	1
55	−1	−1	1	0	−1	−1	0	0	−5	0	3	1	1	−1
55	7	−1	1	0	−1	0	0	0	−5	0	−1	1	1	−1
54	6	2	0	−1	0	0	−1	−1	6	1	−2	0	0	0
66	2	−2	3	1	0	0	0	0	6	1	2	0	0	0
45	−3	−1	0	0	−1	0	1	1	5	0	−1	3	−1	−1
99	3	3	0	−1	1	0	0	0	−1	−1	−1	3	0	−1
120	−8	0	3	0	0	0	−1	−1	0	0	0	0	0	0
144	0	0	0	−1	0	0	1	1	4	−1	0	0	0	0
176	0	0	−4	1	0	0	0	0	−4	1	0	−1	1	0
16	0	0	−2	1	0	0	z	z̄	4	−1	0	1	1	0
16	0	0	−2	1	0	0	z̄	z	4	−1	0	1	1	0

TABLE C-3

M_{22}: $z_1 = (-1 \pm i\sqrt{7})/2$, $z_2 = (-1 \pm i\sqrt{11})/2$

1^{22}	2^8	3^6	5^4	4^42^2	4^42^2	7^3	7^3	$8^2 \cdot 4 \cdot 2$	$6^2 \cdot 3^2 \cdot 2^2$	11^2	11^2
g	384	36	5	16	32	7	7	8	12	11	11
1	1	1	1	1	1	1	1	1	1	1	1
21	5	3	1	1	1	0	0	−1	−1	−1	−1
55	7	1	0	−1	3	−1	−1	−1	−1	0	0
154	10	1	−1	2	−2	0	0	0	1	0	0
210	2	3	0	−2	−2	0	0	0	−1	−1	−1
280	−8	1	0	0	0	0	0	0	−1	z_2	$\bar{z}_2$
231	7	−3	1	−1	−1	0	0	−1	1	0	0
385	1	−2	0	−1	1	0	0	−1	−2	0	0
99	3	0	−1	−1	3	1	1	1	0	0	0
45	−3	0	0	1	1	z_1	$\bar{z}_1$	−1	0	−1	−1
280	−8	1	0	0	0	0	0	0	1	$\bar{z}_2$	z_2
45	−3	0	0	1	1	$\bar{z}_1$	z_1	−1	0	1	1

TABLE C-4

M_{23}: $\alpha = \frac{1}{2}(-1+i\sqrt{7})$, $\beta = \frac{1}{2}(-1+i\sqrt{15})$, $\gamma = \frac{1}{2}(-1+i\sqrt{23})$, $\delta = \frac{1}{2}(-1+i\sqrt{11})$

	1^{24}	$1^8 2^8$	$1^6 3^6$	$1^4 5^4$	$1^4 4^4 2^4$	$1^3 7^3$	$1^3 7^3$	$1^2 8^2 4^2$	$1^2 6^2 3^2 2^2$	$1^2 11^2$	$1^2 11^2$	$1 \cdot 15 \cdot 53$	$1 \cdot 15 \cdot 53$	$1 \cdot 14 \cdot 72$	$1 \cdot 14 \cdot 72$	$1 \cdot 23$	$1 \cdot 23$
g	2688	180	15	32	14	14	8	12	11	11	15	15	14	14	23	23
1	1	1	1	1	1	1	1	1	1	1	1	1	1	1	1	1	
22	6	4	2	2	1	1	0	0	0	0	-1	-1	-1	-1	-1	-1	
45	-3	0	0	1	α	$\bar\alpha$	-1	-1	-1	-1	0	0	-1	-1	-1	-1	
45	-3	0	0	1	$\bar\alpha$	α	-1	-1	-1	-1	0	0	-1	-1	-1	-1	
230	22	5	-1	0	2	-1	-1	0	-1	0	0	0	0	0	0	0	
231	7	6	-1	1	-1	0	0	-1	-2	-1	-1	-1	β	$\bar\beta$	0	0	
-231	7	-3	-1	1	-1	0	0	-1	1	0	0	0	$\bar\beta$	β	0	0	
231	7	-3	-1	1	-1	0	0	-1	1	0	0	0	β	$\bar\beta$	0	0	
253	13	1	1	-2	1	-1	-1	1	1	0	0	0	-1	-1	0	0	
770	-14	5	0	-2	0	0	0	0	0	0	0	0	0	0	γ	$\bar\gamma$	
770	-14	5	0	-2	0	0	0	0	0	0	0	0	0	0	$\bar\gamma$	γ	
896	0	-4	1	0	0	0	0	0	0	δ	$\bar\delta$	-1	0	0	-1	-1	
896	0	-4	1	0	0	0	0	0	0	$\bar\delta$	δ	-1	0	0	-1	-1	
990	-18	0	0	0	2	α	$\bar\alpha$	0	0	0	0	0	α	$\bar\alpha$	1	1	
990	-18	0	0	0	2	$\bar\alpha$	α	0	0	0	0	0	$\bar\alpha$	α	1	1	
1035	27	0	0	-1	-1	-1	-1	1	0	-1	-1	0	-1	-1	0	0	
2024	8	-1	-1	0	0	1	1	0	-1	0	0	-1	1	1	0	0	

TABLE C-5

M_{24}: $\alpha = \tfrac{1}{2}(-1 + \pm i\sqrt{7})$, $\beta = \tfrac{1}{2}(-1 + i\sqrt{15})$, $\gamma = \tfrac{1}{2}(-1 + i\sqrt{23})$

1^{24}	$1^8 2^8$	$1^6 3^6$	$1^4 5^4$	$1^4 4^4 2^2$	$1^3 7^3$	$1^3 7^3$	$1^2 8^2 4^2$	$1^2 6^2 3^2 2^2$	$1^2 11^2$	$1 \cdot 15 \cdot 53$	$1 \cdot 15 \cdot 53$
g	21504	1080	60	128	42	42	16	24	11	15	15
1	1	1	1	1	1	1	1	1	1	1	1
23	7	5	3	3	2	2	1	1	1	0	0
45	−3	0	0	1	α	$\bar{\alpha}$	−1	0	1	0	0
45	−3	0	0	1	$\bar{\alpha}$	α	−1	0	1	0	0
231	7	−3	1	−1	0	0	−1	1	0	β	$\bar{\beta}$
231	7	−3	1	−1	0	0	−1	1	0	$\bar{\beta}$	β
252	28	9	2	4	0	0	0	1	−1	−1	−1
253	13	10	3	1	1	1	−1	−2	0	0	0
483	35	6	−2	3	0	0	−1	2	−1	1	1
770	−14	5	0	−2	0	0	0	1	0	0	0
770	−14	5	0	−2	0	0	0	1	0	0	0
990	−18	0	0	2	α	$\bar{\alpha}$	0	0	0	0	0
990	−18	0	0	2	$\bar{\alpha}$	α	0	0	0	0	0
1035	−21	0	0	3	2α	$2\bar{\alpha}$	−1	0	1	0	0
1035	−21	0	0	3	$2\bar{\alpha}$	2α	−1	0	1	0	0
1035	27	0	0	−1	−1	−1	1	0	1	0	0
1265	49	5	0	1	−2	−2	1	1	0	0	0
1771	−21	16	1	−5	0	0	−1	0	0	1	1
2024	8	−1	−1	0	1	1	0	−1	0	−1	−1
2277	21	0	−3	1	2	2	−1	0	0	0	0
3312	48	0	−3	0	1	1	0	0	1	0	0
3520	64	10	0	0	−1	−1	0	−2	0	0	0
5313	49	−15	3	−3	0	0	−1	1	0	0	0
5544	−56	9	−1	0	0	0	0	1	0	−1	−1
5796	−28	−9	1	4	0	0	0	−1	−1	1	1
10395	−21	0	0	−1	0	0	1	0	0	0	0

TABLE C-5—continued

1^{24}	$1\cdot14\cdot72$	$1\cdot14\cdot72$	$1\cdot23$	$1\cdot23$	12^2	6^4	4^6	3^8	2^{12}	10^22^2	$21\cdot3$	$21\cdot3$	4^42^4	$12\cdot6\cdot4^2$
g	14	14	23	23	12	24	96	504	7680	20	21	21	384	12
1	1	1	1	1	1	1	1	1	1	1	1	1	1	1
23	0	0	0	0	-1	-1	-1	-1	-1	-1	-1	-1	-1	-1
45	$-\alpha$	$-\bar\alpha$	-1	-1	1	-1	1	3	5	0	$\bar\alpha$	α	-3	0
45	$-\bar\alpha$	$-\alpha$	-1	-1	1	-1	1	3	5	0	α	$\bar\alpha$	-3	0
231	0	0	1	1	0	0	3	0	-9	1	0	0	-1	-1
231	0	0	1	1	0	0	3	0	-9	1	0	0	-1	-1
252	0	0	-1	-1	0	0	0	0	12	2	0	0	4	1
253	-1	-1	0	0	1	1	1	1	-11	-1	1	1	-3	0
483	0	0	0	0	0	0	3	0	3	-2	0	0	3	0
770	0	0	γ	$\bar\gamma$	1	1	-2	-7	10	0	0	0	2	-1
770	0	0	$\bar\gamma$	γ	1	1	-2	-7	10	0	0	0	2	-1
990	α	$\bar\alpha$	1	1	1	-1	-2	3	-10	0	$\bar\alpha$	α	6	0
990	$\bar\alpha$	α	1	1	1	-1	-2	3	-10	0	α	$\bar\alpha$	6	0
1035	0	0	0	0	-1	1	-1	-3	-5	0	$-\bar\alpha$	$-\alpha$	3	0
1035	0	0	0	0	-1	1	-1	-3	-5	0	$-\alpha$	$-\bar\alpha$	3	0
1035	-1	-1	0	0	0	2	3	6	35	0	-1	-1	3	0
1265	0	0	0	0	0	0	-3	8	-15	0	1	1	-7	-1
1771	0	0	0	0	-1	-1	-1	7	11	1	0	0	3	0
2024	1	1	0	0	0	0	0	8	24	-1	1	1	8	-1
2277	0	0	0	0	0	2	-3	6	-19	1	-1	-1	-3	0
3312	-1	-1	0	0	0	-2	0	-6	16	1	1	1	0	0
3520	1	1	1	1	0	0	0	-8	0	0	-1	-1	0	0
5313	0	0	0	0	0	0	-3	0	9	-1	0	0	1	1
5544	0	0	1	1	0	0	0	0	24	-1	0	0	-8	1
5796	0	0	0	0	0	0	0	0	36	1	0	0	-4	-1
10395	0	0	-1	-1	0	0	3	0	-45	0	0	0	3	0

References

Alanen, J. D., and Knuth, D. E. (1964). Tables of finite fields. *Sankhyā Ser. A* **26**, 305–328.
Albert, A. (1956). "Fundamental Concepts of Higher Algebra." Univ. of Chicago Press, Chicago, Illinois.
Alltop, W. O. (1969). An infinite class of 4-designs. *J. Combinatorial Theory* **6**, 320–322.
Alltop, W. O. (1971). Some 3-designs and a 4-design. *J. Combinatorial Theory Ser. A* **11**, 190–195.
Alltop, W. O. (1972). An infinite class of 5-designs. *J. Combinatorial Theory Ser. A* **12**, 390–395.
Anderson, D. R. (1968). A new class of cyclic codes. *SIAM J. Appl. Math.* **61**, 181–197.
Artin, E. (1957). "Geometric Algebra." Wiley (Interscience), New York.
Artin, E. (1959). "Galois Theory." Univ. of Notre Dame Press, Notre Dame, Indiana.
Artin, E., Nesbitt, C. J., and Thrall, R. M. (1944). "Rings with Minimum Condition." Univ. of Michigan Press, Ann Arbor.
Assmus, E. F., Jr., and Mattson, H. F., Jr., (1966a). Perfect codes and the Mathieu groups. *Arch. Math.* (*Basel*) **17**, 121–135.
Assumus, E. F., Jr., and Mattson, H. F., Jr. (1966b). Disjoint Steiner systems associated with the Mathieu groups. *Bull. Amer. Math. Soc.* **72**, 843–845.
Assmus, E. F., Jr., and Mattson, H. F., Jr. (1966c). On the number of inequivalent Steiner triple systems. *J. Combinatorial Theory* **1**, 301–305.
Assmus, E. F., Jr., and Mattson, H. F., Jr. (1966d). Cyclic codes. Final Rep., AFCRL-66-348. Air Force Cambridge Res. Lab. April 28, Cambridge, Massachusetts.
Assmus, E. F., Jr., and Mattson, H. F., Jr. (1967a). On tactical configurations and error-correcting codes. *J. Combinatorial Theory* **2**, 243–257.
Assmus, E. F., Jr., and Mattson, H. F., Jr., (1967b). On the automorphism groups of Paley-Hadamard matrices. *Proc. Conf. Combinatorial Math. Its Appl. Univ. of North Carolina, April,* 1967, pp. 98–103, Univ. of North Carolina Press, Chapel Hill, North Carolina.
Assmus, E. F., Jr., and Mattson, H. F., Jr. (1969). New 5-designs. *J. Combinatorial Theory* **6**, 122–151.

Assmus, E. F., Jr., and Mattson, H. F., Jr. (1970). Algebraic theory of codes II. Final Rep AFCRL-71-0013. Air Force Cambridge Res. Lab. October 15, Cambridge, Massachusetts.

Assmus, E. F., Jr., and Mattson, H. F., Jr. (1972a). Contractions of self-orthogonal codes. *Discrete Math.* **3**, 21–32.

Assmus, E. F., Jr., and Mattson, H. F., Jr. (1972b). On weights in quadratic-residue codes. *Discrete Math.* **3**, 1–20.

Bellman, R. (1970). "Introduction to Matrix Analysis," 2nd ed. McGraw-Hill, New York.

Berlekamp, E. R. (1965). Distribution of cyclic matrices in a finite field. *Duke Math. J.* **33**, 45–48.

Berlekamp, E. R. (1968). "Algebraic Coding Theory." McGraw-Hill, New York.

Berlekamp, E. R. (1971). Coding theory and the Mathieu groups. *Information and Control* **17**, 40–64.

Berlekamp, E. R. MacWilliams, F. J., and Sloane, N. J. A. (1972). Gleason's theorem on self-dual codes. *IEEE Trans. Information Theory*, **IT-18**, 409–414.

Berman, G. (1952). Finite projective geometries. *Canad. J. Math.* **4**, 302–313.

Berman, S. D. (1967a). On the theory of group codes. *Kybernetika* **3**, 31–39.

Berman, S. D. (1967b). Semisimple cyclic and Abelian codes, II. *Kybernetika* **3**, 21–30.

Biggs, N. (1971) "Finite Groups of Automorphisms." Cambridge Univ. Press, Cambridge, London and New York.

Biglieri, E., and Elia, M. (1972). On the existence of group codes for the Gaussian channel. *IEEE Trans. Information Theory* **IT-18**, 399–402.

Blake, I. F. (1972) Distance properties of group codes for the Gaussian channel. *SIAM J. Appl. Math.* **23**, 312–324.

Blake, I. F. (1974). Configuration matrices of group codes. *IEEE. Trans Information Theory* **IT-20**, 95–100.

Boerner, H. (1963). "Representations of Groups." North-Holland Publ., Amsterdam.

Bose, R. C. (1942). A note on the resolvability of balanced incomplete block designs. *Sankhyā* **6**, 105–110.

Bose, R. C. (1961). On some connections between the design of experiments and information theory. *Bull. Inst. Internat. Statist.* **38**, 257–271.

Bose, R. C., and Burton, R. C. (1966). A characterization of flat spaces in a finite geometry and the uniqueness of the Hamming and the MacDonald codes. *J. Combinatorial Theory* **1**, 96–104.

Bose, R. C., and Bush, K. A. (1952). Orthogonal arrays of strength two and three. *Ann. Math. Statist.* **23**, 508–524.

Bose, R. C., and Ray-Chaudhuri, D. K. (1960a). On a class of error correcting binary group codes. *Information and Control* **3**, 68–79.

Bose, R. C., and Ray-Chaudhuri, D. K. (1960b). Further results on error correcting binary group codes. *Information and Control* **3**, 279–290.

Bose, R. C., and Shrikhande, S. S. (1959a). On the falsity of Euler's conjecture about the non-existence of two orthogonal Latin squares of order $4t+2$. *Proc. Nat. Acad. Sci. U.S.A.* **45**, 734–737.

Bose, R. C., and Shrikhande, S. S. (1959b). A note on a result in the theory of code construction. *Information and Control* **2**, 183–194.

Bose, R. C., Shrikhande, S. S., and Parker, E. T. (1960). Further results on the construction of mutually orthogonal Latin squares and the falsity of Euler's conjecture. *Canad. J. Math.* **12**, 189–203.

Burgoyne, N., and Fong, P. (1966). The Schur multipliers of the Mathieu groups. *Nagoya Math. J.* **27**, 733–745.

Burrow, M. (1965). "Representation Theory of Finite Groups." Academic Press, New York.

Camion, P. (1970). Abelian codes. Inst. of Statist. Mimeo Ser. No. 600.32. Univ. of North Carolina, Chapel Hill.
Carmichael, R. D. (1956). "Introduction to the Theory of Groups of Finite Order." Dover, New York.
Chen, C. L. (1971). On majority-logic decoding of finite geometry codes. *IEEE Trans. Information Theory* **IT-17**, 332–336.
Chen, C. L. (1972). Note on majority logic decoding of finite geometry codes. *IEEE Trans. Information Theory* **IT-18**, 539–541.
Chen, C. L., and Lin, S. (1969). Further results on polynomial codes. *Information and Control* **15**, 38–60.
Chen, C. L., Peterson, W. W., and Weldon, E. J., Jr. (1969). Some results on quasi-cyclic codes. *Information and Control* **15**, 407–423.
Conway, J. H. (1968a). Tabulation of some information concerning finite fields. In "Computers in Mathematical Research" (R. F. Churchhouse and J. C. Hertz, eds.). North-Holland Publ. Amsterdam.
Conway, J. H. (1968b). A perfect group of order 8,315,553,613,086,720,000 and the sporadic simple groups. *Proc. Nat. Acad. Sci. U.S.A.* **61**, 398–400.
Conway, J. H. (1969). A group of order 8,315,533,613,086,720,000. *Bull. London Math. Soc.* **1**, 79–88.
Cooper, A. B., and Gore, W. C. (1970). A result concerning the dual of polynomial codes. *IEEE Trans. Information Theory* **IT-16**, 638–640.
Coxeter, H. S. M. (1958). Twelve points in PG(5, 3) with 95,040 selftransformations. *Phil. Trans. Roy. Soc. London Ser. A* **247**, 279–293.
Curtis, C. W., and Reiner, I. (1966). "Representation Theory of Finite Groups and Associative Algebras." Wiley (Interscience), New York.
Davenport, H. (1968). Bases for finite fields. *J. London Math. Soc.* **43**, 21–39.
Delsarte, P. (1969). A geometric approach to a class of cyclic codes. *J. Combinatorial Theory* **6**, 340–358.
Delsarte, P. (1970a). On cyclic codes that are invariant under the general linear group. *IEEE Trans. Information Theory* **IT-16**, 760–769.
Delsarte, P. (1970b). Automorphisms of Abelian codes. *Phillips Res. Rep.* **25**, 389–403.
Delsarte, P. (1971a). Majority logic decodable codes derived from finite inversive planes. *Information and Control* **18**, 319–325.
Delsarte, P. (1971b). Weights of p-ary Abelian codes. *Philips Res. Rep.* **26**, 145–156.
Delsarte, P. (1973a). An algebraic approach to the association schemes of coding theory. *Philips Research Repts. Supp.* No. 10.
Delsarte, P. (1973b). Four fundamental parameters of a code and their combinatorial significance. *Information and Control* **23**, 407–438.
Delsarte, P. Goethals, J. M., and MacWilliams, F. J. (1970). On generalized Reed-Muller codes and their relatives. *Information and Control* **16**, 403–442.
Dembowski, P. (1968). "Finite Geometries." Springer–Verlag, New York and Berlin.
Deza, F. (1973). Une propriete extremale des plans projectifs finis dans une classe de codes equidistants. *Discrete Math.* **6**, 343–352.
Dickson, L. E. (1958). "Linear Groups with an Exposition of the Galois Field Theory." Dover, New York.
Djoković, D. Ž., and Blake, I. F. (1972). An optimization problem for unitary and orthogonal representations of finite groups. *Trans. Amer. Math. Soc.* **164**, 267–274.
Farber, S. (1968). Ph. D. Thesis, Calif. Inst. Technol, Pasadena, California.
Finney, D. J. (1960). "An Introduction to the Theory of Experimental Design." Univ. of Chicago Press, Chicago, Illinois.

Forney, G. D., Jr. (1966). "Concatenated Codes." MIT Press, Cambridge, Massachusetts.
Garbe, D., and Mennicke, J. L. (1964). Some remarks on the Mathieu groups. *Canad. Math. Bull.* **7**, 201–212.
Gilbert, E. N., (1952). A comparison of signaling alphabets. *Bell System Tech. J.* **31**, 504–522.
Gleason, A. M. (1971). Weight polynomials of self dual codes and the MacWilliams identities. *Act. Congr. Int. Math.* **3**, 211–215.
Goethals, J. M. (1969). A polynomial approach to linear codes. *Philips Research Repts.* **24**, 145–159.
Goethals, J. M. (1971). On the Golay perfect binary code. *J. Combinatorial Theory Ser. A* **11**, 178–186.
Goethals, J. M., and Delsarte, P. (1968). On a class of majority logic decodable cyclic codes. *IEEE Trans. Information Theory* **IT-14**, 182–188.
Goethals, J. M., and Snover, S. L. (1972). Nearly perfect binary codes. *Discrete Math.* **3**, 65–88.
Golay, M. J. E. (1949). Notes on digital coding. *Proc. IRE* **37**, 657.
Golay, M. J. E. (1958). Notes on the penny-weighing problem, lossless symbol coding with nonprimes, etc. *IRE Trans. Information Theory* **IT-4**, 103–109.
Goldman, J., and Rota, G. C. (1969). The number of subspaces of a vector space. In "Recent Progress in Combinatorics" (W. Tutte, ed.). Academic Press, New York.
Golomb, S. W., and Posner, E. C. (1964). Rook domains, Latin squares, affine planes and error-distributing codes. *IEEE Trans. Information Theory* **IT-10**, 196–208.
Goppa, V. D. (1971). Rational presentation of codes and (L, g) codes. *Probl. Peredachi Informatsii* **7**, 41–49.
Gore, W. C., and Cooper, A. B. (1970). Comments on polynomial codes. *IEEE Trans. Information Theory* **IT-16**, 635–638.
Gorenstein, D., and Zierler, N. (1961). A class of error correcting codes in p^m symbols. *SIAM J. Appl. Math* **9**, 207–214.
Graham, R. L., and MacWilliams, J. (1966). On the number of information symbols in difference-set cyclic codes. *Bell System Tech. J.* **45**, 1057–1070.
Griesmer, J. H. (1960). A bound for error-correcting codes. *IBM J. Res. Develop.* **4**, 532–542.
Hall, M. (1959). "The Theory of Groups." Macmillan, New York.
Hall, M. (1962). Note on the Mathieu group M_{12}. *Arch. Math. (Basel)* **13**, 334–340.
Hall, M. (1964). Block designs. In "Applied Combinatorial Mathematics" (E. F. Beckenbach, ed.) Chapter 13. Wiley, New York.
Hall, M. (1967a). "Combinatorial Theory." Ginn (Blaisdell), Boston, Massachusetts.
Hall, M. (1967b). Group theory and block designs. *Proc. Internat. Conf. Theory of Groups, Canberra, Australia, 1965*. Gordon & Breach, New York.
Hamada, N. (1968). The rank of the incidence matrix of points and d-flats in finite geometries *J. Sci. Hiroshima Univ. Ser.* A-1 **32**, 381–396.
Hamming, R. W. (1950). Error detecting and error correcting codes. *Bell System Tech. J.* **29**, 147–160.
Herstein, I. N. (1964). "Topics in Algebra." Ginn (Blaisdell), Boston, Massachusetts.
Herzog, M., and Schönheim, J. (1971). Linear and nonlinear single-error-correcting perfect mixed codes. *Information and Control* **18**, 364–368.
Higman, D. G., and Sims, C. C. (1968). A simple group of order 44, 353, 000. *Math. Z.* **105**, 110–113.
Hocquenghem, A. (1959). Codes correcteurs d'erreurs. *Chiffres* **2**, 147–156.

Hoffman, K., and Kunze, R. (1961). "Linear Algebra." Prentice-Hall, Englewood Cliffs, New Jersey.
Hsiao, M. Y., Bossen, D. C., and Chien, R. T. (1970). Orthogonal Latin square codes. *IBM J. Res. Develop.* **18**, 390–394.
Hughes, D. R. (1965). On t-designs and groups. *Amer. J. Math.* **87**, 761–778.
Ingemarsson, I. (1968). Signal sets generated by orthogonal transformations of the signal space. Tech. Rep. No. 21. Royal Inst. of Technol., Stockholm, Sweden.
Ireland, K., and Rosen, M. (1972). "Elements of Number Theory: Including an Introduction to Equations over Finite Fields." Bogden and Quigley, Tarrytown-on-Hudson, New York.
Jansen, L., and Boon, M. (1967). "Theory of Finite Groups." North-Holland Publ. Amsterdam.
Johnson, S. M. (1962). A new upper bound for error-correcting codes. *IRE Trans. Information Theory* **IT-8**, 203–207.
Kaplansky, I. (1969a). "Linear Algebra and Geometry." Allyn & Bacon, Rockleigh, New Jersey.
Kaplansky, I. (1969b). "Fields and Rings." Univ. of Chicago Press, Chicago, Illinois.
Karlin, M. (1969). New binary coding results by circulants. *IEEE Trans. Information Theory* **IT-15**, 81–92.
Kasami, T., and Lin, S. (1966). Some codes which are invariant under doubly-transitive permutation groups and their connection with balanced incomplete block designs AFCRL-66-142, Sci. Rep. No. 6. Air Force Cambridge Res. Lab., Cambridge, Massachusetts.
Kasami, T., and Lin, S. (1971). On majority-logic decoding for duals of primitive polynomial codes. *IEEE Trans. Information Theory* **IT-17**, 322–331.
Kasami, T., Lin, S., and Peterson, W. (1968a). New generalizations of the Reed-Muller codes. Pt. 1. Primitive codes. *IEEE Trans. Information Theory* **IT-14**, 189–199.
Kasami, T., Lin, S., and Peterson, W. (1968b). Polynomial codes. *IEEE Trans. Information Theory* **IT-14**, 708–814.
Kasami, T., Lin, S., and Peterson, W. (1968c). Some results on cyclic codes which are invariant under the affine group and their applications. *Information and Control* **11**, 475–496.
Landau, H. J., and Slepian, D. (1966). On the optimality of the regular simplex code. *Bell System Tech. J.* **45**, 1247–1272.
Lang, S. (1971). "Algebra." Addison-Wesley, Reading, Massachusetts.
Leech, J. (1967). Notes on sphere packings. *Canad. J. Math.* **19**, 251–267.
Leech, J., and Sloane, N. J. A. (1970). New sphere packings in dimensions 9-15. *Bull. Amer. Math. Soc.* **76**, 1006–1010.
Lenstra, H. W. Jr. (1972). Two theorems on perfect codes. *Discrete Math.* **3**, 125–132.
Lin, S. (1968). On a class of cyclic codes. *In* "Error Correcting Codes" (H. B. Mann, ed.). Wiley, New York.
Lin, S. (1969). Some codes which are invariant under a transitive permutation group and their connection with balanced incomplete block designs. *In* "Combinatorial Mathematics and Its Applications" (R. C. Bose and T. A. Dowling, eds.), Chapter 24. Univ. of North Carolina Press, Chapel Hill.
Lin, S. (1972a). Shortened finite geometry codes. *IEEE Trans. Information Theory* **IT-18**, 692–696.
Lin, S. (1972b). On the number of information symbols in polynomial codes. *IEEE Trans. Information Theory* **IT-18**, 785–794.

Lindström, B. (1969). On group and nongroup perfect codes in q symbols. *Math. Scand.* **25**, 149–158.
Lloyd, S. P. (1957). Binary block coding. *Bell System Tech. J.* **36**, 517–535.
Lomont, J. S. (1959). "Applications of Finite Groups." Academic Press, New York.
McCoy, N. H. (1948). "Rings and Ideals." Math. Assoc. of Amer., Buffalo, New York.
McCoy, N. H. (1964). The Theory of Rings." Macmillan, New York.
McEliece, R. J. (1969). Factorization of polynomials over finite fields. *Math. Comp.* **23**, 861–867.
MacWilliams, F. J. (1961). Error correcting codes for multiple level transmission. *Bell System Tech. J.* **40**, 281–308.
MacWilliams, F. J. (1962). A theorem on the distribution of weights in a systematic code. *Bell System Tech. J.*, **42**, 79–94.
MacWilliams, F. J. (1965). The structure and properties of binary cyclic alphabets. *Bell System Tech. J.* **44**, 303–332.
MacWilliams, F. J. (1969). Codes and ideals in group algebras. *In* "Combinatorial Mathematics and its Applications" (R. C. Bose and T. A. Dowling, eds.), Chapter 18. Univ. of North Carolina Press, Chapel Hill.
MacWilliams, F. J. (1970). Binary codes which are ideals in the group algebra of an Abelian group. *Bell System Tech. J.* **44**, 987–1011.
MacWilliams, F. J. (1971). Orthogonal circulant matrices over a finite field and how to find them. *J. Combinatorial Theory Ser. A* **10**, 1–17.
MacWilliams, F. J., and Mann, H. B. (1968). On the p-rank of the design matrix of a difference set, *Information and Control* **12**, 474–488.
MacWilliams, F. J., Mallows, C. J., and Sloane, N. J. A. (1972a). Generalizations of Gleason's theorem on weight enumerators of self-dual codes. *IEEE Trans. Information Theory* **IT-18**, 794–805.
MacWilliams, F. J., Sloane, N. J. A., and Thompson, J. G. (1972b). Good self dual codes exist. *Discrete Math.* **3**, 153–162.
MacWilliams, F. J., Sloane, N. J. A., and Goethals, J. M. (1972c). The MacWilliams identities for nonlinear codes. *Bell System Tech. J.* **51**, 803–819.
MacWilliams, F. J., Sloane, N. J. A. and Thompson, J. G. (1973), On the existence of a projective plane of order 10. *J. Combinatorial Theory Ser. A* **14**, 66–78 (1973).
Mallows, C. L., and Sloane, N. J. A. (1973). An upper bound for self-dual codes. *Information and Control* **22**, 188–200.
Mann, H. B. (1949). "Analysis and Design of Experiments." Dover, New York.
Massey, J. L. (1963). "Threshold Decoding." MIT Press, Cambridge, Massachusetts.
Mattson, H. F., and Solomon, G. (1961). A new treatment of Bose-Chaudhuri codes. *SIAM J. Appl. Math.* **9**, 654–669.
Mills, W. H., and Zierler, N. (1969). On a conjecture of Golomb. *Pacific J. Math* **28**, 635–640.
Muller, D. E. (1954). Application of Boolean algebra to switching circuit design and to error detection. *IRE Trans. Elect. Comp.* **EC-3**, 6–12.
Murnaghan, F. D. (1963). "The Theory of Group Representations." Dover, New York.
Ottoson, R. (1971). Group codes for phase- and amplitude-modulated signals on a Gaussian channel. *IEEE Trans. Information Theory* **IT-17**, 315–321.
Paige, L. J. (1956). A note on the Mathieu groups. *Canad. J. Math* **9**, 15–18.
Passman, D. S. (1968). "Permutation Groups." Benjamin, New York.
Peterson, W. W. (1961). "Error Correcting Codes." MIT Press, Cambridge Massachusetts.
Peterson, W. W., and Weldon, E. J., Jr. (1972). "Error Correcting Codes," 2nd ed. MIT Press, Cambridge, Massachusetts.

Plackett, S. R. L., and Burman, J. P. (1946). The design of optimum multifactorial experiments. *Biometrika* **33**, 305–325.
Pless, V. (1963). Power moment identities on weight distributions in error correcting codes. *Information and Control* **6**, 147–152.
Pless, V. (1965). The number or isotropic subspaces in a finite geometry. *Acad. Naz. Lincei* [8], **34**, 418–421.
Pless, V. (1968). On the uniqueness of the Golay codes. *J. Combinatorial Theory* **5**, 215–228.
Pless, V. (1969). On a new family of symmetry codes and related new five designs. *Bull. Amer. Math. Soc.* **75**, 1339–1342.
Pless, V. (1970). The weight of the symmetry code for $p = 29$ and the 5-designs contained therein. *Ann. N.Y. Acad. Sci.* **175**, 310–313.
Pless, V. (1972a). Symmetry codes over GF(3) and new five-designs. *J. Combinatorial Theory Ser A* **12**, 119–142.
Pless, V. (1972b). A classification of self-orthogonal codes over GF(2). *Discrete Math.* **3**, 209–246.
Pless, V., and Pierce, J. N. (1973). Self-dual codes over over GF(q) satisfy a modified Varshamov-Gilbert bound. *Information and Control* **23**, 35–40.
Plotkin, M. (1960). Binary codes with specified minimum distance. *IRE Trans. Information Theory* **IT-6**, 445–450.
Reed, I. S. (1954). A class of multiple-error-correcting codes and the decoding scheme. *IRE Trans. Information Theory* **IT-4**, 38–49.
Reed, I. S., and Solomon, G. (1960). Polynomial codes over certain finite fields. *SIAM J. Appl. Math.* **8**, 300–304.
Riordan, J. (1958). "An Introduction to Combinatorial Analysis." Wiley, New York.
Riordan, J. (1968). "Combinatorial Identities." Wiley, New York.
Rudolph, L. (1964). Geometric configuration and majority logic decodable codes. M.E.E. Thesis, Univ. of Oklahoma, Norman.
Rudolph, L. (1967). A class of majority logic decodable codes. *IEEE Trans. Information Theory* **13**, 305–307.
Ryser, H. J. (1963). "Combinatorial Mathematics." Math. Assoc. of Amer., distributed by Wiley, New York.
Schönheim, J. (1968). On linear and nonlinear single-error-correcting q-nary perfect codes. *Information and Control* **12**, 23–26.
Schönheim, J. (1969). Semilinear codes and some combinatorial applications of them. *Information and Control* **15**, 61–66.
Semakov, N. V., and Zinov'ev, V. A. (1968). Equidistant q-ary codes with maximal distance and resolvable balanced incomplete block designs. *Problems of Information Transmission* **4**, 1–7.
Semakov, N. V., and Zinov'ev, V. A. (1969a). Complete and quasi-complete balanced codes. *Problems of Information Transmission* **5**, 11–13.
Semakov, N. V., and Zinov'ev, V. A. (1969b). Balanced codes and tactical configurations. *Problems of Information Transmission* **5**, 22–28.
Semakov, N. V., Zinov'ev, V. A., and Zaitsev, G. V. (1969). A class of maximum equidistant codes. *Problems of Information Transmission* **5**, 65–68.
Semakov, N. V., Zinov'ev, V. A., and Zaitsev, G. V. (1971) Uniformly packed codes. *Problems of Information Transmission* **7**, 30–39.
Serre, J. P. (1971) "Representation Lineares des Groupes Finis," 2nd ed. Hermann, Paris.
Shannon, C. E. (1959). Probability of error for optimal codes in a Gaussian channel. *Bell System Tech. J.* **38**, 611–656.

Shrikhande, S. S. (1953). The nonexistence of certain affine resolvable balanced incomplete block designs. *Canad. J. Math.* **5**, 413–420.
Shrikhande, S. S. (1964). Generalized Hadamard matrices and orthogonal arrays of strength two *Canad. J. Math.* **16**, 736–740.
Singer, J. (1938). A theorem in finite projective geometry and some applications to number theory. *Trans. Amer. Math. Soc.* **43**, 377–385.
Singleton, R. C. (1964). Maximum distance Q-nary codes. *IEEE Trans. Information Theory* **IT-10**, 116–118.
Slepian, D. (1951). Large signaling alphabets generated by groups. Technical Memorandum, Bell Telephone Laboratories, Murray Hill, New Jersey.
Slepian, D. (1956). A class of binary signalling alphabets. *Bell System Tech. J.* **35**, 203–234.
Slepian, D. (1960). Some further theory of group codes .*Bell System Tech. J.* **39**, 1219–1252.
Slepian, D. (1963). Bounds on communication. *Bell System Tech. J.* **42**, 681–707.
Slepian, D. (1965). Permutation modulation. *Proc. IEEE* **53**, 228–236.
Slepian, D. (1968). Group codes for the Gaussian channel. *Bell System Tech. J.* **47**, 575–602.
Slepian, D. (1971). On neighbour distances and symmetry in group codes. *IEEE Trans. Information Theory* **IT-17**, 630–632.
Sloane, N. J. A. (1972). A survey of constructive coding theory and a table of binary codes of highest known rate. *Discrete Math.* **3**, 265–294.
Sloane, N. J. A., and Whitehead, D. S. (1970). New family of single error correcting codes, *IEEE Trans. Information Theory* **IT-16**, 717–719.
Smirnov, V. I. (1961). " Linear Algebra and Group Theory," Dover translation. New York.
Smith, K. J. C. (1969). On the p-rank of the incidence matrix of points and hyperplanes in a finite projective geometry. *J. Combinatorial Theory* **7**, 122–129.
Stanton, R. G. (1951). The Mathieu groups. *Canad. J. Math.* **3**, 164–174.
Stanton, R. G. (1963). Some types of statistical designs. Lectures. *Canad. Math. Congr. Saskatoon*, 1963. Canad. Math. Congr.
Stiffler, J J. (1971). " Theory of Synchronous Communications." Prentice-Hall, Englewood Cliffs, New Jersey.
Swan, R. G. (1962). Factorization of polynomials over finite fields. *Pacific J. Math.* **12**, 1099–1106.
Tarry, G. (1900). Le Problème de 36 Officieurs. *C. R. Acad. Sci. Paris* **1**, 122–123.
Tarry, G. (1901). Le Problème de 36 Officieurs. *C. R. Acad. Sci. Paris* **2**, 170–203.
Tavares, S., Bhargava, V. K., and Shiva, S. G. S. (1974). Some rate-$P/(P+1)$ quasi-cyclic codes. *IEEE Trans. Information Theory* **IT-20**, 133–135.
Tietäväinen, A. (1973). On the nonexistence of perfect codes over finite fields. *SIAM J. Appl. Math.* **24**, 88–96.
Tietäväinen, A., and Perko, A. (1971). There are no unknown perfect binary codes. *Ann. Univ. Turku., Ser. A* **148**, 3–10.
Todd, J. A. (1959). On representations of the Mathieu groups as collineation groups. *J. London Math. Soc.* **34**, 406–416.
Todd, J. A. (1966). A representation of the Mathieu group M_{24} as a collineation group. *Ann. Mat. Pura Appl.* (*IV*) **71**, 199–238.
Townsend, R. L., and Weldon, E. J., Jr. (1967). Self-orthogonal quasi-cyclic codes. *IEEE Trans. Information Theory* **IT-13**, 183–195.
Tutte, W. T. (1965). Lectures on matroids. *J. Res. Nat. Bur. Standards Sec B* **69**, 1–47.
Tutte, W. T. (1966). " Connectivity in Graphs." Univ. of Toronto Press, Toronto.
Tutte, W. T. (1971). " Introduction to the Theory of Matroids." Amer. Elsevier, New York.
van der Waerden, B. L. (1953). " Modern Algebra," Vol. 1. Ungar New York.
van der Waerden, B. L. (1970). "Algebra," Vol. 2. Ungar, New York.

Van Lint, J. H. (1971), "Coding Theory," Lecture Notes in Math., Vol. 201. Springer-Verlag, Berlin and New York.
Van Lint, J. H. (1973). A theorem on equidistant codes. *Discrete Math.* 6, 353–358.
Vanstone, S. A. (1974). The structure of regular pairwise balanced designs. Ph.D. Thesis, Univ. of Waterloo, Ontario, Canada.
Varshamov, R. R. (1957). Estimate of the number of signals in error correcting codes. *Dokl. Akad. Nauk. SSR* **117**, 739–741.
Vasil'ev, Y. L. (1962). On non-group close-packed codes. *Prob. Cybernet.* **8**, 337–339.
Veblen, O., and Bussey, W. H. (1906). Finite projective geometries. *Trans. Amer. Math. Soc.* **7**, 241–259.
Wallis, W. D., Street, A. P., and Wallis, J. S. (1972). "Combinatorics: Room Squares, Sum-Free Sets, Hadamard Matrices," Lecture Notes in Mathematics Vol. 292. Springer-Verlag, Berlin and New York.
Weber, C. L. (1968). "Elements of Detection and Signal Design." McGraw-Hill, New York.
Weldon, E. J., Jr. (1966). Difference-set cyclic codes. *Bell System Tech. J.* **45**, 1045–1055.
Weldon, E. J., Jr.(1968). New generalizations of the Reed–Muller codes. Pt. II. Nonprimitive codes. *IEEE Trans. Information Theory* **IT-14**, 199–205.
Weldon, E. J., Jr. (1969). Euclidean geometry cyclic codes. *In* "Combinatorial Mathematics and its Applications" (R. C. Bose and T. A. Dowing, Univ. eds.), Chapter 23. Univ. of North Carolina Press, Chapel Hill.
Whitelaw, T. A. (1966). On the Mathieu group of degree twelve. *Proc. Cambridge Philos. Soc.* **62**, 351–364.
Witt, E. (1938a). Die 5-fach Transitiven Gruppen von Mathieu. *Abh. Math. Sem. Univ. Hamburg* **12**, 256–264.
Witt. E. (1938b). Über Steinersche Systeme. *Abh. Math. Sem. Univ. Hamburg* **12**, 265–275.
Wozencraft, J. M., and Jacobs, I. M. (1965). "Principles of Communication Engineering." Wiley, New York.
Wyner, A. D. (1965). Capabilities of bounded discrepancy decoding. *Bell System Tech. J.* **44**, 1061–1122.
Zaremba, S. K. (1950). A covering theorem for Abelian groups. *J. London Math. Soc.* **26**, 71–72.
Zaremba, S. K. (1952). Covering problems concerning Abelian groups. *J. London Math. Soc.* **27**, 242–246.
Zetterberg, L. H. (1965). A class of codes for polyphase signals on a bandlimited Gaussian channel. *IEEE Trans. Information Theory* **IT-11**, 383–395.
Zierler, N. (1970). On $x^n + x + 1$ over $GF(2)$. *Information and Control* **16**, 502–505.
Zierler, N., and Brillhart, J. (1968). On primitive trinomials (Mod 2), *Information and Control* **13**, 541–554.

Index

A

Abelian codes, 227–244
Affine group, 62, 75, 93
Affine resolvable balanced incomplete block design, 125–132, 202
  and equidistant codes, 130–132
  parameters of, 202
Affine transformations, 60
Alanen, J. D., 89
Albert, A. A., 37, 40, 89, 90
Algebra, 207
Algebraic closure, 4
Ambivalent conjugacy class, 279
Anderson, D. R., 57–58
Artin, E., 89, 208, 244
Ascending chain condition, 206
Assmus, E. F., Jr., 44, 56, 144, 155, 160, 165, 180, 181, 183, 184, 185, 242, 332

B

Balanced codes, 148
  and equidistant codes, 192–193
  and t-designs, 189–192
Balanced incomplete block designs, 119–132

affine resolvable, 125–132, 202
affine α-resolvable, 202
and codes, 129–132
complement of, 122
derived, 123
and finite geometries, 123
incidence matrix, 121
residual, 123
resolvable, 125–132
α-resolvable, 202
symmetric, 121–122
Balanced vectors, 318
  and full homogeneous components, 319–325
BCH codes, 50–52
  design distance of, 51
  and Goppa codes, 52–53
  invariance under affine group, 65
  parity check matrix, 51
  and polynomial codes, 82
Berlekamp, E. R., 37, 38, 52, 89, 90, 109, 142, 143, 159, 169, 330, 332
Berman, S. D., 227, 231, 240, 241, 242, 243
Biglieri, E., 301, 302
Bilinear form, 10–12
Binary repetition codes, 55, 161
Binomial coefficient identities, 59
Binomial coefficients, Lucas's theorem on, 64

Blake, I. F., 309, 318, 326
Boerner, H., 290
Boon, M., 290
Bose, R. C., 50, 117, 125, 127, 128, 129, 136, 167, 193
Bounds on code dictionaries, 83–89
Brillhart, J., 38, 91
Bruck–Ryser theorem, 99
Burgoyne, N., 332
Burrow, M., 244, 264, 290
Burton, R. C., 117
Bush, K. A., 127, 128, 129, 167
Bussey, W. H., 96, 98

C

Camion, P., 227, 239
Carmichael, R. D., 89, 106, 166
Chain, 175–178
    domain of, 175
    primitive, 201
Chain group, 175
    contraction, 178
    dendroid, 176
    dendroid basis of, 176
    dual, 178
    reduction, 177
    regular, 201
Character table, see Group character table
Chen, C. L., 82, 83, 117, 156, 157
Circulant matrix, 147, 149
    algebra of, 157–158
Class function, 219
Codes for the Gaussian channel, 292–327
Codeword polynomial, 41
Collineation group, 105–106
    and finite geometries, 104–106
Commutator of a group, 264
Complement of balanced incomplete block design, 122
Configuration matrix, 304–314
    decomposition of, 309–313
Conjugate roots, 4
Conway, C. H., 89, 161
Coset leaders, 20
Coxeter, H. S. M., 332
Curtis, C. W., 208, 244, 272, 290
Cycle space of a graph, 179–180
Cyclic codes, 40–58
    equivalence of, 45

    and ideals, 41
    structure, 222–227
Cyclotomic factorization of x^{n-1}, 38

D

Delsarte, P., 27, 65, 66, 69, 73, 76, 89, 92, 111, 112, 114, 117, 244
Dembowski, P., 117
Derived balanced incomplete block design, 123
Desargues's theorem, 98–99
Descending chain condition, 206
Deza, F., 132, 194
Dickson, L. E., 37, 40, 89, 90
Difference sets, 149–152
    and circulant incidence matrices, 149
    Hadamard, 150
    multiplier, 152
    planar, 152
    and quadratic residues, 150–151
Difference set codes, 116, 152
Direct product of codes, see Kronecker product of codes
Direct product of groups, 273
    irreducible representations of, 274–275
Direct sum of group representations, 251
Direct sum of vector spaces, 11
Discriminant, 91
Distance of block code, 17
Djoković, D. Ž., 318
Dual basis, 13–15
Dual,
    of linear code, 18
    of a vector space, 6

E

Elia, M., 301, 302
Equidistant codes, 130–132, 189, 192–200
    and balanced codes, 192–193
    and affine resolvable balanced incomplete block designs, 130–132
    and projective planes, 194–200
Equivalence of codes, 19
Equivalence of cyclic codes, 45
Euclidean algorithm, 3
Euclidean geometry, 101–104
    flat, 99
Euclidean geometry codes, 112–116
    and majority logic decoding, 113–116

Index 351

and polynomial codes, 78–79, 112–116
and projective geometry codes, 117–118
Euler totient function, 9
Expurgated code, 21
Extended code, 21

F

Factor space, 20
Fano geometry, 97, 106–107, 119, 137, 146, 151
Farber, S., 297
Fermat's theorem, 139
Field, 2–6
  algebraically closed, 4
  perfect, 4
Field automorphism, 5
Field characteristic, 2
Field extension, 2
  algebraic, 4
  degree of, 2
  Galois group of, 5, 10
  normal, 4
  separable, 4
  simple, 4
  transcendental, 4
Field isomorphism, 5
Finite fields, 1–94
  computations in, 15
  fundamental properties of, 7–10
  inclusion relations, 9
  polynomials over, 29–40
  primitive element of, 9
  vector spaces over, 10–15
Finite geometries, 95–118
  and balanced incomplete block designs, 123
  collineation groups, 104–106
Fisher's inequality, 121–122, 125
Fong, P., 332
Frobenius reciprocity theorem, 271–272
Full homogeneous component, 287, 317
  and balanced vectors, 319–325
Fundamental region, 326

G

Galois field, *see* Finite field
Galois group of field extension, 5, 10
Garbe, D., 332
Gaussian coefficients, 59

Gauss's lemma, 141
General linear group, 60, 65, 104, 247
  and generalized Reed–Muller codes, 73
General linear nonhomogeneous group, 60, 62, 66
  and generalized Reed–Muller codes, 72–73
Generalized Reed–Muller codes, 68–78
  and the affine group, 75
  and BCH codes, 75–76
  and the general linear group, 73
  and the general linear nonhomogeneous group, 72–73
  generator matrix, 77
  generator polynomial of cyclic, 74
  nonprimitive, 77–78
  and polynomial codes, 82–83
  and Reed–Solomon codes, 76
Generator matrix, 18
  of cyclic code, 42
Gilbert, E. N., 83
Gilbert bound, *see* Varshamov–Gilbert bound
Gleason, A. M., 159
Goethals, J. M., 54, 112, 117, 186
Golay, M. J. E., 138, 161, 162
Golay code,
  characterizations, 165
  and quadratic residue code, 143
  weight enumerator of extended, 160
Goldman, J., 89
Golomb, S. W., 138
Goppa, V. D., 22
Goppa codes, 23
  and BCH codes, 52–53
Gorenstein, D., 50
Graham, R. L., 116, 152
Gram matrix, 304
Gram–Schmidt orthogonalization, 293
Greatest common divisor, 3
Griesmer, J. H., 94
GRM codes, *see* Generalized Reed–Muller codes
Groups,
  affine, 62, 75, 93
  collineation, 105–106
  direct product, 273–275
  general linear, 60, 65, 104, 247
  general linear nonhomogeneous, 60, 62, 66
  linear fractional, 61, 62
  of linear transformations, 60–63
  permutation, 62, 331
  projective, 104–105

projective general linear, 61
projective special linear, 62
of quaternions, 265–266
semidirect product of, 273–274
special linear, 61
structure constants, 289, 291
Group algebra, 217–222
   Abelian, 231–238
   center, 218
   and modules, 284–290
Group character table, 221, 262–264
   of A_4, 262–263
   of Mathieu groups, 333–338
   of quaternions, 265–266
Group characters, 221–222, 253–259
   Abelian, 222, 228–230
   induced, 269–272
   inner product of, 258
   orthogonality relationships, 221
   properties of, 259, 269
   subduced, 269–272
Group codes for the Gaussian channel, 292–327
   distance properties of, 314–317
   planar, 302
Group representations, 219–222, 246–291
   of Abelian groups, 228–230
   completely reducible, 221
   criterion for irreducibility, 258
   of direct product of groups, 274–275
   direct sum of, 251
   equivalent, 220, 247
   full homogeneous components, 287, 317
   and group algebras, 284–290
   homogeneous components, 287, 307, 317
   induced, 269–272
   irreducible, 220, 252
   kind, 276–278
   Kronecker product of, 252–253
   and modules, 284–290
   monomial, 272
   number of nonequivalent irreducible, 260
   number of one dimensional, 264–265
   orthogonality relationships, 257
   reducible, 220
   regular, 248–249
   subduced, 269–272

H

Hadamard difference set, 150
Hadamard matrices, 86, 144–148
   and symmetric balanced incomplete block designs, 145–147
Hall, M., 123, 127, 135, 136, 137, 147, 165, 167, 199, 266, 332
Hamada, N., 112
Hamming, R. W., 46, 85
Hamming codes, 46–50, 161
   and maximum length codes, 49–50, 92
   weight enumerator, 47–50
Hamming distance, 16
Herstein, I. N., 3, 89, 281
Herzog, M., 164
Higman, D. G., 161
Hocquenghem, A., 50
Hoffman, K., 314
Homogeneous component, 287, 307, 317
Hsiao, M. Y., 168

I

Ideals, 205–216
   ascending chain condition for, 206
   and cyclic codes, 41
   descending chain condition for, 206
   generator of, 205
   intersection of, 206
   maximal, 206
   minimal, 45, 206
   nil, 206
   nilpotent, 206, 207–210
   principal, 205
   sum of, 205
Idempotents, 206
   of Abelian group algebras, 235
   central, 213
   generating, 210–212
   orthogonal, 212
   primitive, 212
Induced characters, 269–272
Induced representations, 269–272
Ingemarsson, I., 318
Initial vector problem, 313, 318–325
Ireland, K., 91
Irreducible representations, 252
   criterion for determining kind, 277–278
   kind, 276

J

Jacobs, I. M., 295
Jansen, L., 290

Index

Johnson, S. M., 86
Johnson bound, 87
   specialized, 187

K

Kaplansky, I., 89, 91
Karlin, M., 156, 157
Kasami, T., 63, 64, 68, 75, 76, 78, 79, 80, 93, 94
Knuth, D. E., 89
Krawtchouk polynomial, 27, 92
Krawtchouk transform, 93
Kronecker product,
   of codes, 22
   of group representations, 252, 253
Kunze, R., 314

L

Lagrange interpolation formula, 36
Landau, H. J., 297
Lang, S., 244, 319
Latin squares, 132–139
   and maximum distance separable codes, 138
   pairwise orthogonal, 133–136, 138
   and projective planes, 136
Lee metric, 16
Leech, J., 292
Legendre symbol, 139
Lenstra, H. W., Jr., 164
Lin S., 82, 83, 113, 117, 167
Lindström, B., 162
Linear block codes, 15–29
Linear fractional group, 61, 62
Linear functional, 6
Linear transformation groups, 60–63
Linear transformations of vector spaces, 58–63
Lloyd, S. P., 164
Lomont, J. S., 280, 290
Lucas's theorem, 64, 330

M

McCoy, N. H., 244
McEliece, R. J., 37, 244
MacWilliams, F. J., 19, 24, 27, 28, 116, 117, 152, 153, 156, 158, 160, 161, 236, 238, 239, 244, 245

MacWilliams identities, 24–27
Majority logic decoding, 107–117
   and Euclidean geometry codes, 113–116
   and projective geometry codes, 111–112
Mallows, C. L., 160
Mann, H. B., 117, 152
Maschke's theorem, 250–251
Massey, J. L., 108
Mathieu groups, 331–338
   character tables, 333–338
Matrix ring, 216
Matroid, 173–175
   cographic, 173
   dendroid of a, 174
   girth, 173
   graphic, 173
   rank, 174
   reduction, 178
   separator of, 201
   spanning subset of, 175
Matroid belts, 180
Matroid circuit, 173
   weight of, 173
Mattson, H. F., Jr., 44, 56, 66, 93, 142, 144, 155, 160, 165, 180, 181, 183–185, 242, 332
Mattson–Solomon polynomial, 67, 76, 92
Maximum distance separable codes, 54–57
   and latin squares, 138
   and matroids, 180–181
   and Reed–Solomon codes, 56
   weight enumerator, 54–55
Maximum length code, 48–49
   distance of, 49
   and Hamming codes, 49–50, 92
   weight enumerator of, 48–50
Maximum likelihood decoding, 17, 20, 295, 327
Mennicke, J. L., 332
Mills, W. H., 91
Modules, 285–290
   and group algebras, 284–290
Möbius function, 32
Möbius inversion formula, 32, 38, 329
Monomial representations, 272
Muller, D. E., 107

N

Natural representation of a permutation group, 247–248, 266–269

Nearly perfect codes, 186–189
Newton identities, 37
Nilpotent element, 206
Noetherian rings, 207
Norm of field element, 6
Normal basis, 13, 15

O

One-step orthogonalizable code, 108
Optimal code, see Maximum distance separable code
Orthogonal arrays, 127–132, 167
  and projective planes, 128, 137
  and transversal systems, 199
Orthogonal code for the Gaussian channel, 296
Orthogonal latin squares, see Pairwise orthogonal latin squares
Orthogonal parity check sum, 108
Orthogonality relationships of group characters, 221
Orthogonality relationships of group represenations, 257
Orthogonal vectors, 11

P

Paige, L. J., 170, 332
Pairwise orthgonal latin squares, 133–136, 138
  and maximum distance separable codes, 138
  and parity check matrices, 168–169
Pappus configuration, 98
Pappus property, 98, 99
Parity check code, 19
Parity check matrix, 19
  of cyclic code, 43
Parity check sum, 108
  orthogonal, 108
Pascal triangle, 162
Passman, D. S., 331
Perfect codes, 21, 161–165
  binary repetition codes, 55, 161
  Hamming codes, 47, 161
  and t designs, 183–186
  Vasil'ev codes, 162–163, 185
Permutation codes, 326–327
Permutation groups, 62, 331
  code invariance under, 63–66
  k-transitive, 331

natural representation of, 247, 248, 266–269
  sharply k-transitive, 331
  transitive constituent of, 266
  transitivity of, 62
Peterson, W. W., 21, 52, 89
Pierce, J. N., 160
Planar difference set codes, 116
Planar group code, 302
Pless, V., 24, 27, 153, 155, 160, 161, 165, 168, 183
Pless identities, 27–29
Plotkin, M., 84, 86, 147
Plotkin bound, 49, 84
  and Singleton bound, 53
Polygon matroid, 173
Polynomial,
  codeword, 41
  cyclotomic, 38–40
  example of irreducible, 40
  exponent, 32, 39
  over finite field, 29–40
  Galois group of, 6
  irreducible, 2
  Krawtchouk, 27, 92
  minimal, 4, 30
  monic, 2
  number of monic irreducible, 33
  primitive, 32
  reciprocal, 42
  roots of irreducible, 3
Polynomial approach to coding, 66–83
Polynomial codes, 78–83
  and BCH codes, 82
  bound on distance of, 81
  and Euclidean geometry codes, 78–79, 112–116
  and generalized Reed–Muller codes, 82–83
  generator polynomial of, 79–80
  and projective geometry codes, 78–79, 109–111
  and Reed–Solomon codes, 82
Posner, E. C., 138
Prime subfield, 2
Primitive element, 9
Primitive root codes, 57–58
Principal vectors, 321–325
Projective general linear group, 61
Projective geometry, 96–101
  construction of, 97, 101
  Desarguesian, 99
  example of, 106–107

Index

flats, 99
  non-Desarguesian, 99
Projective geometry codes, 109–112
  and Euclidean geometry codes, 117–118
  and majority logic decoding, 111–112
  and polynomial codes, 78–79, 109–111
Projective group, 104–105
Projective plane, 96, 116
  and equidistant codes, 194–200
  and latin squares, 136
  and orthogonal arrays, 128, 137
Projective special linear group, 62

Q

Quadratic residue codes, 142–144
  and Golay codes, 143
Quadratic equations over a finite field, 90–91
Quadratic reciprocity, law of, 140
Quadratic residues, 139–144
  and difference sets, 150–151
Quasi-cyclic codes, 156–158
Quasi-perfect codes, 21
q-weight of an integer, 70

R

Radical, 209
Ray-Chaudhuri, D. K., 50
Real representations, 276–284
Reed, I. S., 52, 107
Reed–Muller codes, 107, 116, 166–167
Reed–Solomon codes, 52–53, 76
  and maximum distance separable codes, 56
  and polynomial codes, 82
Regular representation, 248–249
Reiner, I., 208, 244, 272, 290
Representation space, 247
  invariance of, 249
Residual balanced incomplete block design, 123
Resolvable balanced incomplete block design, 125–132
  and equidistant codes, 130–132
Rings, 205–216
  Artinian, 207
  of matrices, 216
  with minimum condition, 208
  Noetherian, 207

  semisimple, 210–215
  simple, 210, 213–216
Riordan, J., 72
Rosen, K., 91
Rota, G. C., 89
Rudolph, L., 107, 112
Ryser, H. J., 122, 123, 125, 136, 165

S

Scalar complete code, 181
Schöneim, J., 162, 164
Schur's lemma, 255–256
Self-dual code, 152–161
  weight enumeration, 158–161
Semakov, N. V., 130, 132, 148, 189, 190, 191
Semidirect product of groups, 273–274
Semisimple rings, 210–215
Serre, J. P., 289, 290
Shannon, C. E., 297
Shrikhande, S. S., 193, 202
Simple component, 213–215
Simple rings, 210, 213–216
Simplex code, 297
Sims, C. C., 161
Singer, J., 152
Singleton, R. C., 56, 138
Singleton bound, 53,
  and Plotkin bound, 53
Slepian, D., 20, 297, 299, 301, 302, 304, 314, 326
Sloane, N. J. A., 94, 160, 292
Smith, K. J. C., 117, 152
Snover, S. L., 186
Solomon, G., 52, 66, 93, 142
Solvability by radicals, 5
Special linear group, 61
Sphere, 16
Sphere packing bound, 85, 184
Splitting field, 2
Standard array, 20
Stanton, R. G., 332
Steiner system, 136–137, 148, 172
  kth order, 202
  Mathieu, 165, 172, 185
Steiner triple systems, 136–137
Stiffler, J. J., 51
Stirling number of the second kind, 27
Strongly balanced vectors, 318
Structure constants of group, 289, 291

Subalgebra, 207
Subduced characters, 269–272
Subduced representations, 269–272
Subfield, 2
Subfield subcode, 83, 118
Subgeometry, 106–107
Submodule, 286
  irreducible, 286
Subring, 205
Swan, R. G., 37, 91
Symmetric balanced incomplete block design, 121–122
  and Hadamard matrices, 145–147
  incidence matrix, 123
  and projective geometries, 124
Symmetric function, 36
  elementary, 36
Symmetry codes, 153–156
  and t designs, 155–156
Syndrome, 21
Systematic code, 18

T

Tarry, G., 136, 139
Tavares, S., 157, 158
t Designs, 170–173
  and balanced codes, 189–192
  and nearly perfect codes, 186–189
  and perfect codes, 183–186
  regular, 201
  and symmetry codes, 155–156
Threshold decoding, 108
Tietäväinen, A., 18, 164, 184
Todd, J. A., 332
Townsend, R. L., 156
Trace of an element, 6
Trace function, 13
Transitive permutation group, 62, 331
Transversal system, 127–128
  and orthogonal array, 127, 199
Trinomial, 37, 91
Tutte, W. T., 173, 179

U

Uniformly packed codes, *see* Nearly perfect codes
Unimodular group, *see* Special linear group

V

van der Monde matrix, 34, 51
van der Waerden, B. L., 89, 290
van Lint, J. H., 27, 93, 132, 164, 198
Vanstone, S. A., 200
Varshamov, R. R., 84
Varshamov–Gilbert bound, 84
Vasil'ev, Y. L., 162
Vasil'ev codes, 162–163, 185
Veblen, O., 96, 98
Vector spaces, 10–15
  direct sum of, 11
  dual of, 6
  dual basis of, 13, 14
  linear transformations of, 58–63

W

Wallis, W. D., 147
Weber, C. L., 297
Wedderburn's theorem, 321
Weight enumerator, 24–29
  of extended Golay codes, 160
  of Hamming codes, 47–50
  of maximum distance separable codes, 54–55
  of maximum length codes, 48–50
  of self-dual codes, 158–161
Weldon, E. J., Jr., 52, 107, 116, 117, 152, 156
Whitehead, D. S., 94
Whitelaw, T. A., 332
Wilson's theorem, 69, 140
Witt, E., 165, 172, 332
Wozencraft, J. M., 295
Wyner, A. D., 326

Z

Zaremba, S. K., 170
Zierler, N., 38, 50, 91
Zinov'ev, V. A., 130, 132, 148, 189, 190, 191